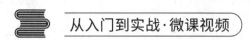

Spring Boot 从入门到实战

微课视频版

陈恒 主编

楼偶俊 巩庆志 董宗然 副主编

清华大学出版社

北京

内 容 简 介

本书从 Spring 和 Spring MVC 的基础知识讲起,让读者无难度地学习 Spring Boot 2。为更好地帮助读者学习,本书以大量实例介绍了 Spring Boot 2 的基本思想、方法和技术。

全书共 12 章,内容涵盖 Spring 基础、Spring MVC 基础、Spring Boot 核心知识、Web 开发、数据访问、安全控制、异步消息、部署与测试、应用监控、电子商务平台的设计与实现(Thymeleaf+MyBatis)、名片管理系统的设计与实现(Vue.js+JPA)等。书中实例侧重实用性、通俗易懂,使读者能够快速掌握 Spring Boot 2 的基础知识、编程技巧以及完整的开发体系,为项目开发打下坚实的基础。

本书提供 1000 分钟的教学视频,还提供教学大纲、教学课件、电子教案、程序源码、习题答案等配套资源。本书可以作为大学计算机及相关专业的教材或教学参考书,也可以作为 Java 技术的培训教材,适合具有 Java 和 Java Web 编程基础的读者,尤其适合广大 Java EE 应用开发人员阅读与使用。

本书封面贴有清华大学出版社防伪标签,无标签者不得销售。
版权所有,侵权必究。举报: 010-62782989, beiqinquan@tup.tsinghua.edu.cn。

图书在版编目(CIP)数据

Spring Boot 从入门到实战: 微课视频版/陈恒主编. —北京: 清华大学出版社,2020.5(2024.1重印)
(从入门到实战·微课视频)
ISBN 978-7-302-55188-1

Ⅰ. ①S… Ⅱ. ①陈… Ⅲ. ①JAVA 语言—程序设计 Ⅳ. ①TP312.8

中国版本图书馆 CIP 数据核字(2020)第 048541 号

策划编辑:	魏江江
责任编辑:	王冰飞
封面设计:	刘 键
责任校对:	李建庄
责任印制:	丛怀宇

出版发行: 清华大学出版社
网　　址: https://www.tup.com.cn, https://www.wqxuetang.com
地　　址: 北京清华大学学研大厦 A 座　　邮　编: 100084
社 总 机: 010-83470000　　邮　购: 010-62786544
投稿与读者服务: 010-62776969, c-service@tup.tsinghua.edu.cn
质量反馈: 010-62772015, zhiliang@tup.tsinghua.edu.cn
课件下载: https://www.tup.com.cn, 010-83470236

印 装 者: 三河市天利华印刷装订有限公司
经　　销: 全国新华书店
开　　本: 185mm×260mm　　印　张: 30.75　　字　数: 748 千字
版　　次: 2020 年 6 月第 1 版　　印　次: 2024 年 1 月第 8 次印刷
印　　数: 16501~18500
定　　价: 79.80 元

产品编号: 083960-01

时至今日,脚本语言和敏捷开发大行其道之时,基于 Spring 框架的 Java EE 开发显得烦琐许多,开发者经常遇到两个非常头疼的问题:①大量的配置文件;②与第三方框架的整合。而 Spring Boot 的出现颠覆了 Java EE 开发,可以说具有划时代的意义。Spring Boot 的目标是帮助开发者编写更少的代码实现所需功能,遵循"约定优于配置"的原则,从而使开发者只需很少的配置,或者使用默认配置就可以快速搭建项目。虽然 Spring Boot 给开发者带来了开发效率,但 Spring Boot 并不是什么新技术,完全是一个基于 Spring 的应用,所以读者在学习 Spring Boot 之前,最好先学习 Spring 与 Spring MVC 的基础知识。

本书系统介绍了 Spring Boot 的主要技术,主要包括三方面内容:①快速开发一个 Web 应用系统(Spring 与 Spring MVC 基础、Thymeleaf 与 Vue.js 视图技术、数据访问技术);②Spring Boot 的高级特性(自动配置、部署、单元测试以及安全机制);③分布式架构技术(REST、MongoDB、Redis、Cache、异步消息以及应用监控)。本书的重点不是简单地介绍基础知识,而是通过精心设计的大量实例,使读者快速地掌握 Spring Boot 的实践应用,提高 Java EE 应用的开发能力。

全书共 12 章,其各章的具体内容如下。

第 1 章介绍 Spring 的基础知识,包括 Spring 的开发环境构建、Spring IoC、Spring AOP、Spring Bean 以及 Spring 的数据库编程等内容。

第 2 章介绍 Spring MVC 的基础知识,包括 Spring MVC 的工作原理、Spring MVC 的工作环境、基于注解的控制器、表单标签库与数据绑定、JSON 数据交互以及 Spring MVC 的基本配置等内容。

第 3 章主要介绍如何快速构建第一个 Spring Boot 应用,包括 Maven 手工构建、http://start.spring.io 快速构建以及 Spring Tool Suite(STS)快速构建等内容。

第 4 章介绍 Spring Boot 的核心,包括核心注解、基本配置、自动配置原理以及条件注解等主要内容。

第 5 章介绍 Spring Boot 的 Web 开发相关技术,包括 Spring Boot 的 Web 开发支持、Thymeleaf 视图模板引擎技术、JSON 数据交互、文件上传与下载、异常统一处理以及对 JSP 的支持。

第 6 章主要讲解 Spring Boot 访问数据库的解决方案,包括 Spring Data JPA、Spring Boot 整合 MyBatis、Spring Boot 的事务管理、Spring Boot 整合 REST、Spring Boot 整合 MongoDB、Spring Boot 整合 Redis、数据缓存 Cache 技术等内容。

第 7 章介绍 Spring Security，包括 Spring Security 快速入门、基于 Spring Data JPA 的 Spring Boot Security 操作实例以及基于 MyBatis 的 Spring Boot Security 操作实例等内容。

第 8 章介绍企业级系统之间的异步消息通信，包括消息模型、JMS 与 AMQP 企业级消息代理、Spring Boot 对异步消息的支持以及异步消息通信实例等内容。

第 9 章介绍 Spring Boot 的部署与单元测试，包括模板引擎的热部署、使用 spring-boot-devtools 热部署以及 Spring Boot 的单元测试等内容。

第 10 章介绍 Spring Boot 应用的监控和管理，包括端点的分类与测试、自定义端点以及自定义 HealthIndicator 等内容。

第 11 章以电子商务平台的设计与实现为综合案例，讲述如何使用 Spring Boot＋Thymeleaf＋MyBatis 开发一个 Spring Boot 应用。

第 12 章以名片系统的设计与实现为综合案例，讲述如何使用 Spring Boot＋Vue.js＋JPA 开发一个前后端分离的应用。

注：为便于教学，本书提供 1000 分钟的教学视频，扫描书中相关章节的二维码可以在线观看、学习；本书还提供教学大纲、教学课件、电子教案、程序源码、习题答案等配套资源，扫描封底的课件二维码可以免费下载。

由于编者水平有限，书中难免存在不足之处，敬请广大读者批评指正。

编　者
2020 年 3 月

目 录

源码下载

第 1 章　Spring 基础 ... 1

学习目的与要求 ... 1
本章主要内容 ... 1
1.1　Spring 概述 ... 1
　　1.1.1　Spring 的由来 ... 1
　　1.1.2　Spring 的体系结构 ... 2
1.2　Spring 开发环境的构建 ... 4
　　1.2.1　使用 Eclipse 开发 Java Web 应用 ... 4
　　1.2.2　Spring 的下载及目录结构 ... 7
　　1.2.3　第一个 Spring 入门程序 ... 8
1.3　Spring IoC ... 10
　　1.3.1　Spring IoC 的基本概念 ... 10
　　1.3.2　Spring 的常用注解 ... 11
　　1.3.3　基于注解的依赖注入 ... 12
　　1.3.4　Java 配置 ... 15
1.4　Spring AOP ... 17
　　1.4.1　Spring AOP 的基本概念 ... 17
　　1.4.2　基于注解开发 AspectJ ... 19
1.5　Spring Bean ... 25
　　1.5.1　Bean 的实例化 ... 25
　　1.5.2　Bean 的作用域 ... 27
　　1.5.3　Bean 的初始化和销毁 ... 29
1.6　Spring 的数据库编程 ... 31

1.6.1　Spring JDBC 的 XML 配置 …………………………………………… 31
　　　1.6.2　Spring JDBC 的 Java 配置 …………………………………………… 32
　　　1.6.3　Spring JdbcTemplate 的常用方法 …………………………………… 33
　　　1.6.4　基于@Transactional 注解的声明式事务管理 ……………………… 38
　　　1.6.5　如何在事务处理中捕获异常 ………………………………………… 42
　1.7　本章小结 ……………………………………………………………………… 43
　习题 1 ……………………………………………………………………………… 44

第 2 章　Spring MVC 基础　45

学习目的与要求 …………………………………………………………………… 45
本章主要内容 ……………………………………………………………………… 45
　2.1　Spring MVC 的工作原理 ……………………………………………………… 45
　2.2　Spring MVC 的工作环境 ……………………………………………………… 47
　　　2.2.1　Spring MVC 所需要的 JAR 包 ……………………………………… 47
　　　2.2.2　使用 Eclipse 开发 Spring MVC 的 Web 应用 ……………………… 47
　　　2.2.3　基于 Java 配置的 Spring MVC 应用 ………………………………… 51
　2.3　基于注解的控制器 …………………………………………………………… 52
　　　2.3.1　Controller 注解类型 …………………………………………………… 53
　　　2.3.2　RequestMapping 注解类型 …………………………………………… 53
　　　2.3.3　编写请求处理方法 …………………………………………………… 54
　　　2.3.4　Controller 接收请求参数的常见方式 ………………………………… 55
　　　2.3.5　重定向与转发 ………………………………………………………… 60
　　　2.3.6　应用@Autowired 进行依赖注入 ……………………………………… 62
　　　2.3.7　@ModelAttribute ……………………………………………………… 64
　2.4　表单标签库与数据绑定 ……………………………………………………… 65
　　　2.4.1　表单标签库 …………………………………………………………… 65
　　　2.4.2　数据绑定 ……………………………………………………………… 69
　2.5　JSON 数据交互 ………………………………………………………………… 77
　　　2.5.1　JSON 数据结构 ………………………………………………………… 77
　　　2.5.2　JSON 数据转换 ………………………………………………………… 78
　2.6　Spring MVC 的基本配置 ……………………………………………………… 82
　　　2.6.1　静态资源配置 ………………………………………………………… 83
　　　2.6.2　拦截器配置 …………………………………………………………… 84
　　　2.6.3　文件上传配置 ………………………………………………………… 85
　2.7　本章小结 ……………………………………………………………………… 90
　习题 2 ……………………………………………………………………………… 90

第3章 Spring Boot 入门 ... 91

学习目的与要求 ... 91
本章主要内容 ... 91
3.1 Spring Boot 概述 ... 91
    3.1.1 什么是 Spring Boot ... 91
    3.1.2 Spring Boot 的优点 ... 92
    3.1.3 Spring Boot 的主要特性 ... 92
3.2 第一个 Spring Boot 应用 ... 93
    3.2.1 Maven 简介 ... 93
    3.2.2 Maven 的 pom.xml ... 93
    3.2.3 在 Eclipse 中创建 Maven Web 项目 ... 94
    3.2.4 Maven 手工构建第一个 Spring Boot 应用 ... 98
3.3 Spring Boot 快速构建 ... 101
    3.3.1 http://start.spring.io ... 101
    3.3.2 Spring Tool Suite ... 104
3.4 本章小结 ... 106
习题3 ... 106

第4章 Spring Boot 核心 ... 107

学习目的与要求 ... 107
本章主要内容 ... 107
4.1 Spring Boot 的基本配置 ... 107
    4.1.1 启动类和核心注解@SpringBootApplication ... 107
    4.1.2 关闭某个特定的自动配置 ... 109
    4.1.3 定制 Banner ... 109
    4.1.4 关闭 banner ... 111
    4.1.5 Spring Boot 的全局配置文件 ... 111
    4.1.6 Spring Boot 的 Starters ... 112
4.2 读取应用配置 ... 116
    4.2.1 Environment ... 116
    4.2.2 @Value ... 117
    4.2.3 @ConfigurationProperties ... 118
    4.2.4 @PropertySource ... 119
4.3 日志配置 ... 120

4.4 Spring Boot 的自动配置原理 …… 122
4.5 Spring Boot 的条件注解 …… 125
    4.5.1 条件注解 …… 125
    4.5.2 实例分析 …… 129
    4.5.3 自定义条件 …… 133
    4.5.4 自定义 Starters …… 135
4.6 本章小结 …… 140
习题 4 …… 141

第 5 章 Spring Boot 的 Web 开发 …… 142

学习目的与要求 …… 142
本章主要内容 …… 142
5.1 Spring Boot 的 Web 开发支持 …… 142
5.2 Thymeleaf 模板引擎 …… 143
    5.2.1 Spring Boot 的 Thymeleaf 支持 …… 144
    5.2.2 Thymeleaf 基础语法 …… 146
    5.2.3 Thymeleaf 的常用属性 …… 151
    5.2.4 Spring Boot 与 Thymeleaf 实现页面信息国际化 …… 156
    5.2.5 Spring Boot 与 Thymeleaf 的表单验证 …… 160
    5.2.6 基于 Thymeleaf 与 BootStrap 的 Web 开发实例 …… 164
5.3 Spring Boot 处理 JSON 数据 …… 169
5.4 Spring Boot 文件上传与下载 …… 174
5.5 Spring Boot 的异常统一处理 …… 180
    5.5.1 自定义 error 页面 …… 180
    5.5.2 @ExceptionHandler 注解 …… 184
    5.5.3 @ControllerAdvice 注解 …… 185
5.6 Spring Boot 对 JSP 的支持 …… 186
5.7 本章小结 …… 190
习题 5 …… 190

第 6 章 Spring Boot 的数据访问 …… 191

学习目的与要求 …… 191
本章主要内容 …… 191
6.1 Spring Data JPA …… 191
    6.1.1 Spring Boot 的支持 …… 193

        6.1.2　简单条件查询 ·· 195
        6.1.3　关联查询 ·· 204
        6.1.4　@Query 和@Modifying 注解 ································ 225
        6.1.5　排序与分页查询 ··· 230
    6.2　Spring Boot 使用 JdbcTemplate ·· 235
    6.3　Spring Boot 整合 MyBatis ··· 241
    6.4　Spring Boot 的事务管理 ··· 245
        6.4.1　Spring Data JPA 的事务支持 ································· 245
        6.4.2　Spring Boot 的事务支持 ····································· 245
    6.5　REST ·· 247
        6.5.1　REST 简介 ··· 247
        6.5.2　Spring Boot 整合 REST ······································ 249
        6.5.3　Spring Data REST ·· 250
        6.5.4　REST 服务测试 ··· 252
    6.6　MongoDB ·· 258
        6.6.1　安装 MongoDB ·· 258
        6.6.2　Spring Boot 整合 MongoDB ·································· 259
        6.6.3　增删改查 ··· 260
    6.7　Redis ·· 265
        6.7.1　安装 Redis ·· 266
        6.7.2　Spring Boot 整合 Redis ······································ 269
        6.7.3　使用 StringRedisTemplate 和 RedisTemplate ··················· 270
    6.8　数据缓存 Cache ··· 276
        6.8.1　Spring 缓存支持 ··· 276
        6.8.2　Spring Boot 缓存支持 ·· 278
        6.8.3　使用 Redis Cache ·· 284
    6.9　本章小结 ··· 285
    习题 6 ·· 285

第 7 章　Spring Boot 的安全控制　286

学习目的与要求 ·· 286
本章主要内容 ·· 286
7.1　Spring Security 快速入门 ··· 286
    7.1.1　什么是 Spring Security ·· 286
    7.1.2　Spring Security 的适配器 ······································· 287
    7.1.3　Spring Security 的用户认证 ····································· 287

		7.1.4	Spring Security 的请求授权	289

        7.1.4 Spring Security 的请求授权 …………………………………… 289
        7.1.5 Spring Security 的核心类 …………………………………… 291
        7.1.6 Spring Security 的验证机制 ………………………………… 292
  7.2 Spring Boot 的支持 ▶ ……………………………………………… 293
  7.3 实际开发中的 Spring Security 操作实例 ▶ ……………………… 293
        7.3.1 基于 Spring Data JPA 的 Spring Boot Security
               操作实例 ……………………………………………………… 293
        7.3.2 基于 MyBatis 的 Spring Boot Security 操作实例 …… 312
  7.4 本章小结 ………………………………………………………………… 317
  习题 7 ………………………………………………………………………… 317

第 8 章　异步消息　318

学习目的与要求 ……………………………………………………………… 318
本章主要内容 ………………………………………………………………… 318
  8.1 消息模型 ▶ …………………………………………………………… 318
        8.1.1 点对点式 ……………………………………………………… 319
        8.1.2 发布/订阅式 ………………………………………………… 319
  8.2 企业级消息代理 ▶ …………………………………………………… 319
        8.2.1 JMS ……………………………………………………………… 319
        8.2.2 AMQP …………………………………………………………… 321
  8.3 Spring Boot 的支持 ▶ ……………………………………………… 323
        8.3.1 JMS 的自动配置 …………………………………………… 323
        8.3.2 AMQP 的自动配置 ………………………………………… 324
  8.4 异步消息通信实例 ▶ ………………………………………………… 324
        8.4.1 JMS 实例 ……………………………………………………… 324
        8.4.2 AMQP 实例 …………………………………………………… 329
  8.5 本章小结 ………………………………………………………………… 334
  习题 8 ………………………………………………………………………… 334

第 9 章　Spring Boot 的热部署与单元测试　335

学习目的与要求 ……………………………………………………………… 335
本章主要内容 ………………………………………………………………… 335
  9.1 开发的热部署 ▶ ……………………………………………………… 335
        9.1.1 模板引擎的热部署 ………………………………………… 335
        9.1.2 使用 spring-boot-devtools 进行热部署 ………… 336

9.2 Spring Boot 的单元测试 ·· 337
 9.2.1 Spring Boot 单元测试程序模板 ···················· 338
 9.2.2 测试 Service ·· 338
 9.2.3 测试 Controller ·· 339
 9.2.4 模拟 Controller 请求 ·································· 340
 9.2.5 比较 Controller 请求返回的结果 ················· 341
 9.2.6 测试实例 ·· 342
9.3 本章小结 ··· 348
习题 9 ··· 348

第 10 章 监控 Spring Boot 应用   349

学习目的与要求 ··· 349
本章主要内容 ·· 349
10.1 端点的分类与测试 ··· 350
 10.1.1 端点的开启与暴露 ····································· 350
 10.1.2 应用配置端点的测试 ································· 352
 10.1.3 度量指标端点的测试 ································· 355
 10.1.4 操作控制端点的测试 ································· 358
10.2 自定义端点 ··· 359
10.3 自定义 HealthIndicator ·································· 362
10.4 本章小结 ··· 363
习题 10 ·· 363

第 11 章 电子商务平台的设计与实现（Thymeleaf＋MyBatis）   364

学习目的与要求 ··· 364
本章主要内容 ·· 364
11.1 系统设计 ··· 364
 11.1.1 系统功能需求 ·· 365
 11.1.2 系统模块划分 ·· 365
11.2 数据库设计 ·· 366
 11.2.1 数据库概念结构设计 ································· 366
 11.2.2 数据逻辑结构设计 ···································· 367
 11.2.3 创建数据表 ··· 369
11.3 系统管理 ··· 369

　　　　11.3.1　添加相关依赖 …………………………………………………… 369
　　　　11.3.2　HTML页面及静态资源管理 ……………………………………… 370
　　　　11.3.3　应用的目录结构 …………………………………………………… 371
　　　　11.3.4　配置文件 …………………………………………………………… 372
　　11.4　组件设计 ……………………………………………………………………… 373
　　　　11.4.1　管理员登录权限验证 ……………………………………………… 373
　　　　11.4.2　前台用户登录权限验证 …………………………………………… 373
　　　　11.4.3　验证码 ……………………………………………………………… 374
　　　　11.4.4　统一异常处理 ……………………………………………………… 375
　　　　11.4.5　工具类 ……………………………………………………………… 376
　　11.5　后台管理子系统的实现 ……………………………………………………… 377
　　　　11.5.1　管理员登录 ………………………………………………………… 377
　　　　11.5.2　类型管理 …………………………………………………………… 380
　　　　11.5.3　添加商品 …………………………………………………………… 386
　　　　11.5.4　查询商品 …………………………………………………………… 390
　　　　11.5.5　修改商品 …………………………………………………………… 396
　　　　11.5.6　删除商品 …………………………………………………………… 398
　　　　11.5.7　查询订单 …………………………………………………………… 400
　　　　11.5.8　用户管理 …………………………………………………………… 402
　　　　11.5.9　安全退出 …………………………………………………………… 403
　　11.6　前台电子商务子系统的实现 ………………………………………………… 403
　　　　11.6.1　导航栏及首页搜索 ………………………………………………… 403
　　　　11.6.2　推荐商品及最新商品 ……………………………………………… 409
　　　　11.6.3　用户注册 …………………………………………………………… 411
　　　　11.6.4　用户登录 …………………………………………………………… 413
　　　　11.6.5　商品详情 …………………………………………………………… 414
　　　　11.6.6　收藏商品 …………………………………………………………… 416
　　　　11.6.7　购物车 ……………………………………………………………… 417
　　　　11.6.8　下单 ………………………………………………………………… 422
　　　　11.6.9　个人信息 …………………………………………………………… 425
　　　　11.6.10　我的收藏 ………………………………………………………… 426
　　　　11.6.11　我的订单 ………………………………………………………… 428
　　11.7　本章小结 ……………………………………………………………………… 433

第12章　名片系统的设计与实现（Vue.js＋JPA） 🎬 ……… 434

　　学习目的与要求 ……………………………………………………………………… 434

本章主要内容 …… 434
12.1 名片系统功能介绍 …… 434
12.2 使用 IntelliJ IDEA 构建名片后端系统 cardmis …… 435
    12.2.1 构建基于 JPA 的 Spring Boot Web 应用 …… 435
    12.2.2 修改 pom.xml …… 435
    12.2.3 配置数据源等信息 …… 437
    12.2.4 创建持久化实体类 …… 438
    12.2.5 创建 Repository 持久层 …… 439
    12.2.6 创建业务层 …… 440
    12.2.7 创建控制器层 …… 442
    12.2.8 创建跨域响应头设置过滤器 …… 444
    12.2.9 创建工具类 …… 445
12.3 使用 IntelliJ IDEA 构建名片前端系统 cardmis-vue …… 445
    12.3.1 安装 Node.js …… 445
    12.3.2 安装 Vue CLI …… 446
    12.3.3 构建前端 Vue 项目 cardmis-vue …… 447
    12.3.4 分析 Vue 项目结构 …… 451
    12.3.5 设置 IntelliJ IDEA 支持创建 *.vue 文件及打开 *.vue 文件 …… 454
    12.3.6 开发前端页面 …… 454
    12.3.7 配置页面路由 …… 469
    12.3.8 设置反向代理 …… 470
    12.3.9 设置跨域支持 …… 472
12.4 Vuex 与前端路由拦截器 …… 472
    12.4.1 引入 Vuex …… 472
    12.4.2 修改路由配置 …… 473
    12.4.3 使用钩子函数判断是否登录 …… 473
    12.4.4 解决跨域请求 session 失效的问题 …… 474
12.5 测试运行 …… 474
12.6 小结 …… 475

参考文献 …… 476

第1章 Spring 基础

学习目的与要求

本章重点讲解 Spring 的基础知识。通过本章的学习，了解 Spring 的体系结构，理解 Spring IoC 与 AOP 的基本原理，了解 Spring Bean 的生命周期、实例化以及作用域，掌握 Spring 的事务管理。

本章主要内容

- Spring 开发环境的构建。
- Spring IoC。
- Spring AOP。
- Spring Bean。
- Spring 的数据库编程。

Spring 是当前主流的 Java 开发框架，为企业级应用开发提供了丰富的功能。掌握 Spring 框架的使用，已是 Java 开发者必备的技能之一。本章将学习如何使用 Eclipse 开发 Spring 程序，不过在此之前需要构建 Spring 的开发环境。

1.1 Spring 概述

1.1.1 Spring 的由来

视频讲解

Spring 是一个轻量级 Java 开发框架，最早由 Rod Johnson 创建，目的是为了解决企业级应用开发的业务逻辑层和其他各层的耦合问题。它是一个分层的 JavaSE/EEfull-stack（一站式）轻量级开源框架，为开发 Java 应用程序提供全面的基础架构支持。Spring 负责基础架构，因此 Java 开发者可以专注于应用程序的开发。

1.1.2　Spring 的体系结构

Spring 的功能模块被有组织地分散到约 20 个模块中，这些模块分布在核心容器、数据访问/集成(Data Access/Integration)、Web、AOP(Aspect Oriented Programming，面向切面的编程)、植入(Instrumentation)、消息传输(Messaging)和测试(Test)中，如图 1.1 所示。

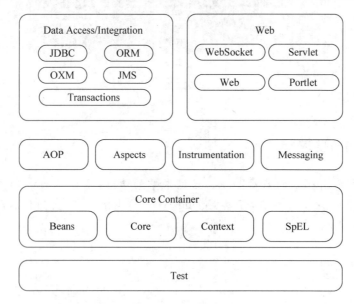

图 1.1　Spring 的体系结构

1. 核心容器

Spring 的核心容器是其他模块建立的基础，由 Spring-core、Spring-beans、Spring-context、Spring-context-support 和 Spring-expression(Spring 表达式语言)等模块组成。

Spring-core 模块：提供了框架的基本组成部分，包括控制反转(Inversion of Control，IoC)和依赖注入(Dependency Injection，DI)功能。

Spring-beans 模块：提供了 BeanFactory，是工厂模式的一个经典实现，Spring 将管理对象称为 Bean。

Spring-context 模块：建立在 Core 和 Beans 模块基础上，提供一个框架式的对象访问方式，是访问定义和配置的任何对象的媒介。ApplicationContext 接口是 Context 模块的焦点。

Spring-context-support 模块：支持整合第三方库到 Spring 应用程序上下文，特别是用于高速缓存(EhCache、JCache)和任务调度(CommonJ、Quartz)的支持。

Spring-expression 模块：提供了强大的表达式语言去支持运行时查询和操作对象图。这是对 JSP 2.1 规范中规定的统一表达式语言(Unified EL)的扩展。该语言支持设置和获取属性值、属性分配、方法调用、访问数组、集合和索引器的内容、逻辑和算术运算、变量命名以及从 Spring 的 IoC 容器中以名称检索对象。它还支持列表投影、选择以及常见的列表聚合。

2. AOP 和 Instrumentation

Spring 框架中与 AOP 和 Instrumentation 相关的模块有 Spring-aop 模块、Spring-aspects 模块以及 Spring-instrument 模块。

Spring-aop 模块：提供了一个符合 AOP 要求的面向切面的编程实现，允许定义方法拦截器和切入点，将代码按照功能进行分离，以便干净地解耦。

Spring-aspects 模块：提供了与 AspectJ 的集成功能，AspectJ 是一个功能强大且成熟的 AOP 框架。

Spring-instrument 模块：提供了类植入（Instrumentation）支持和类加载器的实现，可以在特定的应用服务器中使用。Instrumentation 提供了一种虚拟机级别支持的 AOP 实现方式，使得开发者无须对 JDK 做任何升级和改动，就可以实现某些 AOP 的功能。

3. 消息

Spring 4.0 以后新增了消息（Spring-messaging）模块，该模块提供了对消息传递体系结构和协议的支持。

4. 数据访问/集成

数据访问/集成由 JDBC、ORM、OXM、JMS 和事务模块组成。

Spring-jdbc 模块：提供了一个 JDBC 的抽象层，消除了烦琐的 JDBC 编码和数据库厂商特有的错误代码解析。

Spring-tx 模块（事务模块）：支持用于实现特殊接口和所有 POJO（普通 Java 对象）类的编程和声明式事务管理。

Spring-orm 模块：为流行的对象关系映射（Object-Relational Mapping）API 提供集成层，包括 JPA 和 Hibernate。使用 Spring-orm 模块，可以将这些 O/R 映射框架与 Spring 提供的所有其他功能结合使用，例如声明式事务管理功能。

Spring-oxm 模块：提供了一个支持对象/XML 映射的抽象层实现，如 JAXB、Castor、JiBX 和 XStream。

Spring-jms 模块（Java Messaging Service）：指 Java 消息传递服务，包含用于生产和使用消息的功能。自 Spring 4.1 后，提供了与 Spring-messaging 模块的集成。

5. Web

Web 层由 Spring-web、Spring-webmvc、Spring-websocket 和 Portlet 模块组成。

Spring-web 模块：提供了基本的 Web 开发集成功能。例如多文件上传功能、使用 Servlet 监听器初始化一个 IoC 容器以及 Web 应用上下文。

Spring-webmvc 模块：也称为 Web-Servlet 模块，包含用于 Web 应用程序的 Spring MVC 和 REST Web Services 实现。Spring MVC 框架提供了领域模型代码和 Web 表单之

间的清晰分离,并与Spring Framework的所有其他功能集成,本书后续章节将会详细讲解Spring MVC框架。

Spring-websocket模块:Spring 4.0后新增的模块,它提供了WebSocket和SockJS的实现,主要是与Web前端的全双工通信的协议。

Spring-webmvc-portlet模块(也称为Web-Portlet模块)提供在Portlet环境中使用的MVC实现,并反映基于Servlet的Spring-webmvc模块功能。(该模块在Spring5中已移除)。

Spring-webflux是一个新的非堵塞函数式Reactive Web框架,可以用来建立异步的、非阻塞、事件驱动的服务,并且扩展性好。(该模块是Spring5的新增模块。)

6. 测试

Spring-test模块:支持使用JUnit或TestNG对Spring组件进行单元测试和集成测试。

1.2 Spring开发环境的构建

使用Spring框架开发应用前,应先搭建其开发环境。本书前两章(Spring、Spring MVC)的开发环境都是基于Eclipse平台的Java Web应用的开发环境。

1.2.1 使用Eclipse开发Java Web应用

为了提高开发效率,通常需要安装IDE(集成开发环境)工具。Eclipse是一个可用于开发Web应用的IDE工具。登录http://www.eclipse.org/ide,选择Java EE,根据操作系统的位数,下载相应的Eclipse。本书采用的是"eclipse-jee-2018-12-R-win32-x86_64.zip"。

使用Eclipse之前,需要对JDK、Web服务器和Eclipse进行一些必要的配置。因此,在安装Eclipse之前,应事先安装JDK和Web服务器。

1. 安装JDK

安装并配置JDK(本书采用的JDK是jdk-11.0.1_windows-x64_bin.exe),按照提示安装完成JDK后,需要配置"环境变量"的"系统变量"Java_Home和Path。在Windows 10系统下,系统变量示例如图1.2和图1.3所示。

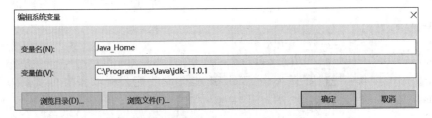

图1.2 新建系统变量Java_Home

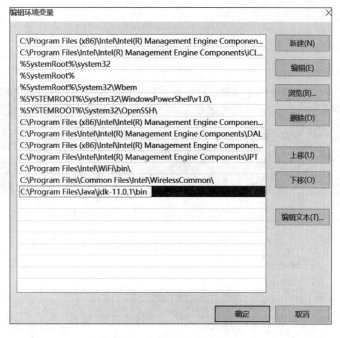

图 1.3　编辑系统变量 Path

2．Web 服务器

目前，比较常用的 Web 服务器包括 Tomcat、JRun、Resin、WebSphere、WebLogic 等，本书采用的是 Tomcat 9.0。

登录 Apache 软件基金会的官方网站 http://jakarta.Apache.org/tomcat，下载 Tomcat 9.0 的免安装版（本书采用 apache-tomcat-9.0.2-windows-x64.zip）。登录网站后，首先在 Download 里选择 Tomcat 9，然后在 Binary Distributions 的 Core 中选择相应版本即可。

安装 Tomcat 之前需要事先安装 JDK 并配置系统环境变量 Java_Home。将下载的 apache-tomcat-9.0.2-windows-x64.zip 解压到某个目录下，例如解压到 E:\Java soft，解压缩后将出现如图 1.4 所示的目录结构。

图 1.4　Tomcat 目录结构

执行 Tomcat 根目录下 bin 文件夹中的 startup.bat 来启动 Tomcat 服务器。执行 startup.bat 启动 Tomcat 服务器会占用一个 MS-DOS 窗口，出现如图 1.5 所示的界面，如果关闭当前 MS-DOS 窗口将关闭 Tomcat 服务器。

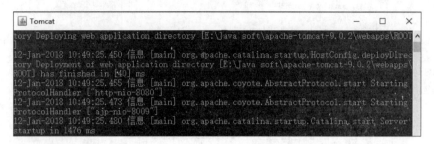

图 1.5　执行 startup.bat 启动 Tomcat 服务器

Tomcat 服务器启动后，在浏览器的地址栏中输入"http://localhost:8080"，将出现如图 1.6 所示的 Tomcat 测试页面。

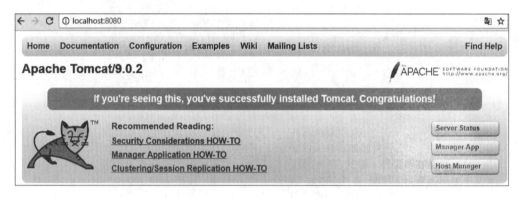

图 1.6　Tomcat 测试页面

3. 安装 Eclipse

Eclipse 下载完成后，解压到自己设置的路径下，即可完成安装。Eclipse 安装后，双击 Eclipse 安装目录下的 eclipse.exe 文件，启动 Eclipse。

4. 集成 Tomcat

启动 Eclipse，选择 Window | Preferences 菜单项，在弹出的对话框中选择 Server | Runtime Environments 命令。在弹出的窗口中，单击 Add 按钮，弹出如图 1.7 所示的 New Server Runtime Environment 界面，在此可以配置各种版本的 Web 服务器。

在图 1.7 中选择 Apache Tomcat v9.0 服务器版本，单击 Next 按钮，进入如图 1.8 所示的界面。

在图 1.8 中单击 Browse 按钮，选择 Tomcat 的安装目录，单击 Finish 按钮即可完成 Tomcat 配置。

至此，可以使用 Eclipse 创建 Dynamic Web project，并在 Tomcat 下运行。

第1章 Spring 基础

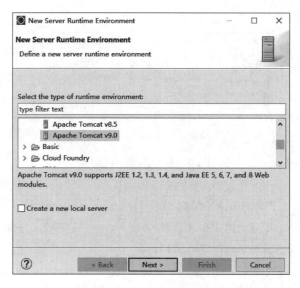

图 1.7 Tomcat 配置界面

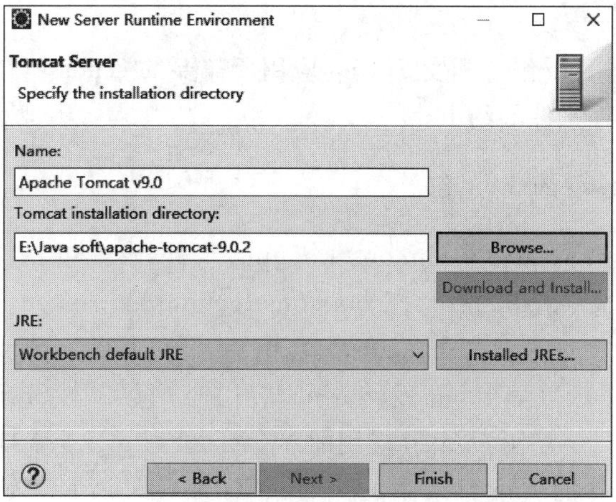

图 1.8 选择 Tomcat 目录

1.2.2 Spring 的下载及目录结构

使用 Spring 框架开发应用程序时,除了引用 Spring 自身的 JAR 包外,还需要引用 commons.logging 的 JAR 包。

1. Spring 的 JAR 包

Spring 官方网站升级后,建议通过 Maven 和 Gradle 下载。对于不使用 Maven 和 Gradle 下载的开发者,本书给出一个 Spring Framework jar 官方直接下载路径：http://repo.springsource.org/libs-release-local/org/springframework/spring/。本书采用的是 spring-framework-5.1.4.RELEASE-dist.zip。将下载到的 ZIP 文件解压缩,解压缩后的目

录结构如图 1.9 所示。

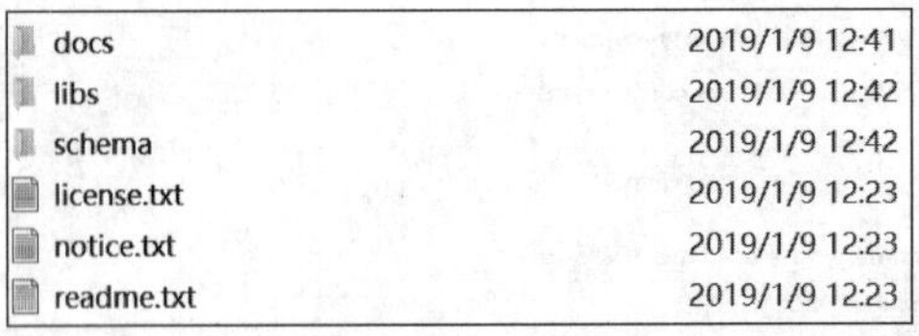

图 1.9　spring-framework-5.1.4 的目录结构

图 1.9 中,docs 目录包含 Spring 的 API 文档和开发规范。

图 1.9 中,libs 目录包含开发 Spring 应用所需要的 JAR 包和源代码。该目录下有三类 JAR 文件,其中,以 RELEASE.jar 结尾的文件是 Spring 框架 class 的 JAR 包,即开发 Spring 应用所需要的 JAR 包;以 RELEASE-javadoc.jar 结尾的文件是 Spring 框架 API 文档的压缩包;以 RELEASE-sources.jar 结尾的文件是 Spring 框架源文件的压缩包。在 libs 目录中,有 4 个基础包:spring-core-5.1.4.RELEASE.jar、spring-beans-5.1.4.RELEASE.jar、spring-context-5.1.4.RELEASE.jar 和 spring-expression-5.1.4.RELEASE.jar,分别对应 Spring 核心容器的 4 个模块:Spring-core 模块、Spring-beans 模块、Spring-context 模块和 Spring-expression 模块。

图 1.9 中,schema 目录包含开发 Spring 应用所需要的 schema 文件,这些 schema 文件定义了 Spring 相关配置文件的约束。

2. commons.logging 的 JAR 包

Spring 框架依赖于 Apache Commons Logging 组件,该组件的 JAR 包可以通过网址 "http://commons.apache.org/proper/commons-logging/download_logging.cgi" 下载,本书下载的是 commons-logging-1.2-bin.zip,解压缩后,即可找到 commons-logging-1.2.jar。

对于 Spring 框架的初学者,开发 Spring 应用时,只需将 Spring 的 4 个基础包和 commons-logging-1.2.jar 复制到 Web 应用的 WEB-INF/lib 目录下即可。当不明白需要哪些 JAR 包时,将 Spring 的 libs 目录中的 spring-XXX-5.1.4.RELEASE.jar 全部复制到 WEB-INF/lib 目录下即可。

1.2.3　第一个 Spring 入门程序

本节通过一个简单的入门程序向读者演示 Spring 框架的使用过程。

【例 1-1】　Spring 框架的使用过程。

具体实现步骤如下。

1. 使用 Eclipse 创建 Web 应用并导入 JAR 包

使用 Eclipse 创建一个名为 ch1_1 的 Dynamic Web project,并将 Spring 的 4 个基础包和第三方依赖包 commons-logging-1.2.jar 复制到 ch1_1 的 WEB-INF/lib 目录中,如图 1.10 所示。

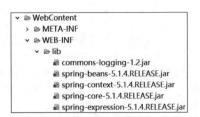

图 1.10　应用 ch1_1 导入的 JAR 包

注意：在讲解 Spring MVC 框架前，本书的实例并没有真正运行 Web 应用。创建 Web 应用的目的是方便添加相关 JAR 包。

2．创建接口 TestDao

Spring 解决的是业务逻辑层和其他各层的耦合问题，因此它将面向接口的编程思想贯穿于整个系统应用。

在 src 目录下，创建一个 dao 包，并在 dao 包中创建接口 TestDao，接口中定义一个 sayHello()方法，代码如下：

```
package dao;
public interface TestDao {
    public void sayHello();
}
```

3．创建接口 TestDao 的实现类 TestDaoImpl

在包 dao 下创建 TestDao 的实现类 TestDaoImpl，具体代码如下：

```
package dao;
public class TestDaoImpl implements TestDao{
    @Override
    public void sayHello() {
        System.out.println("Hello, Study hard!");
    }
}
```

4．创建配置文件 applicationContext.xml

在 src 目录下，创建 Spring 的配置文件 applicationContext.xml，并在该文件中使用实现类 TestDaoImpl 创建一个 id 为 test 的 Bean，具体代码如下：

```xml
<?xml version="1.0" encoding="UTF-8"?>
<beans xmlns="http://www.springframework.org/schema/beans"
    xmlns:xsi="http://www.w3.org/2001/XMLSchema-instance"
    xsi:schemaLocation="http://www.springframework.org/schema/beans
        http://www.springframework.org/schema/beans/spring-beans.xsd">
    <!-- 将指定类 TestDaoImpl 配置给 Spring，让 Spring 创建其实例 -->
    <bean id="test" class="dao.TestDaoImpl" />
</beans>
```

注：配置文件的名称可以自定义，但习惯上命名为 applicationContext.xml，有时候也命名为 beans.xml。配置文件信息不需要读者手写，可以从 Spring 的帮助文档中复制（首先使用浏览器打开\spring-framework-5.1.4.RELEASE\docs\spring-framework-reference\index.html，在页面中单击超链接 Core，在 1.2.1 Configuration metadata 小节下即可找到配置文件的约束信息）。

5．创建测试类

在 src 目录下，创建一个 test 包，并在 test 包中创建 Test 类，具体代码如下：

```
package test;
import org.springframework.context.ApplicationContext;
import org.springframework.context.support.ClassPathXmlApplicationContext;
import dao.TestDao;
public class Test {
    public static void main(String[] args) {
        //初始化Spring容器ApplicationContext,加载配置文件
        //@SuppressWarnings抑制警告的关键字,有泛型未指定类型
        @SuppressWarnings("resource")
        ApplicationContext appCon = new ClassPathXmlApplicationContext("applicationContext.xml");
        //通过容器获取test实例
        TestDao tt = (TestDao)appCon.getBean("test");    //test为配置文件中的id
        tt.sayHello();
    }
}
```

执行上述main()方法后,将在控制台输出"Hello,Study hard!"。上述main()方法中并没有使用new运算符创建TestDaoImpl类的对象,而是通过Spring容器获取的实现类对象,这就是Spring IoC的工作机制。

1.3 Spring IoC

视频讲解

1.3.1 Spring IoC的基本概念

控制反转(Inversion of Control,IoC)是一个比较抽象的概念,是Spring框架的核心,用来消减计算机程序的耦合问题。依赖注入(Dependency Injection,DI)是IoC的另外一种说法,只是从不同的角度,描述相同的概念。下面通过实际生活中的一个例子解释IoC和DI。

当人们需要一件东西时,第一反应就是找东西,例如想吃面包。在没有面包店和有面包店两种情况下,你会怎么做? 在没有面包店时,最直观的做法可能是按照自己的口味制作面包,也就是需要主动制作一个面包。然而,时至今日,各种面包店盛行,当你不想自己制作面包时,可以去面包店,选择你喜欢的面包。注意你并没有制作面包,而是由店家制作,但是完全符合你的口味。

上面只是列举了一个非常简单的例子,但包含了控制反转的思想,即把制作面包的主动权交给店家。下面通过面向对象编程思想,继续探讨这两个概念。

当某个Java对象(调用者,例如你)需要调用另一个Java对象(被调用者,即被依赖对象,例如面包)时,在传统编程模式下,调用者通常会采用"new 被调用者"的代码方式来创建对象(例如你自己制作面包)。这种方式会增加调用者与被调用者之间的耦合性,不利于后期代码的升级与维护。

当Spring框架出现后,对象的实例不再由调用者来创建,而是由Spring容器(例如面包店)来创建。Spring容器会负责控制程序之间的关系(例如面包店负责控制你与面包的关系),而不是由调用者的程序代码直接控制。这样,控制权由调用者转移到Spring容器,控

制权发生了反转,这就是 Spring 的控制反转。

从 Spring 容器角度来看,Spring 容器负责将被依赖对象赋值给调用者的成员变量,相当于为调用者注入它所依赖的实例,这就是 Spring 的依赖注入,主要目的是为了解耦,体现一种"组合"的理念。

综上所述,控制反转是一种通过描述(在 Spring 中可以是 XML 或注解)并通过第三方产生或获取特定对象的方式。在 Spring 中实现控制反转的是 IoC 容器,其实现方法是依赖注入。

1.3.2　Spring 的常用注解

在 Spring 框架中,尽管使用 XML 配置文件可以很简单地装配 Bean(如例 1-1),但如果应用中有大量的 Bean 需要装配时,会导致 XML 配置文件过于庞大,不方便以后的升级维护。因此,更多的时候推荐开发者使用注解(annotation)的方式去装配 Bean。

在 Spring 框架中定义了一系列的注解,常用注解如下所示。

1. 声明 Bean 的注解

1) @Component

该注解是一个泛化的概念,仅仅表示一个组件对象(Bean),可以作用在任何层次上,没有明确的角色。

2) @Repository

该注解用于将数据访问层(DAO)的类标识为 Bean,即注解数据访问层 Bean,其功能与 @Component()相同。

3) @Service

该注解用于标注一个业务逻辑组件类(Service 层),其功能与 @Component()相同。

4) @Controller

该注解用于标注一个控制器组件类(Spring MVC 的 Controller),其功能与 @Component()相同。

2. 注入 Bean 的注解

1) @Autowired

该注解可以对类成员变量、方法及构造方法进行标注,完成自动装配的工作。通过 @Autowired 的使用来消除 setter 和 getter 方法。默认按照 Bean 的类型进行装配。

2) @Resource

该注解与 @Autowired 功能一样。区别在于,该注解默认是按照名称来装配注入的,只有当找不到与名称匹配的 Bean 时才会按照类型来装配注入;而 @Autowired 默认按照 Bean 的类型进行装配,如果想按照名称来装配注入,则需要结合 @Qualifier 注解一起使用。

@Resource 注解有两个属性:name 和 type。name 属性指定 Bean 实例名称,即按照名称来装配注入;type 属性指定 Bean 类型,即按照 Bean 的类型进行装配。

3）@Qualifier

该注解与@Autowired注解配合使用。当@Autowired注解需要按照名称来装配注入时，则需要结合该注解一起使用，Bean的实例名称由@Qualifier注解的参数指定。

1.3.3 基于注解的依赖注入

Spring IoC容器（ApplicationContext）负责创建和注入Bean。Spring提供使用XML配置、注解、Java配置以及groovy配置实现Bean的创建和注入。本书尽量使用注解（@Component、@Repository、@Service以及@Controller等业务Bean的配置）和Java配置（全局配置如数据库、MVC等相关配置）完全代替XML配置，这也是Spring Boot推荐的配置方式。

下面通过一个简单实例向读者演示基于注解的依赖注入的使用过程。

【例1-2】 基于注解的依赖注入的使用过程。

具体实现步骤如下。

1. 使用Eclipse创建Web应用并导入JAR包

使用Eclipse创建一个名为ch1_2的Dynamic Web project，并将Spring的4个基础包、第三方依赖包commons-logging-1.2.jar以及spring-aop-5.1.4.RELEASE.jar（本节扫描注解，需要事先导入Spring AOP的JAR包）复制到ch1_2的WEB-INF/lib目录中，如图1.11所示。

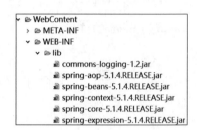

图1.11 ch1_2应用导入的JAR包

2. 创建DAO层

在ch1_2应用的src中，创建annotation.dao包，该包下创建TestDao接口和TestDaoImpl实现类，并将实现类TestDaoImpl使用@Repository注解标注为数据访问层。

TestDao的代码如下：

```
package annotation.dao;
public interface TestDao {
    public void save();
}
```

TestDaoImpl的代码如下：

```
package annotation.dao;
import org.springframework.stereotype.Repository;
@Repository("testDaoImpl")
/** 相当于@Repository，但如果在service层使用@Resource(name = "testDaoImpl")注入Bean,
testDaoImpl不能省略。 **/
public class TestDaoImpl implements TestDao{
    @Override
    public void save() {
```

```
        System.out.println("testDao save");
    }
}
```

3. 创建 Service 层

在 ch1_2 应用的 src 中，创建 annotation.service 包，该包下创建 TestService 接口和 TestSeviceImpl 实现类，并将实现类 TestServiceImpl 使用@Service 注解标注为业务逻辑层。

TestService 的代码如下：

```
package annotation.service;
public interface TestService {
    public void save();
}
```

TestServiceImpl 的代码如下：

```
package annotation.service;
import javax.annotation.Resource;
import org.springframework.stereotype.Service;
import annotation.dao.TestDao;
@Service("testServiceImpl")           //相当于@Service
public class TestSeviceImpl implements TestService{
    @Resource(name = "testDaoImpl")
    /** 相当于@Autowired,@Autowired 默认按照 Bean 类型注入 **/
    private TestDao testDao;
    @Override
    public void save() {
        testDao.save();
        System.out.println("testService save");
    }
}
```

4. 创建 Controller 层

在 ch1_2 应用的 src 中，创建 annotation.controller 包，该包下创建 TestController 类，并将 TestController 类使用@Controller 注解标注为控制器层。

TestController 的代码如下：

```
package annotation.controller;
import org.springframework.beans.factory.annotation.Autowired;
import org.springframework.stereotype.Controller;
import annotation.service.TestService;
@Controller
public class TestController {
    @Autowired
    private TestService testService;
    public void save() {
        testService.save();
        System.out.println("testController save");
```

}
}

5. 创建配置类

本书尽量不使用Spring的XML配置文件,而使用注解和Java配置。因此,在此需要使用@Configuration创建一个Java配置类(相当于一个Spring的XML配置文件),并通过@ComponentScan扫描使用注解的包(相当于在Spring的XML配置文件中使用<context：component-scan base-package="Bean所在的包路径"/>语句)。

在ch1_2应用的annotation包下,创建名为ConfigAnnotation的配置类。ConfigAnnotation的代码如下：

```
package annotation;
import org.springframework.context.annotation.ComponentScan;
import org.springframework.context.annotation.Configuration;
@Configuration//声明当前类是一个配置类(见1.3.4节),相当于一个Spring的XML配置文件
@ComponentScan("annotation")
//自动扫描annotation包下使用的注解,并注册为Bean
/* 相当于在Spring的XML配置文件中使用<context:component - scan base - package = "Bean所在的包路径"/>语句 */
public class ConfigAnnotation {
}
```

6. 创建测试类

在ch1_2应用的annotation包下,创建测试类TestAnnotation,具体代码如下：

```
package annotation;
import org.springframework.context.annotation.AnnotationConfigApplicationContext;
import annotation.controller.TestController;
public class TestAnnotation {
    public static void main(String[] args) {
        //初始化Spring容器ApplicationContext
        AnnotationConfigApplicationContext appCon =
                new AnnotationConfigApplicationContext(ConfigAnnotation.class);
        TestController tc = appCon.getBean(TestController.class);
        tc.save();
        appCon.close();
    }
}
```

7. 运行结果

运行测试类TestAnnotation的main方法,结果如图1.12所示。

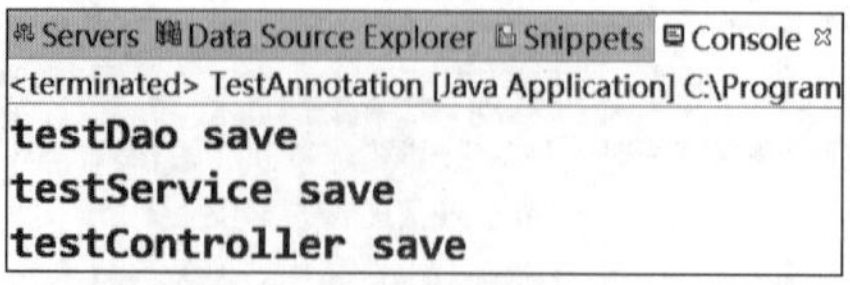

图1.12 ch1_2的运行结果

1.3.4　Java 配置

Java 配置是 Spring4.x 推荐的配置方式，它是通过@Configuration 和@Bean 来实现的。@Configuration 声明当前类是一个配置类，相当于一个 Spring 配置的 XML 文件。@Bean 注解在方法上，声明当前方法的返回值为一个 Bean。下面通过实例演示 Java 配置的使用过程。

【例 1-3】　Java 配置的使用过程。

具体实现步骤如下。

1. 使用 Eclipse 创建 Web 应用并导入 JAR 包

使用 Eclipse 创建一个名为 ch1_3 的 Dynamic Web project，并将与 ch1_2 相同的 JAR 包复制到 WEB-INF/lib 目录中。

2. 创建 DAO 层

在 ch1_3 应用的 src 中，创建 dao 包，该包下创建 TestDao 类。此类中没有使用 @Repository 注解为数据访问层，具体代码如下：

```
package dao;
//此处没有使用@Repository 声明 Bean
public class TestDao {
    public void save() {
        System.out.println("TestDao save");
    }
}
```

3. 创建 Service 层

在 ch1_3 应用的 src 中，创建 service 包，该包下创建 TestService 类。此类中没有使用 @Service 注解为业务逻辑层，具体代码如下：

```
package service;
import dao.TestDao;
//此处没有使用@Service 声明 Bean
public class TestService {
    //此处没有使用@Autowired 注入 testDao
    TestDao testDao;
    public void setTestDao(TestDao testDao) {
        this.testDao = testDao;
    }
    public void save() {
        testDao.save();
    }
}
```

4. 创建 Controller 层

在 ch1_3 应用的 src 中，创建 controller 包，该包下创建 TestController 类。此类中没有使用@Controller 注解为控制器层，具体代码如下：

```
package controller;
import service.TestService;
//此处没有使用@Controller 声明 Bean
public class TestController {
    //此处没有使用@Autowired 注入 testService
    TestService testService;
    public void setTestService(TestService testService) {
        this.testService = testService;
    }
    public void save() {
        testService.save();
    }
}
```

5. 创建配置类

在 ch1_3 应用的 src 中，创建 javaConfig 包，该包下创建 JavaConfig 配置类。此类中使用@Configuration 注解该类为一个配置类，相当于一个 Spring 配置的 XML 文件。在配置类中使用@Bean 注解定义 0 个或多个 Bean，具体代码如下：

```
package javaConfig;
import org.springframework.context.annotation.Bean;
import org.springframework.context.annotation.Configuration;
import controller.TestController;
import dao.TestDao;
import service.TestService;
@Configuration
//一个配置类，相当于一个 Spring 配置的 XML 文件；
//此处没有使用包扫描，是因为所有 Bean 都在此类中定义了
public class JavaConfig {
    @Bean
    public TestDao getTestDao() {
        return new TestDao();
    }
    @Bean
    public TestService getTestService() {
        TestService ts = new TestService();
        //使用 set 方法注入 testDao
        ts.setTestDao(getTestDao());
        return ts;
    }
    @Bean
    public TestController getTestController() {
        TestController tc = new TestController();
        //使用 set 方法注入 testService
```

```
            tc.setTestService(getTestService());
            return tc;
        }
    }
```

6. 创建测试类

在 ch1_3 应用的 javaConfig 包下，创建测试类 TestConfig，具体代码如下：

```
package javaConfig;
import org.springframework.context.annotation.AnnotationConfigApplicationContext;
import controller.TestController;
public class TestConfig {
    public static void main(String[] args) {
        //初始化 Spring 容器 ApplicationContext
        AnnotationConfigApplicationContext appCon =
            new AnnotationConfigApplicationContext(JavaConfig.class);
        TestController tc = appCon.getBean(TestController.class);
        tc.save();
        appCon.close();
    }
}
```

7. 运行结果

运行测试类 TestConfig 的 main 方法，结果如图 1.13 所示。

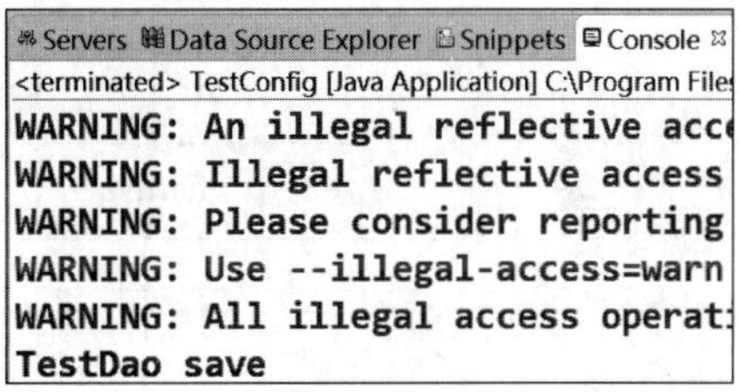

从 ch1_2 应用与 ch1_3 应用对比可以看出，有时候使用 Java 配置反而更加烦琐。何时使用 Java 配置？何时使用注解配置？作者的观点是：全局配置尽量使用 Java 配置，如数据库相关的配置；业务 Bean 的配置尽量使用注解配置，如数据访问层、业务逻辑层、控制器层等相关的配置。

图 1.13　ch1_3 的运行结果

1.4　Spring AOP

视频讲解

Spring AOP 是 Spring 框架体系结构中非常重要的功能模块之一，该模块提供了面向切面编程实现。面向切面编程在事务处理、日志记录、安全控制等操作中被广泛使用。

1.4.1　Spring AOP 的基本概念

1．AOP 的概念

AOP(Aspect-Oriented Programming)，即面向切面编程。它与 OOP(Object-Oriented Programming，面向对象编程)相辅相成，提供了与 OOP 不同的抽象软件结构的视角。在

OOP 中，以类作为程序的基本单元，而 AOP 中的基本单元是 Aspect（切面）。Struts 2 的拦截器设计就是基于 AOP 的思想，是个比较经典的应用。

在业务处理代码中，通常都有日志记录、性能统计、安全控制、事务处理、异常处理等操作。尽管使用 OOP 可以通过封装或继承的方式达到代码的重用，但仍然存在同样的代码分散到各个方法中。因此，采用 OOP 处理日志记录等操作，不仅增加了开发者的工作量，而且提高了升级维护的困难。为了解决此类问题，AOP 思想应运而生。AOP 采取横向抽取机制，即将分散在各个方法中的重复代码提取出来，然后在程序编译或运行阶段，再将这些抽取出来的代码应用到需要执行的地方。这种横向抽取机制，采用传统的 OOP 是无法办到的，因为 OOP 实现的是父子关系的纵向重用。但是 AOP 不是 OOP 的替代品，而是 OOP 的补充，它们相辅相成。

在 AOP 中，横向抽取机制的类与切面的关系如图 1.14 所示。

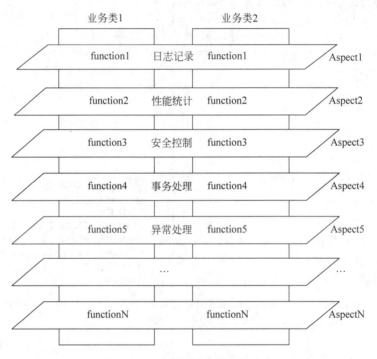

图 1.14　AOP 中类与切面的关系

从图 1.14 可以看出，通过切面 Aspect 分别在业务类 1 和业务类 2 中加入了日志记录、性能统计、安全控制、事务处理、异常处理等操作。

2. AOP 的术语

在 Spring AOP 框架中，涉及以下常用术语。

1）切面

切面（Aspect）是指封装横切到系统功能（如事务处理）的类。

2）连接点

连接点（Joinpoint）是指程序运行中的一些时间点，如方法的调用或异常的抛出。

3）切入点

切入点（Pointcut）是指那些需要处理的连接点。在 Spring AOP 中，所有的方法执行都是连接点，而切入点是一个描述信息，它修饰的是连接点，通过切入点确定哪些连接点需要被处理。切面、连接点和切入点的关系如图 1.15 所示。

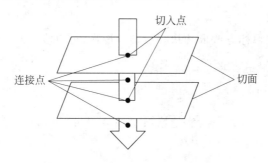

图 1.15　切面、连接点和切入点的关系

4）通知（增强处理）

由切面添加到特定的连接点（满足切入点规则）的一段代码，即在定义好的切入点处所要执行的程序代码。可以将其理解为切面开启后，切面的方法。因此，通知是切面的具体实现。

5）引入

引入（Introduction）允许在现有的实现类中添加自定义的方法和属性。

6）目标对象

目标对象（Target Object）是指所有被通知的对象。如果 AOP 框架使用运行时代理的方式（动态的 AOP）来实现切面，那么通知对象总是一个代理对象。

7）代理

代理（Proxy）是通知应用到目标对象之后，被动态创建的对象。

8）织入

织入（Weaving）是将切面代码插入目标对象上，从而生成代理对象的过程。根据不同的实现技术，AOP 织入有 3 种方式：编译器织入，需要有特殊的 Java 编译器；类装载器织入，需要有特殊的类装载器；动态代理织入，在运行期为目标类添加通知生成子类的方式。Spring AOP 框架默认采用动态代理织入，而 AspectJ（基于 Java 语言的 AOP 框架）采用编译器织入和类装载器织入。

1.4.2　基于注解开发 AspectJ

基于注解开发 AspectJ 要比基于 XML 配置开发 AspectJ 便捷许多，所以在实际开发中推荐使用注解方式。在讲解 AspectJ 之前，先了解一下 Spring 的通知类型。根据 Spring 中通知在目标类方法的连接点位置，可以分为以下 6 种类型。

1. 环绕通知

环绕通知是在目标方法执行前和执行后实施增强，可以应用于日志记录、事务处理等功能。

2. 前置通知

前置通知是在目标方法执行前实施增强,可以应用于权限管理等功能。

3. 后置返回通知

后置返回通知是在目标方法成功执行后实施增强,可以应用于关闭流、删除临时文件等功能。

4. 后置(最终)通知

后置通知是在目标方法执行后实施增强,与后置返回通知不同的是,不管是否发生异常都要执行该通知,可以应用于释放资源。

5. 异常通知

异常通知是在方法抛出异常后实施增强,可以应用于处理异常、记录日志等功能。

6. 引入通知

引入通知是在目标类中添加一些新的方法和属性,可以应用于修改目标类(增强类)。有关 AspectJ 注解如表 1.1 所示。

表 1.1 AspectJ 注解

注 解 名 称	描 述
@Aspect	用于定义一个切面,注解在切面类上
@Pointcut	用于定义切入点表达式。在使用时,需要定义一个切入点方法。该方法是一个返回值 void,且方法体为空的普通方法
@Before	用于定义前置通知。在使用时,通常为其指定 value 属性值,该值可以是已有的切入点,也可以直接定义切入点表达式
@AfterReturning	用于定义后置返回通知。在使用时,通常为其指定 value 属性值,该值可以是已有的切入点,也可以直接定义切入点表达式
@Around	用于定义环绕通知。在使用时,通常为其指定 value 属性值,该值可以是已有的切入点,也可以直接定义切入点表达式
@AfterThrowing	用于定义异常通知。在使用时,通常为其指定 value 属性值,该值可以是已有的切入点,也可以直接定义切入点表达式。另外,还有一个 throwing 属性用于访问目标方法抛出的异常,该属性值与异常通知方法中同名的形参一致
@After	用于定义后置(最终)通知。在使用时,通常为其指定 value 属性值,该值可以是已有的切入点,也可以直接定义切入点表达式

下面通过一个实例讲解基于注解开发 AspectJ 的过程。

【例 1-4】 基于注解开发 AspectJ 的过程。

具体实现步骤如下。

1. 使用 Eclipse 创建 Web 应用并导入 JAR 包

使用 Eclipse 创建一个名为 ch1_4 的 Dynamic Web project,并将 Spring 的 4 个基础包

和第三方依赖包 commons-logging-1.2.jar 复制到 ch1_4 的 WEB-INF/lib 目录中。除了导入上述 5 个 JAR 包,还需要再向 ch1_4 应用导入 spring-aspects-5.1.4.RELEASE.jar、aspectjweaver-1.9.2.jar 和 spring-aop-5.1.4.RELEASE.jar 等 JAR 包。ch1_4 应用导入的 JAR 包如图 1.16 所示。

spring-aspects-5.1.4.RELEASE.jar 是 Spring 为 AspectJ 提供的实现,Spring 的包中已提供。aspectjweaver-1.9.2.jar 是 AspectJ 框架所提供的规范包,可以通过地址 http://mvnrepository.com/artifact/org.aspectj/aspectjweaver 下载。

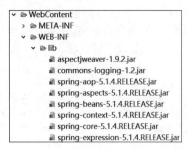

图 1.16 ch1_4 应用导入的 JAR 包

2. 创建接口及实现类

在 ch1_4 的 src 目录下,创建一个 aspectj.dao 包,在该包中创建接口 TestDao 和接口实现类 TestDaoImpl。该实现类作为目标类,在切面类中对其方法进行增强处理。使用注解 @Repository 将目标类 aspectj.dao.TestDaoImpl 注解为目标对象。

TestDao 的代码如下:

```
package aspectj.dao;
public interface TestDao {
    public void save();
    public void modify();
    public void delete();
}
```

TestDaoImpl 的代码如下:

```
package aspectj.dao;
@Repository("testDao")
public class TestDaoImpl implements TestDao{
    @Override
    public void save() {
        System.out.println("保存");
    }
    @Override
    public void modify() {
        System.out.println("修改");
    }
    @Override
    public void delete() {
        System.out.println("删除");
    }
}
```

3. 创建切面类

在 ch1_4 应用的 src 目录下,创建 aspectj.annotation 包,在该包中创建切面类 MyAspect。在该类中,首先使用@Aspect 注解定义一个切面类,由于该类在 Spring 中是作为组件使用的,所以还需要使用@Component 注解。然后,使用@Pointcut 注解切入点表达式,并通过定义方法来表示切入点名称。最后在每个通知方法上添加相应的注解,并将切入点名称作为参数传递给需要执行增强的通知方法。

MyAspect 的代码如下:

```java
package aspectj.annotation;
import org.aspectj.lang.JoinPoint;
import org.aspectj.lang.ProceedingJoinPoint;
import org.aspectj.lang.annotation.After;
import org.aspectj.lang.annotation.AfterReturning;
import org.aspectj.lang.annotation.AfterThrowing;
import org.aspectj.lang.annotation.Around;
import org.aspectj.lang.annotation.Aspect;
import org.aspectj.lang.annotation.Before;
import org.aspectj.lang.annotation.Pointcut;
import org.springframework.stereotype.Component;
/**
 * 切面类,在此类中编写各种类型通知
 */
@Aspect                              //@Aspect 声明一个切面
@Component                           //@Component 让此切面成为 Spring 容器管理的 Bean
public class MyAspect {
    /**
     * 定义切入点,通知增强哪些方法.
"execution( * aspectj.dao.*.*(..))" 是定义切入点表达式,
该切入点表达式的意思是匹配 aspectj.dao 包中任意类的任意方法的执行。
其中 execution()是表达式的主体,第一个 * 表示的是返回类型,使用 * 代表所有类型;
aspectj.dao 表示的是需要匹配的包名,后面第二个 * 表示的是类名,使用 * 代表匹配包中所有的类;
第三个 * 表示的是方法名,使用 * 表示所有方法;后面(..)表示方法的参数,其中".."表示任意参数。
另外,注意第一个 * 与包名之间有一个空格
     */
    @Pointcut("execution( * aspectj.dao.*.*(..))")
    private void myPointCut() {
    }
    /**
     * 前置通知,使用 Joinpoint 接口作为参数获得目标对象信息
     */
    @Before("myPointCut()")              //myPointCut()是切入点的定义方法
    public void before(JoinPoint jp) {
        System.out.print("前置通知:模拟权限控制");
        System.out.println(",目标类对象:" + jp.getTarget()
            + ",被增强处理的方法:" + jp.getSignature().getName());
    }
    /**
     * 后置返回通知
     */
    @AfterReturning("myPointCut()")
    public void afterReturning(JoinPoint jp) {
        System.out.print("后置返回通知:" + "模拟删除临时文件");
        System.out.println(",被增强处理的方法:" + jp.getSignature().getName());
```

```java
    }
    /**
     * 环绕通知
     * ProceedingJoinPoint 是 JoinPoint 子接口,代表可以执行的目标方法
     * 返回值类型必须是 Object
     * 必须一个参数是 ProceedingJoinPoint 类型
     * 必须 throws Throwable
     */
    @Around("myPointCut()")
    public Object around(ProceedingJoinPoint pjp) throws Throwable{
        //开始
        System.out.println("环绕开始:执行目标方法前,模拟开启事务");
        //执行当前目标方法
        Object obj = pjp.proceed();
        //结束
        System.out.println("环绕结束:执行目标方法后,模拟关闭事务");
        return obj;
    }
    /**
     * 异常通知
     */
    @AfterThrowing(value = "myPointCut()",throwing = "e")
    public void except(Throwable e) {
        System.out.println("异常通知:" + "程序执行异常" + e.getMessage());
    }
    /**
     * 后置(最终)通知
     */
    @After("myPointCut()")
    public void after() {
        System.out.println("最终通知:模拟释放资源");
    }
}
```

4. 创建配置类

在 ch1_4 应用的 src 目录下,创建 aspectj.config 包,在该包中创建配置类 AspectjAOPConfig。在该类中使用@Configuration 注解声明此类为配置类;使用@ComponentScan("aspectj")注解自动扫描 aspectj 包下使用的注解;使用@EnableAspectJAutoProxy 注解开启 Spring 对 AspectJ 的支持。

AspectjAOPConfig 的代码如下:

```java
package aspectj.config;
import org.springframework.context.annotation.ComponentScan;
import org.springframework.context.annotation.Configuration;
import org.springframework.context.annotation.EnableAspectJAutoProxy;
@Configuration                          //声明一个配置类
@ComponentScan("aspectj")               //自动扫描 aspectj 包下使用的注解
@EnableAspectJAutoProxy                 //开启 Spring 对 AspectJ 的支持
public class AspectjAOPConfig {
}
```

5. 创建测试类

在 ch1_4 应用的 aspectj. config 包中创建测试类 AOPTest。

AOPTest 的代码如下：

```
package aspectj.config;
import org.springframework.context.annotation.AnnotationConfigApplicationContext;
import aspectj.dao.TestDao;
public class AOPTest {
    public static void main(String[] args) {
        //初始化 Spring 容器 ApplicationContext
        AnnotationConfigApplicationContext appCon =
            new AnnotationConfigApplicationContext(AspectjAOPConfig.class);
        //从容器中获取增强后的目标对象
        TestDao testDaoAdvice = appCon.getBean(TestDao.class);
        //执行方法
        testDaoAdvice.save();
        System.out.println("================");
        testDaoAdvice.modify();
        System.out.println("================");
        testDaoAdvice.delete();
        appCon.close();
    }
}
```

6. 运行测试类

运行测试类 AOPTest 的 main 方法，运行结果如图 1.17 所示。

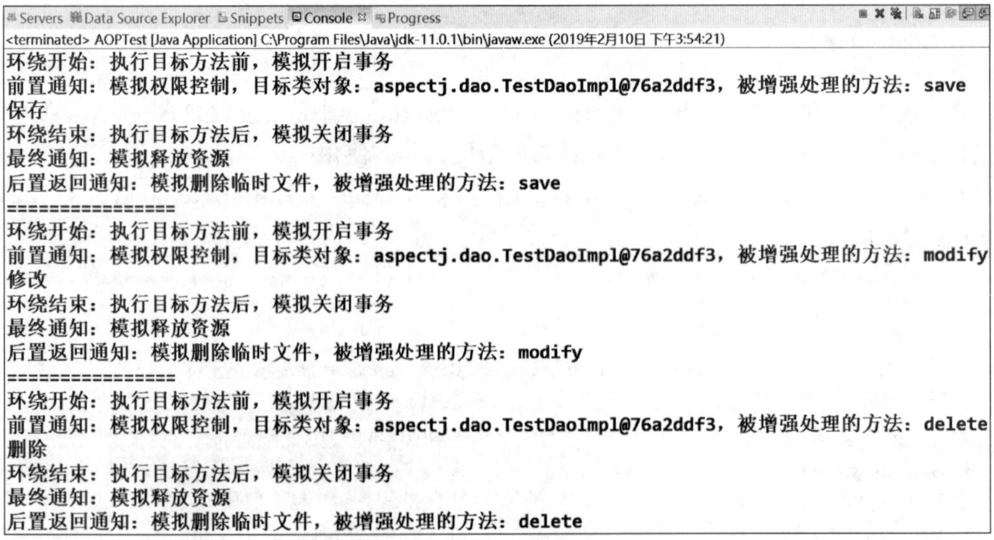
Servers Data Source Explorer Snippets Console Progress
<terminated> AOPTest [Java Application] C:\Program Files\Java\jdk-11.0.1\bin\javaw.exe (2019年2月10日 下午3:54:21)
环绕开始：执行目标方法前，模拟开启事务
前置通知：模拟权限控制，目标类对象：aspectj.dao.TestDaoImpl@76a2ddf3，被增强处理的方法：save
保存
环绕结束：执行目标方法后，模拟关闭事务
最终通知：模拟释放资源
后置返回通知：模拟删除临时文件，被增强处理的方法：save
================
环绕开始：执行目标方法前，模拟开启事务
前置通知：模拟权限控制，目标类对象：aspectj.dao.TestDaoImpl@76a2ddf3，被增强处理的方法：modify
修改
环绕结束：执行目标方法后，模拟关闭事务
最终通知：模拟释放资源
后置返回通知：模拟删除临时文件，被增强处理的方法：modify
================
环绕开始：执行目标方法前，模拟开启事务
前置通知：模拟权限控制，目标类对象：aspectj.dao.TestDaoImpl@76a2ddf3，被增强处理的方法：delete
删除
环绕结束：执行目标方法后，模拟关闭事务
最终通知：模拟释放资源
后置返回通知：模拟删除临时文件，被增强处理的方法：delete

图 1.17　ch1_4 应用的运行结果

1.5　Spring Bean

视频讲解

在 Spring 的应用中，Spring IoC 容器可以创建、装配和配置应用组件对象，这里的组件对象称为 Bean。

1.5.1　Bean 的实例化

在面向对象编程中，想使用某个对象时，需要事先实例化该对象。同样，在 Spring 框架中，想使用 Spring 容器中的 Bean，也需要实例化 Bean。Spring 框架实例化 Bean 有 3 种方式：构造方法实例化、静态工厂实例化和实例工厂实例化，其中，最常用的实例方法是构造方法实例化。

下面通过一个实例 ch1_5 来演示 Bean 的实例化过程。

【例 1-5】　Bean 的实例化过程。

具体实现步骤如下。

1. 使用 Eclipse 创建 Web 应用并导入 JAR 包

使用 Eclipse 创建一个名为 ch1_5 的 Dynamic Web project，并将 Spring 的 4 个基础包、第三方依赖包 commons-logging-1.2.jar 以及 spring-aop-5.1.4.RELEASE.jar 等 JAR 包复制到 ch1_5 的 WEB-INF/lib 目录中。

2. 创建实例化 Bean 的类

在 ch1_5 应用的 src 目录下，创建 instance 包，并在该包中创建 BeanClass、BeanInstanceFactory 以及 BeanStaticFactory 等实例化 Bean 的类。

BeanClass 的代码如下：

```
package instance;
public class BeanClass {
    public String message;
    public BeanClass() {
        message = "构造方法实例化 Bean";
    }
    public BeanClass(String s) {
        message = s;
    }
}
```

BeanInstanceFactory 的代码如下：

```
package instance;
public class BeanInstanceFactory {
    public BeanClass createBeanClassInstance() {
        return new BeanClass("调用实例工厂方法实例化 Bean");
    }
}
```

BeanStaticFactory 的代码如下：

```java
package instance;
public class BeanStaticFactory {
    private static BeanClass beanInstance = new BeanClass("调用静态工厂方法实例化 Bean");
    public static BeanClass createInstance() {
        return beanInstance;
    }
}
```

3. 创建配置类

在 ch1_5 应用的 src 目录下，创建 config 包，并在该包中创建配置类 JavaConfig。在该配置类中使用 @Bean 定义 3 个 Bean，具体代码如下：

```java
package config;
import org.springframework.context.annotation.Bean;
import org.springframework.context.annotation.Configuration;
import instance.BeanClass;
import instance.BeanInstanceFactory;
import instance.BeanStaticFactory;
@Configuration
public class JavaConfig {
    /**
     * 构造方法实例化
     */
    @Bean(value = "beanClass")              //value 可以省略
    public BeanClass getBeanClass() {
        return new BeanClass();
    }
    /**
     * 静态工厂实例化
     */
    @Bean(value = "beanStaticFactory")
    public BeanClass getBeanStaticFactory() {
        return BeanStaticFactory.createInstance();
    }
    /**
     * 实例工厂实例化
     */
    @Bean(value = "beanInstanceFactory")
    public BeanClass getBeanInstanceFactory() {
        BeanInstanceFactory bi = new BeanInstanceFactory();
        return bi.createBeanClassInstance();
    }
}
```

4. 创建测试类

在 ch1_5 应用的 config 包中，创建测试类 TestBean，在该类中测试配置类定义的 Bean，具体代码如下：

```java
package config;
import org.springframework.context.annotation.AnnotationConfigApplicationContext;
import instance.BeanClass;
public class TestBean {
    public static void main(String[] args) {
        //初始化 Spring 容器 ApplicationContext
        AnnotationConfigApplicationContext appCon =
            new AnnotationConfigApplicationContext(JavaConfig.class);
        BeanClass b1 = (BeanClass)appCon.getBean("beanClass");
        System.out.println(b1 + b1.message);
        BeanClass b2 = (BeanClass)appCon.getBean("beanStaticFactory");
        System.out.println(b2 + b2.message);
        BeanClass b3 = (BeanClass)appCon.getBean("beanInstanceFactory");
        System.out.println(b3 + b3.message);
        appCon.close();
    }
}
```

5．运行测试类

运行测试类 TestBean 的 main 方法，运行结果如图 1.18 所示。

图 1.18　ch1_5 应用的运行结果

1.5.2　Bean 的作用域

在 Spring 中，不仅可以完成 Bean 的实例化，还可以为 Bean 指定作用域。在 Spring 中为 Bean 的实例定义了如表 1.2 所示的作用域，通过 @Scope 注解来实现。

表 1.2　Bean 的作用域

作用域名称	描述
singleton	默认的作用域，使用 singleton 定义的 Bean 在 Spring 容器中只有一个 Bean 实例
prototype	Spring 容器每次获取 prototype 定义的 Bean，容器都将创建一个新的 Bean 实例
request	在一次 HTTP 请求中容器将返回一个 Bean 实例，不同的 HTTP 请求返回不同的 Bean 实例。仅在 Web Spring 应用程序上下文中使用
session	在一个 HTTP Session 中，容器将返回同一个 Bean 实例。仅在 Web Spring 应用程序上下文中使用
application	为每个 ServletContext 对象创建一个实例，即同一个应用共享一个 Bean 实例。仅在 Web Spring 应用程序上下文中使用
websocket	为每个 WebSocket 对象创建一个 Bean 实例。仅在 Web Spring 应用程序上下文中使用

在表 1.2 的 6 种作用域中，singleton 和 prototype 是最常用的两种，后面 4 种作用域仅使用在 Web Spring 应用程序上下文中。下面通过一个实例来演示 Bean 的作用域。

【例 1-6】 Bean 的作用域。

具体实现步骤如下。

1. 使用 Eclipse 创建 Web 应用并导入 JAR 包

使用 Eclipse 创建一个名为 ch1_6 的 Dynamic Web project，并将 Spring 的 4 个基础包、第三方依赖包 commons-logging-1.2.jar 以及 spring-aop-5.1.4.RELEASE.jar 等 JAR 包复制到 ch1_6 的 WEB-INF/lib 目录中。

2. 编写不同作用域的 Bean

在 ch1_6 应用的 src 目录下，创建 service 包，并在该包中创建 SingletonService 和 PrototypeService 类。在 SingletonService 类中，Bean 的作用域为默认作用域 singleton；在 PrototypeService 类中，Bean 的作用域为 prototype。

SingletonService 的代码如下：

```java
package service;
import org.springframework.stereotype.Service;
@Service                    //默认为 singleton 相当于@Scope("singleton")
public class SingletonService {
}
```

PrototypeService 的代码如下：

```java
package service;
import org.springframework.context.annotation.Scope;
import org.springframework.stereotype.Service;
@Service
@Scope("prototype")
public class PrototypeService {
}
```

3. 创建配置类

在 ch1_6 应用的 src 目录下，创建包 config，并在该包中创建配置类 ScopeConfig，具体代码如下：

```java
package config;
import org.springframework.context.annotation.ComponentScan;
import org.springframework.context.annotation.Configuration;
@Configuration
@ComponentScan("service")
public class ScopeConfig {
}
```

4. 创建测试类

在 ch1_6 应用的 config 包中，创建测试类 TestScope，在该测试类中分别获得

SingletonService 和 PrototypeService 的两个 Bean 实例，具体代码如下：

```java
package config;
import org.springframework.context.annotation.AnnotationConfigApplicationContext;
import service.PrototypeService;
import service.SingletonService;
public class TestScope {
    public static void main(String[] args) {
        //初始化 Spring 容器 ApplicationContext
        AnnotationConfigApplicationContext appCon =
            new AnnotationConfigApplicationContext(ScopeConfig.class);
        SingletonService ss1 = appCon.getBean(SingletonService.class);
        SingletonService ss2 = appCon.getBean(SingletonService.class);
        System.out.println(ss1);
        System.out.println(ss2);
        PrototypeService ps1 = appCon.getBean(PrototypeService.class);
        PrototypeService ps2 = appCon.getBean(PrototypeService.class);
        System.out.println(ps1);
        System.out.println(ps2);
        appCon.close();
    }
}
```

5．运行测试类

运行测试类 TestScope 的 main 方法，运行结果如图 1.19 所示。

从图 1.19 运行结果可以得知，两次获取 SingletonService 的 Bean 实例时，IoC 容器返回两个相同的 Bean 实例；而两次获取 PrototypeService 的 Bean 实例时，IoC 容器返回两个不同的 Bean 实例。

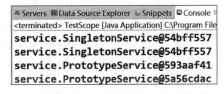

图 1.19　ch1_6 应用的运行结果

1.5.3　Bean 的初始化和销毁

在实际工程应用中，经常需要在 Bean 使用之前或之后做些必要的操作，Spring 对 Bean 的生命周期的操作提供了支持。可以使用@Bean 注解的 initMethod 和 destroyMethod 属性（相当于 XML 配置的 init-method 和 destroy-method）对 Bean 进行初始化和销毁。下面通过一个实例来演示 Bean 的初始化和销毁。

【例 1-7】　Bean 的初始化和销毁。
具体实现步骤如下。

1．使用 Eclipse 创建 Web 应用并导入 JAR 包

使用 Eclipse 创建一个名为 ch1_7 的 Dynamic Web project，并将 Spring 的 4 个基础包、第三方依赖包 commons-logging-1.2.jar 以及 spring-aop-5.1.4.RELEASE.jar 等 JAR 包复制到 ch1_7 的 WEB-INF/lib 目录中。

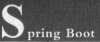

2. 创建 Bean 的类

在 ch1_7 应用的 src 目录下,创建 service 包,并在该包中创建 MyService 类,具体代码如下:

```java
package service;
public class MyService {
    public void initService() {
        System.out.println("initMethod");
    }
    public MyService() {
        System.out.println("构造方法");
    }
    public void destroyService() {
        System.out.println("destroyMethod");
    }
}
```

3. 创建配置类

在 ch1_7 应用的 src 目录下,创建 config 包,并在该包中创建配置类 JavaConfig,具体代码如下:

```java
package config;
import org.springframework.context.annotation.Bean;
import org.springframework.context.annotation.ComponentScan;
import org.springframework.context.annotation.Configuration;
import service.MyService;
@Configuration
public class JavaConfig {
    //initMethod 和 destroyMethod 指定 MyService 类的 initService 和 destroyService 方法
    //在构造方法之后、销毁之前执行
    @Bean(initMethod = "initService", destroyMethod = "destroyService")
    public MyService getMyService() {
        return new MyService();
    }
}
```

4. 创建测试类

在 ch1_7 应用的 config 包中,创建测试类 TestInitAndDestroy,具体代码如下:

```java
package config;
import org.springframework.context.annotation.AnnotationConfigApplicationContext;
import service.MyService;
public class TestInitAndDestroy {
    public static void main(String[] args) {
        //初始化 Spring 容器 ApplicationContext
        AnnotationConfigApplicationContext appCon =
            new AnnotationConfigApplicationContext(JavaConfig.class);
```

```
            MyService ms =        appCon.getBean(MyService.class);
            appCon.close();
    }
}
```

5．运行测试类

运行测试类 TestInitAndDestroy 的 main 方法，运行结果如图 1.20 所示。

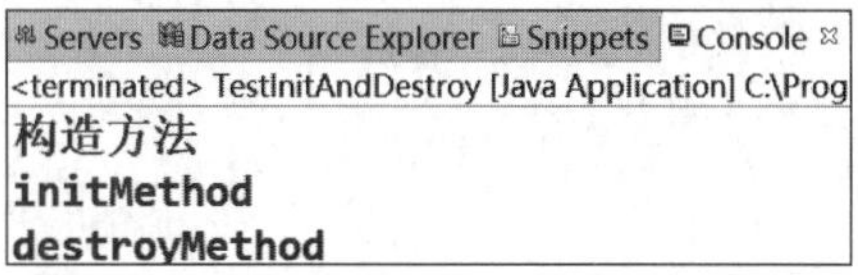

图 1.20 ch1_7 应用的运行结果

1.6 Spring 的数据库编程

视频讲解

数据库编程是互联网编程的基础，Spring 框架为开发者提供了 JDBC 模板模式，即 jdbcTemplate，它可以简化许多代码，但在实际应用中 jdbcTemplate 并不常用。工作更多的时候，用的是 Hibernate 框架和 MyBatis 框架进行数据库编程。本节仅简要介绍 Spring jdbcTemplate 的使用方法，而 Hibernate 框架和 MyBatis 框架的相关内容不属于本节的内容。

1.6.1 Spring JDBC 的 XML 配置

本节主要使用 Spring JDBC 模块的 core 和 dataSource 包进行 Spring 数据库编程。core 包是 JDBC 的核心功能包，包括常用的 JdbcTemplate 类；dataSource 包是访问数据源的工具类包。使用 Spring JDBC 操作数据库，需要对其进行配置。XML 配置文件示例代码如下：

```
<!-- 配置数据源 -->
<bean id = "dataSource" class = "org.springframework.jdbc.datasource.DriverManagerDataSource">
    <!-- MySQL 数据库驱动 -->
    <property name = "driverClassName" value = "com.mysql.jdbc.Driver"/>
    <!-- 连接数据库的 URL -->
    <property name = " url"  value = " jdbc: mysql://localhost: 3306/springtest?characterEncoding = utf8"/>
    <!-- 连接数据库的用户名 -->
    <property name = "username" value = "root"/>
    <!-- 连接数据库的密码 -->
    <property name = "password" value = "root"/>
</bean>
<!-- 配置 JDBC 模板 -->
<bean id = "jdbcTemplate" class = "org.springframework.jdbc.core.JdbcTemplate">
    <property name = "dataSource" ref = "dataSource"/>
</bean>
```

上述示例代码中，配置 JDBC 模板时，需要将 dataSource 注入 jdbcTemplate，而在数据访问层（如 Dao 类）使用 jdbcTemplate 时，也需要将 jdbcTemplate 注入对应的 Bean 中。代码如下：

```
…
@Repository
public class TestDaoImpl implements TestDao{
    @Autowired
    //使用配置文件中的 JDBC 模板
    private JdbcTemplate jdbcTemplate;
    …
}
```

1.6.2　Spring JDBC 的 Java 配置

与 1.6.1 节 XML 配置文件内容等价的 Java 配置示例代码如下：

```
package config;
import org.springframework.beans.factory.annotation.Value;
import org.springframework.context.annotation.Bean;
import org.springframework.context.annotation.ComponentScan;
import org.springframework.context.annotation.Configuration;
import org.springframework.context.annotation.PropertySource;
import org.springframework.jdbc.core.JdbcTemplate;
import org.springframework.jdbc.datasource.DriverManagerDataSource;
@Configuration        //通过该注解来表明该类是一个 Spring 的配置,相当于一个 xml 文件
@ComponentScan(basePackages = "dao")      //配置扫描包
@PropertySource(value = {"classpath:jdbc.properties"},ignoreResourceNotFound = true)
//配置多个属性文件时 value = {"classpath:jdbc.properties","xx","xxx"}
public class SpringJDBCConfig {
    @Value("${jdbc.url}")              //注入属性文件 jdbc.properties 中的 jdbc.url
    private String jdbcUrl;
    @Value("${jdbc.driverClassName}")
    private String jdbcDriverClassName;
    @Value("${jdbc.username}")
    private String jdbcUsername;
    @Value("${jdbc.password}")
    private String jdbcPassword;
    /**
     * 配置数据源
     */
    @Bean
    public DriverManagerDataSource dataSource() {
        DriverManagerDataSource myDataSource = new DriverManagerDataSource();
        //数据库驱动
        myDataSource.setDriverClassName(jdbcDriverClassName);;
        //相应驱动的 jdbcUrl
        myDataSource.setUrl(jdbcUrl);
        //数据库的用户名
```

```
            myDataSource.setUsername(jdbcUsername);
            //数据库的密码
            myDataSource.setPassword(jdbcUsername);
            return myDataSource;
    }
    /**
     * 配置 JdbcTemplate
     */
    @Bean(value = "jdbcTemplate")
    public JdbcTemplate getJdbcTemplate() {
    return new JdbcTemplate(dataSource());
    }
}
```

上述 Java 配置示例中,需要事先在 classpath 目录(如应用的 src 目录)下创建属性文件,代码如下:

```
jdbc.driverClassName = com.mysql.jdbc.Driver
jdbc.url = jdbc:mysql://localhost:3306/springtest?characterEncoding = utf8
jdbc.username = root
jdbc.password = root
```

另外,在数据访问层(如 Dao 类)使用 jdbcTemplate 时,也需要将 jdbcTemplate 注入对应的 Bean 中,代码如下:

```
…
@Repository
public class TestDaoImpl implements TestDao{
        @Autowired
        //使用配置文件中的 JDBC 模板
        private JdbcTemplate jdbcTemplate;
        …
}
```

1.6.3 Spring JdbcTemplate 的常用方法

获取 JDBC 模板后,如何使用它是本节将要讲述的内容。首先,需要了解 JdbcTemplate 类的常用方法——update()和 query()方法。

- public int update(String sql,Object args[])

该方法可以对数据表进行增加、修改、删除等操作。使用 args[]设置 SQL 语句中的参数,并返回更新的行数。代码如下:

```
String insertSql = "insert into user values(null,?,?)";
Object param1[] = {"chenheng1", "男"};
jdbcTemplate.update(sql, param1);
```

- public List < T > query (String sql, RowMapper < T > rowMapper, Object args[])

该方法可以对数据表进行查询操作。rowMapper 将结果集映射到用户自定义的类中(前提是自定义类中的属性要与数据表的字段对应)。代码如下:

```
String selectSql = "select * from user";
RowMapper<MyUser> rowMapper = new BeanPropertyRowMapper<MyUser>(MyUser.class);
List<MyUser> list = jdbcTemplate.query(sql, rowMapper, null);
```

下面通过一个实例演示 Spring JDBC 的使用过程。

【例 1-8】 Spring JDBC 的使用过程。

具体实现步骤如下。

1. 使用 Eclipse 创建 Web 应用并导入 JAR 包

使用 Eclipse 创建一个名为 ch1_8 的 Dynamic Web project，并将 Spring 的 4 个基础包、第三方依赖包 commons-logging-1.2.jar、spring-aop-5.1.4.RELEASE.jar、mysql-connector-java-5.1.45-bin.jar（MySQL 数据库的驱动 JAR 包）、spring-jdbc-5.1.4.RELEASE.jar（Spring JDBC 的 JAR 包）以及 spring-tx-5.1.4.RELEASE.jar（Spring 事务处理的 JAR 包）等 JAR 包复制到 ch1_8 的 WEB-INF/lib 目录中。ch1_8 应用导入的 JAR 包如图 1.21 所示。

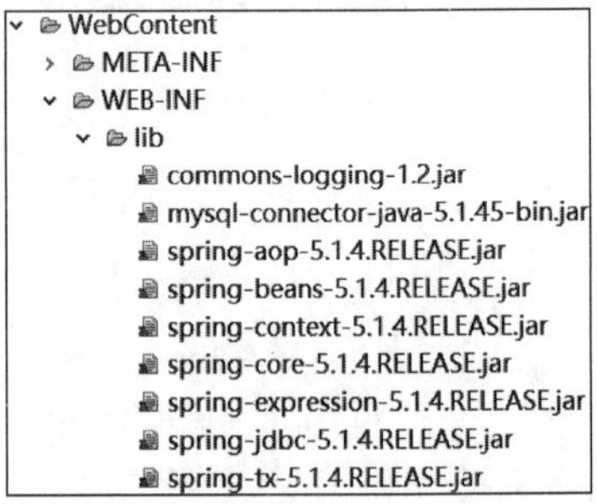

图 1.21　ch1_8 应用导入的 JAR 包

2. 创建属性文件与配置类

本书使用 MySQL 数据库演示有关数据库访问的内容，因此，需要在 ch1_8 应用的 src 目录下，创建数据库配置的属性文件 jdbc.properties，具体内容如下：

```
jdbc.driverClassName=com.mysql.jdbc.Driver
jdbc.url=jdbc:mysql://localhost:3306/springtest?characterEncoding=utf8
jdbc.username=root
jdbc.password=root
```

在 ch1_8 应用的 src 目录下，创建 config 包，并在该包中创建配置类 SpringJDBCConfig。在该配置类中使用@PropertySource 注解读取属性文件 jdbc.properties，并配置数据源和 JdbcTemplate，具体代码如下：

```
package config;
import org.springframework.beans.factory.annotation.Value;
import org.springframework.context.annotation.Bean;
import org.springframework.context.annotation.ComponentScan;
import org.springframework.context.annotation.Configuration;
import org.springframework.context.annotation.PropertySource;
import org.springframework.jdbc.core.JdbcTemplate;
import org.springframework.jdbc.datasource.DriverManagerDataSource;
@Configuration    //通过该注解来表明该类是一个 Spring 的配置,相当于一个 xml 文件
@ComponentScan(basePackages = {"dao","service"})    //配置扫描包
@PropertySource(value = {"classpath:jdbc.properties"},ignoreResourceNotFound = true)
//配置多个配置文件 value = {"classpath:jdbc.properties","xx","xxx"}
public class SpringJDBCConfig {
    @Value("${jdbc.url}")            //注入属性文件 jdbc.properties 中的 jdbc.url
    private String jdbcUrl;
```

```
    @Value(" ${jdbc.driverClassName}")
    private String jdbcDriverClassName;
    @Value(" ${jdbc.username}")
    private String jdbcUsername;
    @Value(" ${jdbc.password}")
    private String jdbcPassword;
    /**
     * 配置数据源
     */
    @Bean
    public DriverManagerDataSource dataSource() {
        DriverManagerDataSource myDataSource = new DriverManagerDataSource();
        //数据库驱动
        myDataSource.setDriverClassName(jdbcDriverClassName);;
        //相应驱动的 jdbcUrl
        myDataSource.setUrl(jdbcUrl);
        //数据库的用户名
        myDataSource.setUsername(jdbcUsername);
        //数据库的密码
        myDataSource.setPassword(jdbcUsername);
        return myDataSource;
    }
    /**
     * 配置 JdbcTemplate
     */
    @Bean(value = "jdbcTemplate")
    public JdbcTemplate getJdbcTemplate() {
      return new JdbcTemplate(dataSource());
    }
}
```

3. 创建数据表与实体类

使用 Navicat for MySQL 创建数据库 springtest，并在该数据库中创建数据表 user，数据表 user 的结构如图 1.22 所示。

名	类型	长度	小数点	允许空值 (	
uid	int	10	0	☐	1
uname	varchar	20	0	☑	
usex	varchar	10	0	☑	

图 1.22　user 表的结构

在 ch1_8 应用的 src 目录下，创建包 entity，在该包中创建实体类 MyUser，具体代码如下：

```
package entity;
public class MyUser {
    private Integer uid;
    private String uname;
    private String usex;
    //省略 set 和 get 方法
    public String toString() {
        return "myUser [uid = " + uid +", uname = " + uname + ", usex = " + usex + "]";
    }
}
```

4. 创建数据访问层

在 ch1_8 应用的 src 目录下，创建 dao 包，在该包中创建数据访问接口 TestDao 和接口实现类 TestDaoImpl。在实现类 TestDaoImpl 中使用@Repository 注解标注此类为数据访问层，并使用@Autowired 注解依赖注入 JdbcTemplate。

TestDao 的代码如下：

```java
package dao;
import java.util.List;
import entity.MyUser;
public interface TestDao {
    public int update(String sql, Object[] param);
    public List<MyUser> query(String sql, Object[] param);
}
```

TestDaoImpl 的代码如下：

```java
package dao;
import java.util.List;
import org.springframework.beans.factory.annotation.Autowired;
import org.springframework.jdbc.core.BeanPropertyRowMapper;
import org.springframework.jdbc.core.JdbcTemplate;
import org.springframework.jdbc.core.RowMapper;
import org.springframework.stereotype.Repository;
import entity.MyUser;
@Repository
public class TestDaoImpl implements TestDao{
    @Autowired
    //使用配置类中的 JDBC 模板
    private JdbcTemplate jdbcTemplate;
    /**
     * 更新方法,包括添加、修改、删除
     * param 为 sql 中的参数,如通配符?
     */
    @Override
    public int update(String sql, Object[] param) {
        return jdbcTemplate.update(sql, param);
    }
    /**
     * 查询方法
     * param 为 sql 中的参数,如通配符?
     */
    @Override
    public List<MyUser> query(String sql, Object[] param) {
        RowMapper<MyUser> rowMapper = new BeanPropertyRowMapper<MyUser>(MyUser.class);
        return jdbcTemplate.query(sql, rowMapper);
    }
}
```

5. 创建业务逻辑层

在 ch1_8 应用的 src 目录下，创建 service 包，在该包中创建数据访问接口 TestService 和接口实现类 TestServiceImpl。在实现类 TestServiceImpl 中使用@Service 注解标注此类为业务逻辑层，并使用@Autowired 注解依赖注入 TestDao。

TestService 的代码如下：

```
package service;
public interface TestService {
    public void testJDBC();
}
```

TestServiceImpl 的代码如下：

```
package service;
import java.util.List;
import org.springframework.beans.factory.annotation.Autowired;
import org.springframework.stereotype.Service;
import dao.TestDao;
import entity.MyUser;
@Service
public class TestServiceImpl implements TestService{
    @Autowired
    public TestDao testDao;
    @Override
    public void testJDBC() {
        String insertSql = "insert into user values(null,?,?)";
        //数组 param 的值与 insertSql 语句中的?一一对应
        Object param1[] = {"chenheng1","男"};
        Object param2[] = {"chenheng2","女"};
        Object param3[] = {"chenheng3","男"};
        Object param4[] = {"chenheng4","女"};
        //添加用户
        testDao.update(insertSql, param1);
        testDao.update(insertSql, param2);
        testDao.update(insertSql, param3);
        testDao.update(insertSql, param4);
        //查询用户
        String selectSql = "select * from user";
        List<MyUser> list = testDao.query(selectSql, null);
        for(MyUser mu : list) {
            System.out.println(mu);
        }
    }
}
```

6. 创建测试类

在 ch1_8 应用的 config 包中，创建测试类 TestJDBC，具体代码如下：

```
package config;
import org.springframework.context.annotation.AnnotationConfigApplicationContext;
import service.TestService;
public class TestJDBC {
    public static void main(String[] args) {
        //初始化 Spring 容器 ApplicationContext
        AnnotationConfigApplicationContext appCon =
            new AnnotationConfigApplicationContext(SpringJDBCConfig.class);
        TestService ts = appCon.getBean(TestService.class);
        ts.testJDBC();
        appCon.close();
    }
}
```

7. 运行测试类

运行测试类 TestJDBC 的 main 方法，运行结果如图 1.23 所示。

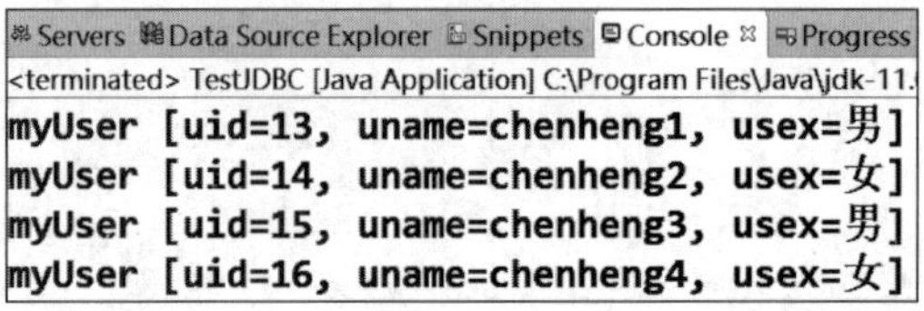

```
myUser [uid=13, uname=chenheng1, usex=男]
myUser [uid=14, uname=chenheng2, usex=女]
myUser [uid=15, uname=chenheng3, usex=男]
myUser [uid=16, uname=chenheng4, usex=女]
```

图 1.23　ch1_8 应用的运行结果

1.6.4　基于@Transactional 注解的声明式事务管理

Spring 的声明式事务管理，是通过 AOP 技术实现的事务管理，其本质是对方法前后进行拦截，然后在目标方法开始之前创建或者加入一个事务，在执行完目标方法之后根据执行情况提交或者回滚事务。

声明式事务管理最大的优点是不需要通过编程的方式管理事务，因而不需要在业务逻辑代码中掺杂事务处理的代码，只需相关的事务规则声明，便可以将事务规则应用到业务逻辑中。通常情况下，在开发中使用声明式事务处理，不仅因为其简单，更主要是因为这样使得纯业务代码不被污染，极大地方便了后期的代码维护。

和编程式事务管理相比，声明式事务管理唯一不足的地方是，最细粒度只能作用到方法级别，无法做到像编程式事务管理那样可以作用到代码块级别。但即便有这样的需求，也可以通过变通的方法进行解决，例如，可以将需要进行事务处理的代码块独立为方法。Spring 的声明式事务管理可以通过两种方式来实现，一是基于 XML 的方式，一是基于@Transactional 注解的方式。

@Transactional 注解可以作用于接口、接口方法、类以及类方法上。当作用于类上时，该类的所有 public 方法都将具有该类型的事务属性，同时，也可以在方法级别使用该注解来覆盖类级别的定义。虽然@Transactional 注解可以作用于接口、接口方法、类以及类方法上，但是 Spring 小组建议不要在接口或者接口方法上使用该注解，因为只有在使用基于接口的代理时它才会生效。可以使用@Transactional 注解的属性定制事务行为，具体属性如表1.3所示。

表1.3　@Transactional 的属性

属　　性	属性值含义	默　认　值
propagation	Propagation 定义了事务的生命周期,主要有以下选项。 ① Propagation.REQUIRED：需要事务支持的方法 A 被调用时,没有事务新建一个事务。当在方法 A 中调用另一个方法 B 时,方法 B 将使用相同的事务。如果方法 B 发生异常需要数据回滚时,整个事务数据回滚。 ② Propagation.REQUIRES_NEW：对于方法 A 和 B,在方法调用时,无论是否有事务都开启一个新的事务；方法 B 有异常不会导致方法 A 的数据回滚。 ③ Propagation.NESTED：和 Propagation.REQUIRES_NEW 类似,仅支持 JDBC,不支持 JPA 或 Hibernate。 ④ Propagation.SUPPORTS：方法调用时有事务就使用事务,没有事务就不创建事务。 ⑤ Propagation.NOT_SUPPORTED：强制方法在事务中执行,若有事务,在方法调用到结束阶段事务都将会被挂起。 ⑥ Propagation.NEVER：强制方法不在事务中执行,若有事务则抛出异常。 ⑦ Propagation.MANDATORY：强制方法在事务中执行,若无事务则抛出异常	Propagation.REQUIRED
isolation	Isolation(隔离)决定了事务的完整性,处理在多事务对相同数据下的处理机制,主要包含以下隔离级别(前提是当前数据库是否支持)： ① Isolation.READ_UNCOMMITTED：对于在事务 A 中修改了一条记录但没有提交的事务,在事务 B 中可以读取到修改后的记录。可导致脏读、不可重复读以及幻读。 ② Isolation.READ_COMMITTED：只有当在事务 A 中修改了一条记录且提交事务后,事务 B 才可以读取到提交后的记录,防止脏读,但可能导致不可重复读和幻读。 ③ Isolation.REPEATABLE_READ：不仅能实现 Isolation.READ_COMMITTED 的功能,而且还能阻止当事务 A 读取了一条记录,事务 B 将不允许修改该条记录；阻止脏读和不可重复读,但可出现幻读。 ④ Isolation.SERIALIZABLE：此级别下事务是顺序执行的,可以避免上述级别的缺陷,但开销较大。 ⑤ Isolation.DEFAULT：使用当前数据库的默认隔离级别。如 Oracle 和 SQL Server 是 READ_COMMITTED；MySQL 是 REPEATABLE_READ	Isolation.DEFAULT

续表

属　　性	属性值含义	默　认　值
timeout	timeout 指定事务过期时间,默认为当前数据库的事务过期时间	
readOnly	指定当前事务是否是只读事务	false
rollbackFor	指定哪个或哪些异常可以引起事务回滚(Class 对象数组,必须继承自 Throwable)	Throwable 的子类
rollbackForClassName	指定哪个或哪些异常可以引起事务回滚(类名数组,必须继承自 Throwable)	Throwable 的子类
noRollbackFor	指定哪个或哪些异常不可以引起事务回滚(Class 对象数组,必须继承自 Throwable)	Throwable 的子类
noRollbackForClassName	指定哪个或哪些异常不可以引起事务回滚(类名数组,必须继承自 Throwable)	Throwable 的子类

本节通过实例来演示基于@Transactional 注解的声明式事务管理。

【例 1-9】 基于@Transactional 注解的声明式事务管理。例 1-9 是通过修改例 1-8 中的代码实现的。

具体实现步骤如下。

1. 修改配置类

在配置类中,使用@EnableTransactionManagement 注解开启声明式事务的支持,同时为数据源添加事务管理器。修改后的配置类代码如下:

```
package config;
import org.springframework.beans.factory.annotation.Value;
import org.springframework.context.annotation.Bean;
import org.springframework.context.annotation.ComponentScan;
import org.springframework.context.annotation.Configuration;
import org.springframework.context.annotation.PropertySource;
import org.springframework.jdbc.core.JdbcTemplate;
import org.springframework.jdbc.datasource.DataSourceTransactionManager;
import org.springframework.jdbc.datasource.DriverManagerDataSource;
import org.springframework.transaction.annotation.EnableTransactionManagement;
@Configuration        //通过该注解来表明该类是一个 Spring 的配置,相当于一个 xml 文件
@ComponentScan(basePackages = {"dao","service"})        //配置扫描包
@PropertySource(value = {"classpath:jdbc.properties"},ignoreResourceNotFound = true)
@EnableTransactionManagement        //开启声明式事务的支持
//配置多个配置文件 value = {"classpath:jdbc.properties","xx","xxx"}
public class SpringJDBCConfig {
    @Value("${jdbc.url}")                //注入属性文件 jdbc.properties 中的 jdbc.url
    private String jdbcUrl;
    @Value("${jdbc.driverClassName}")
    private String jdbcDriverClassName;
    @Value("${jdbc.username}")
```

```java
    private String jdbcUsername;
    @Value("${jdbc.password}")
    private String jdbcPassword;
    /**
     * 配置数据源
     */
    @Bean
    public DriverManagerDataSource dataSource() {
        DriverManagerDataSource myDataSource = new DriverManagerDataSource();
        //数据库驱动
        myDataSource.setDriverClassName(jdbcDriverClassName);;
        //相应驱动的jdbcUrl
        myDataSource.setUrl(jdbcUrl);
        //数据库的用户名
        myDataSource.setUsername(jdbcUsername);
        //数据库的密码
        myDataSource.setPassword(jdbcUsername);
        return myDataSource;
    }
    /**
     * 配置JdbcTemplate
     */
    @Bean(value = "jdbcTemplate")
    public JdbcTemplate getJdbcTemplate() {
        return new JdbcTemplate(dataSource());
    }
    /**
     * 为数据源添加事务管理器
     */
    @Bean
    public DataSourceTransactionManager transactionManager() {
        DataSourceTransactionManager dt = new DataSourceTransactionManager();
        dt.setDataSource(dataSource());
        return dt;
    }
}
```

2. 修改业务逻辑层

在实际开发中，通常通过 Service 层进行事务管理，因此需要为 Service 层添加 @Transactional 注解。

添加 @Transactional 注解后的 TestServiceImpl 类的代码如下：

```java
package service;
import java.util.List;
import org.springframework.beans.factory.annotation.Autowired;
import org.springframework.stereotype.Service;
import org.springframework.transaction.annotation.Transactional;
import dao.TestDao;
import entity.MyUser;
```

```java
@Service
@Transactional
public class TestServiceImpl implements TestService{
    @Autowired
    public TestDao testDao;
    @Override
    public void testJDBC() {
        String insertSql = "insert into user values(null,?,?)";
        //数组 param 的值与 insertSql 语句中的?一一对应
        Object param1[] = {"chenheng1", "男"};
        Object param2[] = {"chenheng2", "女"};
        Object param3[] = {"chenheng3", "男"};
        Object param4[] = {"chenheng4", "女"};
        String insertSql1 = "insert into user values(?,?,?)";
        Object param5[] = {1,"chenheng5", "女"};
        Object param6[] = {1,"chenheng6", "女"};
        //添加用户
        testDao.update(insertSql, param1);
        testDao.update(insertSql, param2);
        testDao.update(insertSql, param3);
        testDao.update(insertSql, param4);
        //添加两个 ID 相同的用户,出现唯一性约束异常,使事务回滚
        testDao.update(insertSql1, param5);
        testDao.update(insertSql1, param6);
        //查询用户
        String selectSql = "select * from user";
        List<MyUser> list = testDao.query(selectSql, null);
        for(MyUser mu : list) {
            System.out.println(mu);
        }
    }
}
```

1.6.5 如何在事务处理中捕获异常

声明式事务处理的流程是:

(1) Spring 根据配置完成事务定义,设置事务属性。

(2) 执行开发者的代码逻辑。

(3) 如果开发者的代码产生异常(如主键重复)并且满足事务回滚的配置条件,则事务回滚;否则,事务提交。

(4) 事务资源释放。

现在的问题是,如果开发者在代码逻辑中加入了 try…catch…语句,Spring 还能不能在声明式事务处理中正常得到事务回滚的异常信息? 答案是不能。例如,我们将 1.6.4 节中 TestServiceImpl 实现类的 testJDBC 方法的代码修改如下:

```java
@Override
public void testJDBC () {
```

```
        String insertSql = "insert into user values(null,?,?)";
        //数组 param 的值与 insertSql 语句中的?一一对应
        Object param1[] = {"chenheng1", "男"};
        Object param2[] = {"chenheng2", "女"};
        Object param3[] = {"chenheng3", "男"};
        Object param4[] = {"chenheng4", "女"};
        String insertSql1 = "insert into user values(?,?,?)";
        Object param5[] = {1,"chenheng5", "女"};
        Object param6[] = {1,"chenheng6", "女"};
            try {
            //添加用户
            testDao.update(insertSql, param1);
            testDao.update(insertSql, param2);
            testDao.update(insertSql, param3);
            testDao.update(insertSql, param4);
            //添加两个 ID 相同的用户,出现唯一性约束异常,使事务回滚
            testDao.update(insertSql1, param5);
            testDao.update(insertSql1, param6);
            //查询用户
            String selectSql = "select * from user";
            List<MyUser> list = testDao.query(selectSql, null);
            for(MyUser mu : list) {
                System.out.println(mu);
            };
        } catch (Exception e) {
            System.out.println("主键重复,事务回滚。");
        }
    }
```

这时,再运行测试类,发现主键重复但事务并没有回滚。这是因为默认情况下,Spring 只在发生未被捕获的 RuntimeExcetpion 时才回滚事务。现在,如何在事务处理中捕获异常呢? 具体修改如下:

(1) 修改@Transactional 注解。

需要将 TestServiceImpl 类中的@Transactional 注解修改为:

```
@Transactional(rollbackFor = {Exception.class})
//rollbackFor 指定回滚生效的异常类,多个异常类逗号分隔;
//noRollbackFor 指定回滚失效的异常类
```

(2) 在 catch 语句中添加"throw new RuntimeException();"语句。

注意:在实际工程应用中,在 catch 语句中添加"TransactionAspectSupport. currentTransactionStatus().setRollbackOnly();"语句即可。也就是说,不需要在@Transaction 注解中添加 rollbackFor 属性。

1.7 本章小结

本章讲解了 Spring IoC、AOP、Bean 以及事务管理等基础知识,目的是让读者在学习 Spring Boot 之前,对 Spring 有个简要了解。

习题 1

1. Spring 的核心容器由哪些模块组成？
2. 如何找到 Spring 框架的官方 API？
3. 什么是 Spring IoC？什么是依赖注入？
4. 在 Java 配置类中如何开启 Spring 对 AspectJ 的支持？又如何开启 Spring 对声明式事务的支持？
5. 什么是 Spring AOP？它与 OOP 是什么关系？

第 2 章 Spring MVC 基础

学习目的与要求

本章重点讲解 Spring MVC 的工作原理、控制器以及数据绑定。通过本章的学习，了解 Spring MVC 的工作原理，掌握 Spring MVC 应用的开发步骤。

本章主要内容

- Spring MVC 的工作原理。
- Spring MVC 的工作环境。
- 基于注解的控制器。
- 表单标签库与数据绑定。
- JSON 数据交互。
- Spring MVC 的基本配置。

MVC 思想将一个应用分成 3 个基本部分：Model（模型）、View（视图）和 Controller（控制器），这 3 个部分以最低的耦合进行协同工作，从而提高应用的可扩展性及可维护性。Spring MVC 是一款优秀的基于 MVC 思想的应用框架，它是 Spring 提供的一个实现了 Web MVC 设计模式的轻量级 Web 框架。

2.1 Spring MVC 的工作原理

Spring MVC 框架是高度可配置的，包含多种视图技术，如 JSP 技术、Velocity、Tiles、iText 和 POI。Spring MVC 框架并不关心使用的视图技术，也不会强迫开发者只使用 JSP

技术，但本书使用的视图是JSP。

Spring MVC框架主要由DispatcherServlet、处理器映射、控制器、视图解析器、视图组成，其工作原理如图2.1所示。

从图2.1可总结出Spring MVC的工作流程如下：

（1）客户端请求提交到DispatcherServlet。

（2）由DispatcherServlet控制器寻找一个或多个HandlerMapping，找到处理请求的Controller。

（3）DispatcherServlet将请求提交到Controller。

（4）Controller调用业务逻辑处理后，返回ModelAndView。

（5）DispatcherServlet寻找一个或多个ViewResoler视图解析器，找到ModelAndView指定的视图。

（6）视图负责将结果显示到客户端。

图2.1中包含4个Spring MVC接口：DispatcherServlet、HandlerMapping、Controller和ViewResoler。

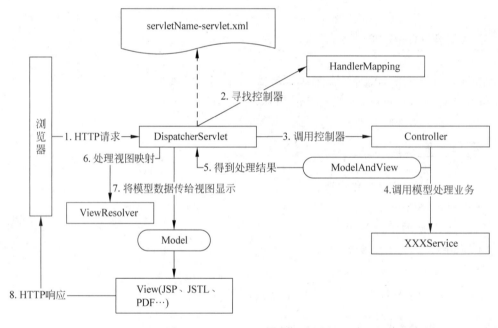

图2.1 Spring MVC工作原理图

Spring MVC所有的请求都经过DispatcherServlet来统一分发。DispatcherServlet将请求分发给Controller之前，需要借助Spring MVC提供的HandlerMapping定位到具体的Controller。

HandlerMapping接口负责完成客户请求到Controller映射。

Controller接口将处理用户请求，这和Java Servlet扮演的角色是一致的。一旦Controller处理完用户请求，则返回ModelAndView对象给DispatcherServlet前端控制器，ModelAndView中包含了模型（Model）和视图（View）。从宏观角度考虑，DispatcherServlet是整个Web应用的控制器；从微观角度考虑，Controller是单个Http请求处理过程中的控制器,而 ModelAndView 是 Http 请求过程中返回的模型(Model)和视图(View)。

ViewResolver 接口(视图解析器)在 Web 应用中负责查找 View 对象,从而将相应结果渲染给客户。

2.2 Spring MVC 的工作环境

2.2.1 Spring MVC 所需要的 JAR 包

在第 1 章 Java Web 开发环境的基础上,导入 Spring MVC 的相关 JAR 包,即可开发 Spring MVC 应用。

对于 Spring MVC 框架的初学者,开发 Spring MVC 应用时,只需要将 Spring 的 4 个基础包、commons-logging-1.2.jar、注解时需要的 JAR 包 spring-aop-5.1.4.RELEASE.jar 和 Spring MVC 相关的 JAR 包(spring-web-5.1.4.RELEASE.jar 和 spring-webmvc-5.1.4.RELEASE.jar)复制到 Web 应用的 WEB-INF/lib 目录下即可。添加后的 JAR 包如图 2.2 所示。

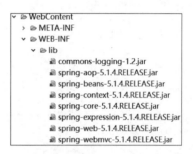

图 2.2 Spring MVC 所需要的 JAR 包

2.2.2 使用 Eclipse 开发 Spring MVC 的 Web 应用

本节通过一个实例来演示 Spring MVC 入门程序的实现过程。

【例 2-1】 Spring MVC 入门程序的实现过程。

具体实现步骤如下。

1. 创建 Web 应用 ch2_1 并导入 JAR 包

创建 Web 应用 ch2_1,导入如图 2.2 所示的 JAR 包。

2. 在 web.xml 文件中部署 DispatcherServlet

在开发 Spring MVC 应用时,需要在 web.xml 中部署 DispatcherServlet,具体代码如下:

```
<?xml version="1.0" encoding="UTF-8"?>
<web-app
xmlns:xsi="http://www.w3.org/2001/XMLSchema-instance"
xmlns="http://xmlns.jcp.org/xml/ns/javaee"
xsi:schemaLocation=" http://xmlns.jcp.org/xml/ns/javaee http://xmlns.jcp.org/xml/ns/javaee/web-app_4_0.xsd"
id="WebApp_ID"
version="4.0">
```

```xml
<!-- 配置 springmvcDispatcherServlet -->
<servlet>
    <servlet-name>springmvc</servlet-name>
    <servlet-class>org.springframework.web.servlet.DispatcherServlet</servlet-class>
    <load-on-startup>1</load-on-startup>
</servlet>
<servlet-mapping>
    <servlet-name>springmvc</servlet-name>
    <url-pattern>/</url-pattern>
</servlet-mapping>
</web-app>
```

上述 DispatcherServlet 的 servlet 对象 springmvc 初始化时，将在应用程序的 WEB-INF 目录下查找一个配置文件，该配置文件的命名规则是"servletName-servlet.xml"，例如 springmvc-servlet.xml。

3. 创建 Web 应用首页

在 ch2_1 应用的 WebContent 目录下，有个应用首页 index.jsp。index.jsp 的代码如下：

```jsp
<%@ page language="java" contentType="text/html; charset=UTF-8" pageEncoding="UTF-8"%>
<!DOCTYPE html>
<html>
<head>
<meta charset="UTF-8">
<title>Insert title here</title>
</head>
<body>
    没注册的用户，请<a href="index/register">注册</a>!<br>
    已注册的用户，去<a href="index/login">登录</a>!
</body>
</html>
```

4. 创建 Controller 类

在 ch2_1 应用的 src 目录下，创建 controller 包，并在该包中创建基于注解的名为 IndexController 的控制器类，该类中有两个处理请求方法，分别处理首页中"注册"和"登录"超链接请求。

```java
package controller;
import org.springframework.stereotype.Controller;
import org.springframework.web.bind.annotation.RequestMapping;
/** "@Controller"表示 IndexController 的实例是一个控制器
 * @Controller 相当于@Controller("indexController")
 * 或@Controller(value = "indexController")
 */
@Controller
@RequestMapping("/index")
public class IndexController {
```

```java
    @RequestMapping("/login")
    public String login() {
        /** login 代表逻辑视图名称,需要根据 Spring MVC 配置文件中
         * internalResourceViewResolver 的前缀和后缀找到对应的物理视图
         */
        return "login";
    }
    @RequestMapping("/register")
    public String register() {
        return "register";
    }
}
```

5. 创建 Spring MVC 的配置文件

在 Spring MVC 中,使用扫描机制找到应用中所有基于注解的控制器类。所以,为了让控制器类被 Spring MVC 框架扫描到,需要在配置文件中声明 spring-context,并使用<context:component-scan/>元素指定控制器类的基本包(请确保所有控制器类都在基本包及其子包下)。另外,需要在配置文件中定义 Spring MVC 的视图解析器(ViewResolver),代码如下:

```xml
<bean class="org.springframework.web.servlet.view.InternalResourceViewResolver"
        id="internalResourceViewResolver">
    <!-- 前缀 -->
    <property name="prefix" value="/WEB-INF/jsp/"/>
    <!-- 后缀 -->
    <property name="suffix" value=".jsp"/>
</bean>
```

上述视图解析器配置了前缀和后缀两个属性。因此,控制器类中视图路径仅需提供 register 和 login,视图解析器将会自动添加前缀和后缀。

因此,在 ch2_1 应用的 WEB-INF 目录下创建名为 springmvc-servlet.xml 的配置文件,代码如下:

```xml
<?xml version="1.0" encoding="UTF-8"?>
<beans xmlns="http://www.springframework.org/schema/beans"
    xmlns:xsi="http://www.w3.org/2001/XMLSchema-instance"
    xmlns:context="http://www.springframework.org/schema/context"
    xmlns:mvc="http://www.springframework.org/schema/mvc"
    xsi:schemaLocation="
        http://www.springframework.org/schema/beans
        http://www.springframework.org/schema/beans/spring-beans.xsd
        http://www.springframework.org/schema/context
        http://www.springframework.org/schema/context/spring-context.xsd
        http://www.springframework.org/schema/mvc
        http://www.springframework.org/schema/mvc/spring-mvc.xsd">
    <!-- 使用扫描机制,扫描控制器类 -->
    <context:component-scan base-package="controller"/>
    <mvc:annotation-driven/>
```

```xml
<!-- 使用resources过滤掉不需要dispatcher servlet的资源(即静态资源,如css、js、html、images)。
     使用resources时,必须使用annotation-driven,否则resources元素会阻止任意控制器被调用。
 -->
<!-- 允许css目录下所有文件可见 -->
<mvc:resources location="/css/" mapping="/css/**"></mvc:resources>
    <!-- 配置视图解析器 -->
    <bean class="org.springframework.web.servlet.view.InternalResourceViewResolver"
        id="internalResourceViewResolver">
     <!-- 前缀 -->
     <property name="prefix" value="/WEB-INF/jsp/" />
     <!-- 后缀 -->
     <property name="suffix" value=".jsp" />
    </bean>
</beans>
```

6. 应用的其他页面

IndexController 控制器的 register 方法处理成功后,跳转到/WEB-INF/jsp/register.jsp 视图；IndexController 控制器的 login 方法处理成功后,跳转到/WEB-INF/jsp/login.jsp 视图。因此,应用的/WEB-INF/jsp 目录下应有 register.jsp 和 login.jsp 页面,此两个 JSP 页面代码略。

7. 发布并运行 Spring MVC 应用

在 Eclipse 中第一次运行 Spring MVC 应用时,需要将应用发布到 Tomcat。例如,运行 ch2_1 应用时,可以右击应用名称 ch2_1,选择 Run As|Run on Server 命令,打开如图 2.3 所示的对话框,在对话框中单击 Finish 按钮即可完成发布并运行。

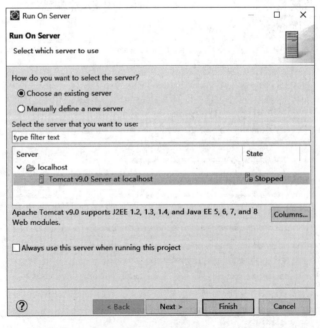

图 2.3 在 Eclipse 中发布并运行 Spring MVC 应用

2.2.3 基于 Java 配置的 Spring MVC 应用

在例 2-1 中，我们使用 web.xml 和 springmvc-servlet.xml 配置文件进行 Web 配置和 Spring MVC 配置。但 Spring Boot 推荐使用 Java 配置的方式进行项目配置，因此，本节通过一个实例来演示 Spring MVC 应用的 Java 配置。

【例 2-2】 Spring MVC 应用的 Java 配置。

具体实现步骤如下。

1. 创建 Web 应用 ch2_2 并导入 JAR 包

创建 Web 应用 ch2_2，导入如图 2.2 所示的 JAR 包。

2. 复制 JSP 和 Java 文件

将 Web 应用 ch2_1 中的 JSP 和 Java 文件按照相同目录复制到应用 ch2_2 中。

3. 创建 Spring MVC 的 Java 配置（相当于 springmvc-servlet.xml 文件）

在 ch2_2 应用的 src 目录中，创建名为 config 的包，在该包中创建 Spring MVC 的 Java 配置类 SpringMVCConfig。在该配置类中使用 @Configuration 注解声明该类为 Java 配置类；使用 @EnableWebMvc 注解开启默认配置，如 ViewResolver；使用 @ComponentScan 注解扫描注解的类；使用 @Bean 注解配置视图解析器。该类需要实现 WebMvcConfigurer 接口来配置 Spring MVC。具体代码如下：

```java
package config;
import org.springframework.context.annotation.Bean;
import org.springframework.context.annotation.ComponentScan;
import org.springframework.context.annotation.Configuration;
import org.springframework.web.servlet.config.annotation.EnableWebMvc;
import org.springframework.web.servlet.config.annotation.ResourceHandlerRegistry;
import org.springframework.web.servlet.config.annotation.WebMvcConfigurer;
import org.springframework.web.servlet.view.InternalResourceViewResolver;
@Configuration
@EnableWebMvc
@ComponentScan("controller")
public class SpringMVCConfig implements WebMvcConfigurer {
    /**
     * 配置视图解析器
     */
    @Bean
    public InternalResourceViewResolver getViewResolver() {
        InternalResourceViewResolver viewResolver = new InternalResourceViewResolver();
        viewResolver.setPrefix("/WEB-INF/jsp/");
        viewResolver.setSuffix(".jsp");
        return viewResolver;
    }
    /**
```

```
 * 配置静态资源
 */
@Override
public void addResourceHandlers(ResourceHandlerRegistry registry) {
    registry.addResourceHandler("/html/**").addResourceLocations("/html/");
}
}
```

4. 创建 Web 的 Java 配置（相当于 web.xml 文件）

在 ch2_2 应用的 config 包中，创建 Web 的 Java 类 WebConfig。该类需要实现 WebApplicationInitializer 接口替代 web.xml 文件的配置。实现该接口将会自动启动 Servlet 容器。在 WebConfig 类中需要使用 AnnotationConfigWebApplicationContext 注册 Spring MVC 的 Java 配置类 SpringMVCConfig，并和当前 ServletContext 关联。最后，在该类中需要注册 Spring MVC 的 DispatcherServlet。具体代码如下：

```
package config;
import javax.servlet.ServletContext;
import javax.servlet.ServletException;
import javax.servlet.ServletRegistration.Dynamic;
import org.springframework.web.WebApplicationInitializer;
import org.springframework.web.context.support.AnnotationConfigWebApplicationContext;
import org.springframework.web.servlet.DispatcherServlet;
public class WebConfig implements WebApplicationInitializer{
    @Override
    public void onStartup(ServletContext arg0) throws ServletException {
        AnnotationConfigWebApplicationContext ctx
            = new AnnotationConfigWebApplicationContext();
        ctx.register(SpringMVCConfig.class);      //注册 Spring MVC 的 Java 配置类 SpringMVCConfig
        ctx.setServletContext(arg0);              //和当前 ServletContext 关联
        /**
         * 注册 Spring MVC 的 DispatcherServlet
         */
        Dynamic servlet = arg0.addServlet("dispatcher", new DispatcherServlet(ctx));
        servlet.addMapping("/");
        servlet.setLoadOnStartup(1);
    }
}
```

5. 发布并运行 Spring MVC 应用

右击 ch2_2 应用，选择 Run As|Run on Server 命令发布并运行应用。

2.3 基于注解的控制器

视频讲解

在使用 Spring MVC 进行 Web 应用开发时，Controller 是 Web 应用的核心。Controller 实现类包含了对用户请求的处理逻辑，是用户请求和业务逻辑之间的"桥梁"，是 Spring MVC 框架的核心部分，负责具体的业务逻辑处理。

2.3.1 Controller 注解类型

在 Spring MVC 中，使用 org.springframework.stereotype.Controller 注解类型声明某类的实例是一个控制器，例如，2.2.2 节中的 IndexController 控制器类。别忘了在 Spring MVC 的配置文件中使用<context:component-scan/>元素（见例 2-1）或在 Spring MVC 配置类中使用@ComponentScan（见例 2-2）指定控制器类的基本包，进而扫描所有注解的控制器类。

2.3.2 RequestMapping 注解类型

在基于注解的控制器类中，可以为每个请求编写对应的处理方法。如何将请求与处理方法一一对应呢？需要使用 org.springframework.web.bind.annotation.RequestMapping 注解类型。

1. 方法级别注解

方法级别注解示例代码如下：

```
package controller;
import org.springframework.stereotype.Controller;
import org.springframework.web.bind.annotation.RequestMapping;
@Controller
public class IndexController {
    @RequestMapping(value = "/index/login")
    public String login() {
        /** login 代表逻辑视图名称，需要根据 Spring MVC 配置中
         * internalResourceViewResolver 的前缀和后缀找到对应的物理视图
         */
        return "login";
    }
    @RequestMapping(value = "/index/register")
    public String register() {
        return "register";
    }
}
```

上述示例中有两个 RequestMapping 注解语句，它们都作用在处理方法上。注解的 value 属性将请求 URI 映射到方法，value 属性是 RequestMapping 注解的默认属性，如果就一个 value 属性，则可省略该属性。可以使用如下 URL 访问 login 方法（请求处理方法）：

http://localhost:8080/ch10/index/login

在访问 login 方法之前，需要事先在/WEB-INF/jsp/目录下创建 login.jsp。

2. 类级别注解

类级别注解示例代码如下：

```
package controller;
import org.springframework.stereotype.Controller;
import org.springframework.web.bind.annotation.RequestMapping;
@Controller
@RequestMapping("/index")
public class IndexController {
    @RequestMapping("/login")
    public String login() {
        return "login";
    }
    @RequestMapping("/register")
    public String register() {
        return "register";
    }
}
```

在类级别注解的情况下,控制器类中的所有方法都将映射为类级别的请求。可以使用如下 URL 访问 login 方法:

http://localhost:8080/ch10/index/login

为了方便程序维护,建议开发者采用类级别注解,将相关处理放在同一个控制器类中,例如,对商品的增、删、改、查处理方法都可以放在一个名为 GoodsOperate 的控制类中。

2.3.3 编写请求处理方法

在控制类中每个请求处理方法可以有多个不同类型的参数,以及一个多种类型的返回结果。

1. 请求处理方法中常出现的参数类型

如果需要在请求处理方法中使用 Servlet API 类型,那么可以将这些类型作为请求处理方法的参数类型。Servlet API 参数类型示例代码如下:

```
package controller;
import javax.servlet.http.HttpServletRequest;
import javax.servlet.http.HttpSession;
import org.springframework.stereotype.Controller;
import org.springframework.web.bind.annotation.RequestMapping;
@Controller
@RequestMapping("/index")
public class IndexController {
    @RequestMapping("/login")
    public String login(HttpSession session, HttpServletRequest request) {
        session.setAttribute("skey", "session 范围的值");
        request.setAttribute("rkey", "request 范围的值");
        return "login";
    }
}
```

除了Servlet API参数类型外,还有输入输出流、表单实体类、注解类型、与Spring框架相关的类型等,这些类型在后续章节中使用时再详细介绍。但特别重要的类型是org.springframework.ui.Model类型,该类型是一个包含Map的Spring框架类型。每次调用请求处理方法时,Spring MVC都将创建org.springframework.ui.Model对象。Model参数类型示例代码如下:

```
package controller;
import org.springframework.stereotype.Controller;
import org.springframework.ui.Model;
import org.springframework.web.bind.annotation.RequestMapping;
@Controller
@RequestMapping("/index")
public class IndexController {
    @RequestMapping("/register")
    public String register(Model model) {
    /*在视图中可以使用EL表达式${success}取出model中的值。*/
        model.addAttribute("success", "注册成功");
        return "register";
    }
}
```

2. 请求处理方法常见的返回类型

最常见的返回类型,就是代表逻辑视图名称的String类型,如前面章节中的请求处理方法。除了String类型外,还有Model、View以及其他任意的Java类型。

2.3.4　Controller接收请求参数的常见方式

Controller接收请求参数的方式有很多种,有的适合get请求方式,有的适合post请求方式,有的二者都适合。下面介绍几个常用的方式,读者可根据实际情况选择合适的接收方式。

1. 通过实体bean接收请求参数

通过一个实体bean来接收请求参数,适用于get和post提交请求方式。需要注意的是,bean的属性名称必须与请求参数名称相同。下面通过一个实例,讲解"通过实体bean接收请求参数"。

【例2-3】　通过实体bean接收请求参数。
具体实现步骤如下。
1) 创建Web应用并导入JAR包
创建Web应用ch2_3,导入如图2.2所示的JAR包。
2) 创建视图文件
在应用ch2_3的/WEB-INF/jsp/目录下有register.jsp、login.jsp和main.jsp文件,main.jsp的代码略。

register.jsp 的代码如下：

```jsp
<%@ page language="java" contentType="text/html; charset=UTF-8" pageEncoding="UTF-8"%>
<!DOCTYPE html>
<html>
<head>
<meta charset="UTF-8">
<title>Insert title here</title>
</head>
<body>
<form action="${pageContext.request.contextPath}/user/register" method="post" name="registForm">
    <table border=1>
        <tr>
            <td>姓名:</td>
            <td>
                <input type="text" name="uname" value="${user.uname}"/>
            </td>
        </tr>
        <tr>
            <td>密码:</td>
            <td><input type="password" name="upass"/></td>
        </tr>
        <tr>
            <td>确认密码:</td>
            <td><input type="password" name="reupass"/></td>
        </tr>
        <tr>
            <td colspan="2" align="center">
                <input type="submit" value="注册"/>
            </td>
        </tr>
    </table>
</form>
</body>
</html>
```

login.jsp 的代码如下：

```jsp
<%@ page language="java" contentType="text/html; charset=UTF-8" pageEncoding="UTF-8"%>
<!DOCTYPE html>
<html>
<head>
<meta charset="UTF-8">
<title>Insert title here</title>
</head>
<body>
    <form action="${pageContext.request.contextPath}/user/login" method="post">
    <table>
        <tr>
            <td align="center" colspan="2">登录</td>
```

```
            </tr>
            <tr>
                <td>姓名:</td>
                <td><input type="text" name="uname"></td>
            </tr>
            <tr>
                <td>密码:</td>
                <td><input type="password" name="upass"></td>
            </tr>
            <tr>
                <td colspan="2">
                    <input type="submit" value="提交">
                    <input type="reset" value="重置">
                </td>
            </tr>
        </table>
        ${messageError}
    </form>
</body>
</html>
```

在 ch2_3 应用的/WebContent/目录下有 index.jsp,具体代码如下:

```
<%@ page language="java" contentType="text/html; charset=UTF-8" pageEncoding="UTF-8"%>
<!DOCTYPE html>
<html>
<head>
<meta charset="UTF-8">
<title>Insert title here</title>
</head>
<body>
    没注册的用户,请<a href="user/register">注册</a>!<br>
    已注册的用户,去<a href="user/login">登录</a>!
</body>
</html>
```

3) 创建 POJO 实体类

在 ch2_3 应用的 src 目录下,创建 pojo 包,并在该包中创建实体类 UserForm,具体代码如下:

```
package pojo;
public class UserForm {
    private String uname;              //与请求参数名称相同
    private String upass;
    private String reupass;
    //省略 getter 和 setter 方法
}
```

4) 创建控制器类

在 ch2_3 应用的 src 目录下,创建 controller 包,并在该包中创建控制器类 UserController。

UserController 的代码如下：

```java
package controller;
import javax.servlet.http.HttpSession;
import org.apache.commons.logging.Log;
import org.apache.commons.logging.LogFactory;
import org.springframework.stereotype.Controller;
import org.springframework.ui.Model;
import org.springframework.web.bind.annotation.RequestMapping;
import pojo.UserForm;
@Controller
@RequestMapping("/user")
public class UserController {
    //得到一个用来记录日志的对象,这样打印信息的时候能够标记打印的是哪个类的信息
    private static final Log logger = LogFactory.getLog(UserController.class);
    /**
     * 处理登录
     * 使用 UserForm 对象(实体 bean)user 接收登录页面提交的请求参数
     */
    @RequestMapping("/login")
    public String login(UserForm user, HttpSession session, Model model) {
        if("zhangsan".equals(user.getUname())
                && "123456".equals(user.getUpass())) {
            session.setAttribute("u", user);
            logger.info("成功");
            return "main";            //登录成功,跳转到 main.jsp
        }else{
            logger.info("失败");
            model.addAttribute("messageError", "用户名或密码错误");
            return "login";
        }
    }

    /**
     * 处理注册
     * 使用 UserForm 对象(实体 bean)user 接收注册页面提交的请求参数
     */
    @RequestMapping("/register")
    public String register(UserForm user, Model model) {
        if("zhangsan".equals(user.getUname())
                && "123456".equals(user.getUpass())) {
            logger.info("成功");
            return "login";           //注册成功,跳转到 login.jsp
        }else{
            logger.info("失败");
            //在 register.jsp 页面上可以使用 EL 表达式取出 model 的 uname 值
            model.addAttribute("uname", user.getUname());
            return "register";        //返回 register.jsp
        }
    }
}
```

5)创建 Web 与 Spring MVC 的配置类

将例 2-2 的 ch2_2 应用的 Web 配置类 WebConfig 和 Spring MVC 配置类 SpringMVCConfig 复制到 ch2_3 应用的相同位置。

6)发布并运行应用

右击 ch2_3 应用,选择 Run As|Run on Server 命令发布并运行应用。

2. 通过处理方法的形参接收请求参数

通过处理方法的形参接收请求参数,也就是直接把表单参数写在控制器类相应方法的形参中,即形参名称与请求参数名称完全相同。该接收参数方式适用于 get 和 post 提交请求方式。可以将例 2-3 的控制器类 UserController 中 register 方法的代码修改如下:

```
@RequestMapping("/register")
/**
 * 通过形参接收请求参数,形参名称与请求参数名称完全相同
 */
public String register(String uname, String upass, Model model) {
    if("zhangsan".equals(uname)
            && "123456".equals(upass)) {
        logger.info("成功");
        return "login";              //注册成功,跳转到 login.jsp
    }else{
        logger.info("失败");
        //在 register.jsp 页面上可以使用 EL 表达式取出 model 的 uname 值
        model.addAttribute("uname", uname);
        return "register";           //返回 register.jsp
    }
}
```

3. 通过@RequestParam 接收请求参数

通过@RequestParam 接收请求参数,适用于 get 和 post 提交请求方式。可以将例 2-3 的控制器类 UserController 中 register 方法的代码修改如下:

```
@RequestMapping("/register")
/**
 * 通过@RequestParam 接收请求参数
 */
public String register(@RequestParam String uname, @RequestParam String upass, Model model) {
    if("zhangsan".equals(uname)
            && "123456".equals(upass)) {
        logger.info("成功");
        return "login";              //注册成功,跳转到 login.jsp
    }else{
        logger.info("失败");
        //在 register.jsp 页面上可以使用 EL 表达式取出 model 的 uname 值
        model.addAttribute("uname", uname);
```

```
            return "register";              //返回register.jsp
    }
}
```

通过"@RequestParam 接收请求参数"与"通过处理方法的形参接收请求参数"的区别是：当请求参数与接收参数名不一致时，"通过处理方法的形参接收请求参数"不会报 400 错误，而"通过@RequestParam 接收请求参数"会报 400 错误。

4．通过@ModelAttribute 接收请求参数

@ModelAttribute 注解放在处理方法的形参上时，用于将多个请求参数封装到一个实体对象，从而简化数据绑定流程，而且自动暴露为模型数据用于视图页面展示时使用。而"通过实体 bean 接收请求参数"只是将多个请求参数封装到一个实体对象，并不能暴露为模型数据（需要使用 model.addAttribute 语句才能暴露为模型数据，数据绑定与模型数据展示可参考第 2.4 节的内容）。

通过@ModelAttribute 注解接收请求参数适用于 get 和 post 提交请求方式。可以将例 2-3 的控制器类 UserController 中 register 方法的代码修改如下：

```
@RequestMapping("/register")
public String register(@ModelAttribute("user") UserForm user) {
    if("zhangsan".equals(user.getUname())
            && "123456".equals(user.getUpass())){
        logger.info("成功");
        return "login";              //注册成功，跳转到login.jsp
    }else{
        logger.info("失败");
        //使用@ModelAttribute("user")与model.addAttribute("user", user)功能相同
    //在register.jsp页面上可以使用EL表达式${user.uname}取出ModelAttribute的uname值
        return "register";              //返回register.jsp
    }
}
```

2.3.5 重定向与转发

重定向是将用户从当前处理请求定向到另一个视图（如 JSP）或处理请求，以前的请求（request）中存放的信息全部失效，并进入一个新的 request 作用域；转发是将用户对当前处理的请求转发给另一个视图或处理请求，以前的 request 中存放的信息不会失效。

转发是服务器行为，重定向是客户端行为。具体工作流程如下。

转发过程：客户浏览器发送 http 请求，Web 服务器接受此请求，调用内部的一个方法在容器内部完成请求处理和转发动作，将目标资源发送给客户；在这里，转发的路径必须是同一个 Web 容器下的 URL，其不能转向到其他的 Web 路径上去，中间传递的是自己的容器内的 request。在客户浏览器的地址栏中显示的仍然是其第一次访问的路径，也就是说客户是感觉不到服务器做了转发的。转发行为是浏览器只做了一次访问请求。

重定向过程：客户浏览器发送 http 请求，Web 服务器接受后发送 302 状态码响应及对

应新的 location 给客户浏览器，客户浏览器发现是 302 响应，则自动再发送一个新的 http 请求，请求 URL 是新的 location 地址，服务器根据此请求寻找资源并发送给客户。在这里 location 可以重定向到任意 URL，既然是浏览器重新发出了请求，那就没有什么 request 传递的概念了。在客户浏览器的地址栏中显示的是其重定向的路径，客户可以观察到地址的变化。重定向行为是浏览器做了至少两次的访问请求。

在 Spring MVC 框架中，控制器类中处理方法的 return 语句默认就是转发实现，只不过实现的是转发到视图。具体代码如下：

```
@RequestMapping("/register")
public String register() {
    return "register";                //转发到 register.jsp
}
```

在 Spring MVC 框架中，重定向与转发的代码如下：

```
package controller;
import org.springframework.stereotype.Controller;
import org.springframework.web.bind.annotation.RequestMapping;
@Controller
@RequestMapping("/index")
public class IndexController {
    @RequestMapping("/login")
    public String login() {
        //转发到一个请求方法(同一个控制器类中,可省略/index/)
        return "forward:/index/isLogin";
    }
    @RequestMapping("/isLogin")
    public String isLogin() {
        //重定向到一个请求方法
        return "redirect:/index/isRegister";
    }
    @RequestMapping("/isRegister")
    public String isRegister() {
        //转发到一个视图
        return "register";
    }
}
```

在 Spring MVC 框架中，无论是重定向还是转发，都需要符合视图解析器的配置，如果直接重定向到一个不需要 DispatcherServlet 的资源，例如：

```
return "redirect:/html/my.html";
```

在 Spring MVC 配置文件中，需要使用 mvc:resources 配置：

```
<mvc:resources location = "/html/" mapping = "/html/**"></mvc:resources>
```

在 Spring MVC 配置类中，需要实现 WebMvcConfigurer 的接口方法 public void addResourceHandlers(ResourceHandlerRegistry registry)，具体代码如下：

```
@Override
public void addResourceHandlers(ResourceHandlerRegistry registry) {
    registry.addResourceHandler("/html/**").addResourceLocations("/html/");
}
```

2.3.6　应用@Autowired进行依赖注入

在前面学习的控制器中,并没有体现 MVC 的 M 层,这是因为控制器既充当 C 层,又充当 M 层。这样设计程序的系统结构很不合理,应该将 M 层从控制器中分离出来。Spring MVC 框架本身就是一个非常优秀的 MVC 框架,它具有一个依赖注入的优点。可以通过 org.springframework.beans.factory.annotation.Autowired 注解类型将依赖注入一个属性(成员变量)或方法,例如:

```
@Autowired
public UserService userService;
```

在 Spring MVC 中,为了能被作为依赖注入,服务层的类必须使用 org.springframework.stereotype.Service 注解类型注明为@Service(一个服务)。另外,还需要在配置文件中使用< context:component-scan base-package = "基本包"/>元素或者在配置类中使用@ComponentScan("基本包")来扫描依赖基本包。下面将例 2-3 的 ch2_3 应用的"登录"和"注册"的业务逻辑处理分离出来,使用 Service 层实现。

首先,创建 service 包,在包中创建 UserService 接口和 UserServiceImpl 实现类。
UserService 接口的具体代码如下:

```
package service;
import pojo.UserForm;
public interface UserService {
    boolean login(UserForm user);
    boolean register(UserForm user);
}
```

UserServiceImpl 实现类的具体代码如下:

```
package service;
import org.springframework.stereotype.Service;
import pojo.UserForm;
//注解为一个服务
@Service
public class UserServiceImpl implements UserService{
    @Override
    public boolean login(UserForm user) {
        if("zhangsan".equals(user.getUname())
                && "123456".equals(user.getUpass()))
            return true;
        return false;
    }
    @Override
    public boolean register(UserForm user) {
```

```
            if("zhangsan".equals(user.getUname())
                    && "123456".equals(user.getUpass()))
                return true;
            return false;
        }
    }
```

其次,将配置类中的@ComponentScan("controller")修改如下:

@ComponentScan(basePackages = {"controller","service"})   //扫描基本包

最后,修改控制器类 UserController,具体代码如下:

```
package controller;
import javax.servlet.http.HttpSession;
import org.apache.commons.logging.Log;
import org.apache.commons.logging.LogFactory;
import org.springframework.beans.factory.annotation.Autowired;
import org.springframework.stereotype.Controller;
import org.springframework.ui.Model;
import org.springframework.web.bind.annotation.ModelAttribute;
import org.springframework.web.bind.annotation.RequestMapping;
import pojo.UserForm;
import service.UserService;
@Controller
@RequestMapping("/user")
public class UserController {
    //得到一个用来记录日志的对象,这样打印信息的时候能够标记打印的是哪个类的信息
    private static final Log logger = LogFactory.getLog(UserController.class);
    //将服务层依赖注入属性 userService
    @Autowired
     public UserService userService;
    /**
     * 处理登录
     */
    @RequestMapping("/login")
    public String login(UserForm user, HttpSession session, Model model) {
        if(userService.login(user)){
            session.setAttribute("u", user);
            logger.info("成功");
            return "main";              //登录成功,跳转到 main.jsp
        }else{
            logger.info("失败");
            model.addAttribute("messageError", "用户名或密码错误");
            return "login";
        }
    }
    /**
     * 处理注册
     */
    @RequestMapping("/register")
    public String register(@ModelAttribute("user") UserForm user) {
```

```
        if(userService.register(user)){
            logger.info("成功");
            return "login";                //注册成功,跳转到login.jsp
        }else{
            logger.info("失败");
            //使用@ModelAttribute("user")与model.addAttribute("user", user)功能相同
          //在register.jsp页面上可以使用EL表达式${user.uname}取出ModelAttribute的uname值
            return "register";             //返回register.jsp
        }
    }
}
```

2.3.7 @ModelAttribute

通过 org.springframework.web.bind.annotation.ModelAttribute 注解类型可以实现以下两个功能。

1. 绑定请求参数到实体对象（表单的命令对象）

该用法与 2.3.4 节的"通过@ModelAttribute 接收请求参数"一样：

```
@RequestMapping("/register")
public String register(@ModelAttribute("user") UserForm user) {
    if("zhangsan".equals(user.getUname())
            && "123456".equals(user.getUpass())){
        return "login";
    }else{
        return "register";
    }
}
```

上述代码中"@ModelAttribute("user") UserForm user"语句的功能有两个：一是将请求参数的输入封装到 user 对象中；一是创建 UserForm 实例，以 user 为键值存储在 Model 对象中，与"model.addAttribute("user", user)"语句功能一样。如果没有指定键值，即"@ModelAttribute UserForm user"，那么创建 UserForm 实例时，以 userForm 为键值存储在 Model 对象中，与"model.addAttribute("userForm", user)"语句功能一样。

2. 注解一个非请求处理方法

被@ModelAttribute 注解的方法，将在每次调用该控制器类的请求处理方法前被调用。这种特性可以用来控制登录权限,当然控制登录权限的方法很多,例如拦截器、过滤器等。

使用该特性控制登录权限的代码如下：

```
package controller;
import javax.servlet.http.HttpSession;
import org.springframework.web.bind.annotation.ModelAttribute;
public class BaseController {
```

```
        @ModelAttribute
        public void isLogin(HttpSession session) throws Exception {
            if(session.getAttribute("user") == null){
                throw new Exception("没有权限");
            }
        }
    }
    package controller;
    import org.springframework.stereotype.Controller;
    import org.springframework.web.bind.annotation.RequestMapping;
    @Controller
    @RequestMapping("/admin")
    public class ModelAttributeController extends BaseController{
        @RequestMapping("/add")
        public String add(){
            return "addSuccess";
        }
        @RequestMapping("/update")
        public String update(){
            return "updateSuccess";
        }
        @RequestMapping("/delete")
        public String delete(){
            return "deleteSuccess";
        }
    }
```

上述 ModelAttributeController 类中的 add、update、delete 请求处理方法执行时,首先执行父类 BaseController 中的 isLogin 方法判断登录权限。

视频讲解

2.4 表单标签库与数据绑定

数据绑定是将用户参数输入值绑定到领域模型的一种特性。在 Spring MVC 的 Controller 和 View 参数数据传递中,所有 HTTP 请求参数的类型均为字符串。如果模型需要绑定的类型为 double 或 int,则需要手动进行类型转换。而有了数据绑定后,就不再需要手动将 HTTP 请求中的 String 类型转换为模型需要的类型。数据绑定的另一个好处是,当输入验证失败时,会重新生成一个 HTML 表单,无须重新填写输入字段。在 Spring MVC 中,为了方便、高效地使用数据绑定,还需要学习表单标签库。

2.4.1 表单标签库

表单标签库中包含可以用在 JSP 页面中渲染 HTML 元素的标签。JSP 页面使用 Spring 表单标签库时,必须在 JSP 页面开头处声明 taglib 指令,代码如下:

```
<%@ taglib prefix="form" uri="http://www.springframework.org/tags/form" %>
```

表单标签库中有 form、input、password、hidden、textarea、checkbox、checkboxes、radiobutton、radiobuttons、select、option、options、errors 等元素。

form：渲染表单元素。

input：渲染<input type="text"/>元素。

password：渲染<input type="password"/>元素。

hidden：渲染<input type="hidden"/>元素。

textarea：渲染 textarea 元素。

checkbox：渲染一个<input type="checkbox"/>元素。

checkboxes：渲染多个<input type="checkbox"/>元素。

radiobutton：渲染一个<input type="radio"/>元素。

radiobuttons：渲染多个<input type="radio"/>元素。

select：渲染一个选择元素。

option：渲染一个选项元素。

options：渲染多个选项元素。

errors：在 span 元素中渲染字段错误。

1. 表单标签

表单标签的语法格式如下：

```
<form:form modelAttribute = "xxx" method = "post" action = "xxx">
    …
</form:form>
```

除了具有 HTML 表单元素属性外，表单标签还具有 acceptCharset、commandName、cssClass、cssStyle、htmlEscape 和 modelAttribute 等属性。各属性含义如下所示。

acceptCharset：定义服务器接受的字符编码列表。

commandName：暴露表单对象的模型属性名称，默认为 command。

cssClass：定义应用到 form 元素的 CSS 类。

cssStyle：定义应用到 form 元素的 CSS 样式。

htmlEscape：true 或 false，表示是否进行 HTML 转义。

modelAttribute：暴露 form backing object 的模型属性名称，默认为 command。

其中，commandName 和 modelAttribute 属性功能基本一致，属性值绑定一个 JavaBean 对象。假设控制器类 UserController 的方法 inputUser，是返回 userAdd.jsp 的请求处理方法。

inputUser 方法的代码如下：

```
@RequestMapping(value = "/input")
public String inputUser(Model model) {
    …
    model.addAttribute("user", new User());
    return "userAdd";
}
```

userAdd.jsp 的表单标签代码如下：

```
<form:form modelAttribute = "user" method = "post" action = "user/save">
    ...
</form:form>
```

注意：在 inputUser 方法中，如果没有 Model 属性 user，userAdd.jsp 页面就会抛出异常，因为表单标签无法找到在其 modelAttribute 属性中指定的 form backing object。

2. input 标签

input 标签的语法格式如下：

```
<form:input path = "xxx"/>
```

该标签除了 cssClass、cssStyle、htmlEscape 属性外，还有一个最重要的属性——path。path 属性将文本框输入值绑定到 form backing object 的一个属性。代码如下：

```
<form:form modelAttribute = "user" method = "post" action = "user/save">
    <form:input path = "userName"/>
</form:form>
```

上述代码将输入值绑定到 user 对象的 userName 属性。

3. password 标签

password 标签的语法格式如下：

```
<form:password path = "xxx"/>
```

该标签与 input 标签用法完全一致，不再赘述。

4. hidden 标签

hidden 标签的语法格式如下：

```
<form:hidden path = "xxx"/>
```

该标签与 input 标签用法基本一致，只不过它不可显示，不支持 cssClass 和 cssStyle 属性。

5. textarea 标签

textarea 基本上就是一个支持多行输入的 input 元素，语法格式如下：

```
<form:textarea path = "xxx"/>
```

该标签与 input 标签用法完全一致，不再赘述。

6. checkbox 标签

checkbox 标签的语法格式如下：

```
<form:checkbox path = "xxx" value = "xxx"/>
```

多个 path 相同的 checkbox 标签，它们是一个选项组，允许多选。选项值绑定到一个数组属性。示例代码如下：

```
<form:checkbox path = "friends" value = "张三"/>张三
<form:checkbox path = "friends" value = "李四"/>李四
<form:checkbox path = "friends" value = "王五"/>王五
<form:checkbox path = "friends" value = "赵六"/>赵六
```

上述示例代码中复选框的值绑定到一个字符串数组属性 friends(String[] friends)。该标签的其他用法与 input 标签基本一致，不再赘述。

7. checkboxes 标签

checkboxes 标签渲染多个复选框，是一个选项组，等价于多个 path 相同的 checkbox 标签。它有 3 个非常重要的属性：items、itemLabel 和 itemValue。

items：用于生成 input 元素的 Collection、Map 或 Array。
itemLabel：items 属性中指定的集合对象的属性，为每个 input 元素提供 label。
itemValue：items 属性中指定的集合对象的属性，为每个 input 元素提供 value。
checkboxes 标签语法格式如下：

```
<form:checkboxes items = "xxx" path = "xxx"/>
```

代码如下：

```
<form:checkboxes items = "${hobbys}" path = "hobby" />
```

上述代码是将 model 属性 hobbys 的内容（集合元素）渲染为复选框。在 itemLabel 和 itemValue 缺省的情况下，如果集合是数组，复选框的 label 和 value 相同；如果是 Map 集合，复选框的 label 是 Map 的值(value)，复选框的 value 是 Map 的关键字(key)。

8. radiobutton 标签

radiobutton 标签的语法格式如下：

```
<form:radiobutton path = "xxx" value = "xxx"/>
```

多个 path 相同的 radiobutton 标签，它们是一个选项组，只允许单选。

9. radiobuttons 标签

radiobuttons 标签渲染多个 radio，是一个选项组，等价于多个 path 相同的 radiobutton 标签。radiobuttons 标签的语法格式如下：

```
<form:radiobuttons path = "xxx" items = "xxx"/>
```

该标签的 itemLabel 和 itemValue 属性与 checkboxes 标签的 itemLabel 和 itemValue 属性完全一样，但只允许单选。

10. select 标签

select 标签的选项可能来自其属性 items 指定的集合，或者来自一个嵌套的 option 标签或 options 标签。select 标签的语法格式如下：

`<form:select path = "xxx" items = "xxx" />`

或

```
<form:select path = "xxx" items = "xxx" >
    <option value = "xxx"> xxx </option>
</form:select>
```

或

```
<form:select path = "xxx">
    <form:options items = "xxx"/>
</form:select>
```

该标签的 itemLabel 和 itemValue 属性与 checkboxes 标签的 itemLabel 和 itemValue 属性完全一样。

11. options 标签

options 标签生成一个 select 标签的选项列表。因此，需要与 select 标签一同使用，具体用法参见 select 标签。

12. errors 标签

errors 标签渲染一个或者多个 span 元素，每个 span 元素包含一个错误消息。它可以用于显示一个特定的错误消息，也可以显示所有错误消息。errors 标签的语法格式如下：

`<form:errors path = " * "/>`

或

`<form:errors path = "xxx"/>`

其中，"*"表示显示所有错误消息；"xxx"表示显示由"xxx"指定的特定错误消息。

2.4.2 数据绑定

为了让读者进一步学习数据绑定和表单标签，本节给出了一个应用实例 ch2_4。ch2_4 应用中实现了 User 类属性和 JSP 页面中表单参数的绑定，同时在 JSP 页面中分别展示了 input、textarea、checkbox、checkboxes、select 等标签。

【例 2-4】 数据绑定和表单标签。

具体实现步骤如下。

1. 创建应用并导入相关的 JAR 包

在 ch2_4 应用中需要使用 JSTL，因此，不仅需要将 Spring MVC 相关 jar 包复制到应用的 WEN-INF/lib 目录下还需要从 Tomcat 的 webapps\examples\WEB-INF\lib 目录下将 JSTL 相关 jar 包复制到应用的 WEN-INF/lib 目录下。ch2_4 应用的 JAR 包如图 2.4 所示。

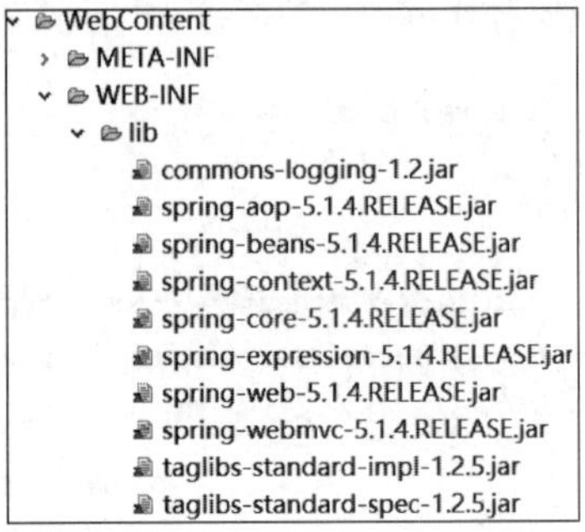

2. 创建 Web 和 Spring MVC 配置类

在 ch2_4 应用的 src 目录下创建名为 config 的包，并在该包中创建 Web 配置类 WebConfig 和 Spring MVC 配置类 SpringMVCConfig。

图 2.4 ch2_4 应用的 JAR 包

为了避免中文乱码问题，需要在 Web 配置类 WebConfig 中注册编码过滤器，同时 JSP 页面编码设置为 UTF-8，form 表单的提交方式为 post。

WebConfig 的代码如下：

```
package config;
import javax.servlet.ServletContext;
import javax.servlet.ServletException;
import org.springframework.web.WebApplicationInitializer;
import org.springframework.web.context.support.AnnotationConfigWebApplicationContext;
import org.springframework.web.filter.CharacterEncodingFilter;
import org.springframework.web.servlet.DispatcherServlet;
public class WebConfig implements WebApplicationInitializer{
    @Override
    public void onStartup(ServletContext arg0) throws ServletException {
        AnnotationConfigWebApplicationContext ctx
          = new AnnotationConfigWebApplicationContext();
        ctx.register(SpringMVCConfig.class);       //注册 Spring MVC 的 Java 配置类 SpringMVCConfig
        ctx.setServletContext(arg0);               //和当前 ServletContext 关联
        /**
         * 注册 Spring MVC 的 DispatcherServlet
         */
        javax.servlet.ServletRegistration.Dynamic servlet =
                arg0.addServlet("dispatcher", new DispatcherServlet(ctx));
        servlet.addMapping("/");
        servlet.setLoadOnStartup(1);
        /**
         * 注册字符编码过滤器
         */
        javax.servlet.FilterRegistration.Dynamic filter =
                arg0.addFilter("characterEncodingFilter", CharacterEncodingFilter.class);
        filter.setInitParameter("encoding", "UTF-8");
        filter.addMappingForUrlPatterns(null, false, "/*");
    }
}
```

SpringMVCConfig 的代码如下：

```java
package config;
import org.springframework.context.annotation.Bean;
import org.springframework.context.annotation.ComponentScan;
import org.springframework.context.annotation.Configuration;
import org.springframework.web.servlet.config.annotation.EnableWebMvc;
import org.springframework.web.servlet.config.annotation.ResourceHandlerRegistry;
import org.springframework.web.servlet.config.annotation.WebMvcConfigurer;
import org.springframework.web.servlet.view.InternalResourceViewResolver;
@Configuration
@EnableWebMvc
@ComponentScan(basePackages = {"controller","service"})
public class SpringMVCConfig implements WebMvcConfigurer {
    /**
     * 配置视图解析器
     */
    @Bean
    public InternalResourceViewResolver getViewResolver() {
        InternalResourceViewResolver viewResolver = new InternalResourceViewResolver();
        viewResolver.setPrefix("/WEB-INF/jsp/");
        viewResolver.setSuffix(".jsp");
        return viewResolver;
    }
    /**
     * 配置静态资源
     */
    @Override
    public void addResourceHandlers(ResourceHandlerRegistry registry) {
        registry.addResourceHandler("/html/**").addResourceLocations("/html/");
    }
}
```

3．创建 View 层

View 层包含两个 JSP 页面，一个是信息输入页面 userAdd.jsp，一个是信息显示页面 userList.jsp。在 ch2_4 应用的 WEB-INF/jsp/ 目录下，创建这两个 JSP 页面。

在 userAdd.jsp 页面中将 Map 类型的 hobbys 绑定到 checkboxes 上，将 String[] 类型的 carrers 和 houseRegisters 绑定到 select 上，实现通过 option 标签对 select 添加选项，同时表单的 method 方法需指定为 post 来避免中文乱码问题。

在 userList.jsp 页面中使用 JSTL 标签遍历集合中的用户信息。

userAdd.jsp 的代码如下：

```jsp
<%@ page language="java" contentType="text/html; charset=UTF-8" pageEncoding="UTF-8"%>
<%@ taglib prefix="form" uri="http://www.springframework.org/tags/form" %>
<!DOCTYPE html>
<html>
<head>
<meta charset="UTF-8">
```

```html
<title>Insert title here</title>
</head>
<body>
<form:form modelAttribute="user" method="post" action="${pageContext.request.contextPath}/user/save">
    <fieldset>
        <legend>添加一个用户</legend>
        <p>
            <label>用户名:</label>
            <form:input path="userName"/>
        </p>
        <p>
            <label>爱好:</label>
            <form:checkboxes items="${hobbys}" path="hobby"/>
        </p>
        <p>
            <label>朋友:</label>
            <form:checkbox path="friends" value="张三"/>张三
            <form:checkbox path="friends" value="李四"/>李四
            <form:checkbox path="friends" value="王五"/>王五
            <form:checkbox path="friends" value="赵六"/>赵六
        </p>
        <p>
            <label>职业:</label>
            <form:select path="carrer">
                <option/>请选择职业
                <form:options items="${carrers}"/>
            </form:select>
        </p>
        <p>
            <label>户籍:</label>
            <form:select path="houseRegister">
                <option/>请选择户籍
                <form:options items="${houseRegisters}"/>
            </form:select>
        </p>
        <p>
            <label>个人描述:</label>
            <form:textarea path="remark" rows="5"/>
        </p>
        <p id="buttons">
            <input id="reset" type="reset">
            <input id="submit" type="submit" value="添加">
        </p>
    </fieldset>
</form:form>
</body>
</html>
```

userList.jsp 的代码如下：

```jsp
<%@ page language="java" contentType="text/html; charset=UTF-8" pageEncoding="UTF-8"%>
<%@ taglib uri="http://java.sun.com/jsp/jstl/core" prefix="c" %>
<!DOCTYPE html>
<html>
<head>
<meta charset="UTF-8">
<title>Insert title here</title>
</head>
<body>
    <h1>用户列表</h1>
    <a href="<c:url value="${pageContext.request.contextPath}/user/input"/>">继续添加</a>
    <table>
        <tr>
            <th>用户名</th>
            <th>兴趣爱好</th>
            <th>朋友</th>
            <th>职业</th>
            <th>户籍</th>
            <th>个人描述</th>
        </tr>
        <!-- JSTL 标签,请参考本书的相关内容 -->
        <c:forEach items="${users}" var="user">
        <tr>
            <td>${user.userName}</td>
            <td>
                <c:forEach items="${user.hobby}" var="hobby">
                    ${hobby} 
                </c:forEach>
            </td>
            <td>
                <c:forEach items="${user.friends}" var="friend">
                    ${friend} 
                </c:forEach>
            </td>
            <td>${user.carrer}</td>
            <td>${user.houseRegister}</td>
            <td>${user.remark}</td>
        </tr>
        </c:forEach>
    </table>
</body>
</html>
```

4. 创建领域模型

在 ch2_4 应用中实现了 User 类属性和 JSP 页面中表单参数的绑定，User 类包含和表单参数名对应的属性，以及属性的 set 和 get 方法。在 ch2_4 应用的 src 目录下，创建 pojo 包，并在该包中创建 User 类。

User 类的代码如下：

```java
package pojo;
public class User {
    private String userName;
    private String[] hobby;              //兴趣爱好
    private String[] friends;            //朋友
    private String carrer;
    private String houseRegister;
    private String remark;
    //省略 setter 和 getter 方法
}
```

5. 创建 Service 层

在 ch2_4 应用中使用了 Service 层，在 Service 层使用静态集合变量 users 模拟数据库存储用户信息，包括添加用户和查询用户两个功能。在 ch2_4 应用的 src 目录下，创建 service 包，并在该包中创建 UserService 接口和 UserServiceImpl 实现类。

UserService 接口的代码如下：

```java
package service;
import java.util.ArrayList;
import pojo.User;
public interface UserService {
    boolean addUser(User u);
    ArrayList<User> getUsers();
}
```

UserServiceImpl 实现类的代码如下：

```java
package service;
import java.util.ArrayList;
import org.springframework.stereotype.Service;
import pojo.User;
@Service
public class UserServiceImpl implements UserService{
    //使用静态集合变量 users 模拟数据库
    private static ArrayList<User> users = new ArrayList<User>();
    @Override
    public boolean addUser(User u) {
        if(!"IT民工".equals(u.getCarrer())){     //不允许添加 IT 民工
            users.add(u);
            return true;
        }
        return false;
    }
    @Override
    public ArrayList<User> getUsers() {
        return users;
    }
}
```

6. 创建 Controller 层

在 Controller 类的 UserController 中定义了请求处理方法，包括处理 user/input 请求的 inputUser 方法和处理 user/save 请求的 addUser 方法，其中在 addUser 方法中用到了重定向。在 UserController 类中，通过 @Autowired 注解在 UserController 对象中主动注入 UserService 对象，实现对 user 对象的添加和查询等操作；通过 model 的 addAttribute 方法将 User 类对象、HashMap 类型的 hobbys 对象、String[]类型的 carrers 对象以及 String[]类型的 houseRegisters 对象传递给 View(userAdd.jsp)。在 ch2_4 应用的 src 目录下，创建 controller 包，并在该包中创建 UserController 控制器类。

UserController 类的代码如下：

```java
package controller;
import java.util.HashMap;
import java.util.List;
import org.apache.commons.logging.Log;
import org.apache.commons.logging.LogFactory;
import org.springframework.beans.factory.annotation.Autowired;
import org.springframework.stereotype.Controller;
import org.springframework.ui.Model;
import org.springframework.web.bind.annotation.ModelAttribute;
import org.springframework.web.bind.annotation.RequestMapping;
import pojo.User;
import service.UserService;
@Controller
@RequestMapping("/user")
public class UserController {
    //得到一个用来记录日志的对象,这样打印信息的时候能够标记打印的是哪个类的信息
    private static final Log logger = LogFactory.getLog(UserController.class);
    @Autowired
    private UserService userService;
    @RequestMapping(value = "/input")
    public String inputUser(Model model) {
        HashMap<String, String> hobbys = new HashMap<String, String>();
        hobbys.put("篮球", "篮球");
        hobbys.put("乒乓球", "乒乓球");
        hobbys.put("电玩", "电玩");
        hobbys.put("游泳", "游泳");
        //如果 model 中没有 user 属性,userAdd.jsp 会抛出异常,因为表单标签无法找到
        //modelAttribute 属性指定的 form backing object
        model.addAttribute("user", new User());
        model.addAttribute("hobbys", hobbys);
        model.addAttribute("carrers", new String[] { "教师", "学生", "coding 搬运工", "IT 民工", "其他" });
        model.addAttribute("houseRegisters", new String[] { "北京", "上海", "广州", "深圳", "其他" });
        return "userAdd";
    }
    @RequestMapping(value = "/save")
    public String addUser(@ModelAttribute User user, Model model) {
        if (userService.addUser(user)) {
            logger.info("成功");
```

```
                return "redirect:/user/list";
            } else {
                logger.info("失败");
                HashMap<String, String> hobbys = new HashMap<String, String>();
                hobbys.put("篮球", "篮球");
                hobbys.put("乒乓球", "乒乓球");
                hobbys.put("电玩", "电玩");
                hobbys.put("游泳", "游泳");
                //这里不需要 model.addAttribute("user", new User(),
                //因为@ModelAttribute 指定 form backing object
                model.addAttribute("hobbys", hobbys);
        model.addAttribute("carrers", new String[] { "教师", "学生", "coding 搬运工", "IT 民工", "其他" });
                model.addAttribute("houseRegisters", new String[] { "北京", "上海", "广州", "深圳", "其他" });
                return "userAdd";
            }
        }
        @RequestMapping(value = "/list")
        public String listUsers(Model model) {
            List<User> users = userService.getUsers();
            model.addAttribute("users", users);
            return "userList";
        }
    }
```

7. 测试应用

通过地址 http://localhost:8080/ch2_4/user/input 测试应用，添加用户信息页面效果如图 2.5 所示。

如果在图 2.5 中，职业选择"IT 民工"，则添加失败。失败后还回到添加页面，输入过的信息不再输入，自动回填（必须结合 form 标签）。自动回填是数据绑定的一个优点。失败页面如图 2.6 所示。

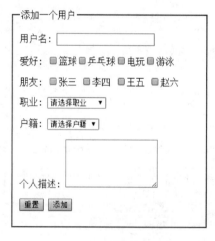 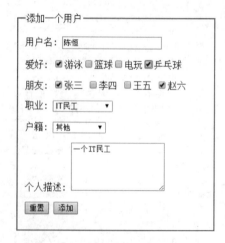

图 2.5　添加用户信息页面　　　　　　图 2.6　添加用户信息失败页面

在图 2.6 中输入正确信息，添加成功后，重定向到信息显示页面，效果如图 2.7 所示。

图 2.7 信息显示页面

视频讲解

2.5 JSON 数据交互

Spring MVC 在数据绑定的过程中，需要对传递数据的格式和类型进行转换，它既可以转换 String 等类型的数据，也可以转换 JSON 等其他类型的数据。本节将针对 Spring MVC 中 JSON 类型的数据交互进行讲解。

2.5.1 JSON 数据结构

JSON(JavaScript Object Notation，JS 对象标记)是一种轻量级的数据交换格式。与 XML 一样，JSON 也是基于纯文本的数据格式。它有以下两种数据结构。

1. 对象结构

对象结构以"{"开始，以"}"结束。中间部分由 0 个或多个以英文","分隔的 key/value 对构成，key 和 value 之间以英文":"分隔。对象结构的语法结构如下：

```
{
    key1:value1,
    key2:value2,
    …
}
```

其中，key 必须为 String 类型，value 可以是 String、Number、Object、Array 等数据类型。例如，一个 person 对象包含姓名、密码、年龄等信息，使用 JSON 的表示形式如下：

```
{
    "pname":"陈恒",
    "password":"123456",
    "page":40
}
```

2. 数组结构

数组结构以"["开始，以"]"结束。中间部分由 0 个或多个以英文","分隔的值的列表组成。数组结构的语法结构如下：

```
[
    value1,
    value2,
    …
]
```

上述两种（对象、数组）数据结构也可以分别组合构成更为复杂的数据结构。例如，一个student对象包含sno、sname、hobby和college对象，其JSON的表示形式如下：

```
{
    "sno":"201802228888",
    "sname":"陈恒",
    "hobby":["篮球","足球"],
    "college":{
        "cname":"清华大学",
        "city":"北京"
    }
}
```

2.5.2 JSON数据转换

为实现浏览器与控制器类之间的JSON数据交互，Spring MVC提供了MappingJackson2HttpMessageConverter实现类默认处理JSON格式请求响应。该实现类利用Jackson开源包读写JSON数据，将Java对象转换为JSON对象和XML文档，同时也可以将JSON对象和XML文档转换为Java对象。

Jackson开源包及其描述如下所示。

- jackson-annotations.jar：JSON转换的注解包。
- jackson-core.jar：JSON转换的核心包。
- jackson-databind.jar：JSON转换的数据绑定包。

以上3个Jackson的开源包的最新版本是2.9.8，读者可通过地址http://mvnrepository.com/artifact/com.fasterxml.jackson.core下载得到。

在使用注解开发时，需要用到两个重要的JSON格式转换注解，分别是@RequestBody和@ResponseBody。

- @RequestBody：用于将请求体中的数据绑定到方法的形参中，该注解应用在方法的形参上。
- @ResponseBody：用于直接返回JSON对象，该注解应用在方法上。

下面通过一个实例来演示JSON数据交互过程。在该实例中，针对返回实体对象、ArrayList集合、Map< String，Object >集合以及List< Map< String，Object >>集合分别处理。

【例2-5】JSON数据交互过程。

具体实现步骤如下。

1. 创建 Web 应用并导入相关的 JAR 包

创建 Web 应用 ch2_5，导入如图 2.8 所示的 JAR 包。

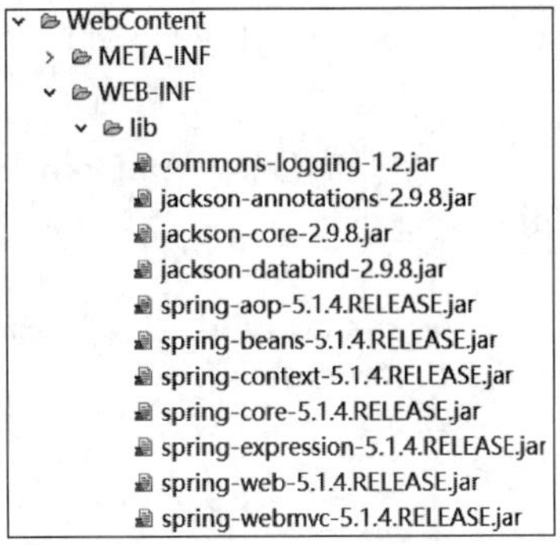

2. 创建 Web 和 Spring MVC 配置类

在 ch2_5 应用的 src 目录下，创建名为 config 的包，在该包中创建 Web 配置类 WebConfig 和 Spring MVC 配置类 SpringMVCConfig。

WebConfig 的代码与 ch2_3 应用的代码相同，不再赘述。

图 2.8　ch2_5 应用的 JAR 包

在 SpringMVCConfig 配置类中，需要配置静态资源，代码如下：

```java
package config;
import org.springframework.context.annotation.Bean;
import org.springframework.context.annotation.ComponentScan;
import org.springframework.context.annotation.Configuration;
import org.springframework.web.servlet.config.annotation.EnableWebMvc;
import org.springframework.web.servlet.config.annotation.ResourceHandlerRegistry;
import org.springframework.web.servlet.config.annotation.WebMvcConfigurer;
import org.springframework.web.servlet.view.InternalResourceViewResolver;
@Configuration
@EnableWebMvc
@ComponentScan("controller")
public class SpringMVCConfig implements WebMvcConfigurer {
    /**
     * 配置视图解析器
     */
    @Bean
    public InternalResourceViewResolver getViewResolver() {
        InternalResourceViewResolver viewResolver = new InternalResourceViewResolver();
        viewResolver.setPrefix("/WEB-INF/jsp/");
        viewResolver.setSuffix(".jsp");
        return viewResolver;
    }
    /**
     * 配置静态资源
     */
    @Override
    public void addResourceHandlers(ResourceHandlerRegistry registry) {
        registry.addResourceHandler("/js/**").addResourceLocations("/js/");
    }
}
```

3. 创建 JSP 页面，并引入 jQuery

首先从 jQuery 官方网站 http://jquery.com/download/下载 jQuery 插件 jquery-3.2.1.min.js，将其复制到 Web 项目开发目录的 WebContent/js 目录下；然后在 JSP 页面中，通过

`<script type="text/javascript" src="js/jquery-3.2.1.min.js"></script>`代码将 jquery-3.2.1.min.js 引入当前页面中。

在 ch2_5 应用的 WebContent 目录下创建 JSP 文件 index.jsp，在该页面中使用 AJAX 向控制器异步提交数据，代码如下：

```jsp
<%@ page language="java" contentType="text/html; charset=UTF-8" pageEncoding="UTF-8"%>
<!DOCTYPE html>
<html>
<head>
<meta charset="UTF-8">
<title>Insert title here</title>
<script type="text/javascript" src="${pageContext.request.contextPath}/js/jquery-3.2.1.min.js"></script>
<script type="text/javascript">
    function testJson(){
        //获取输入的值 pname 为 id
        var pname = $("#pname").val();
        var password = $("#password").val();
        var page = $("#page").val();
        $.ajax({
            //请求路径
            url : "${pageContext.request.contextPath}/testJson",
            //请求类型
            type : "post",
            //data 表示发送的数据
            data : JSON.stringify({pname:pname,password:password,page:page}),
            //定义发送请求的数据格式为 JSON 字符串
            contentType : "application/json;charset=utf-8",
            //定义回调响应的数据格式为 JSON 字符串,该属性可以省略
            dataType : "json",
            //成功响应的结果
            success : function(data){
                if(data != null){
                    //返回一个 Person 对象
                    //alert("输入的用户名:" + data.pname + ",密码:" + data.password + ",年龄:" + data.page);
                    //ArrayList<Person>对象
                    /** for(var i = 0; i < data.length; i++){
                        alert(data[i].pname);
                    } **/
                    //返回一个 Map<String,Object>对象
                    //alert(data.pname);//pname 为 key
                    //返回一个 List<Map<String,Object>>对象
                    for(var i = 0; i < data.length; i++){
                        alert(data[i].pname);
                    }
                }
            }
        });
    }
</script>
```

```
</head>
<body>
    <form action = "">
        用户名:<input type = "text" name = "pname" id = "pname"/><br>
        密码:<input type = "password" name = "password" id = "password"/><br>
        年龄:<input type = "text" name = "page" id = "page"/><br>
        <input type = "button" value = "测试" onclick = "testJson()"/>
    </form>
</body>
</html>
```

4. 创建实体类

在 ch2_5 应用的 src 目录下，创建名为 pojo 的包，在该包中创建 Person 实体类，代码如下：

```
package pojo;
public class Person {
    private String pname;
    private String password;
    private Integer page;
    //省略 set 和 get 方法
}
```

5. 创建控制器类

在 ch2_5 应用的 src 目录下，创建名为 controller 的包，在该包中创建 TestController 控制器类，在处理方法中使用@ResponseBody 和@RequestBody 注解进行 JSON 数据交互，具体代码如下：

```
package controller;
import java.util.ArrayList;
import java.util.HashMap;
import java.util.List;
import java.util.Map;
import org.springframework.stereotype.Controller;
import org.springframework.web.bind.annotation.RequestBody;
import org.springframework.web.bind.annotation.RequestMapping;
import org.springframework.web.bind.annotation.ResponseBody;
import pojo.Person;
@Controller
public class TestController {
    /**
     * 接收页面请求的 JSON 数据，并返回 JSON 格式结果
     */
    @RequestMapping("/testJson")
    @ResponseBody
    public List<Map<String, Object>> testJson(@RequestBody Person user) {
        //打印接收的 JSON 格式数据
        System.out.println("pname = " + user.getPname() +
```

```
            ", password = " + user.getPassword() + ",page = " + user.getPage());
        //返回 Person 对象
        //return user;
        /** ArrayList<Person> allp = new ArrayList<Person>();
        Person p1 = new Person();
        p1.setPname("陈恒 1");
        p1.setPassword("123456");
        p1.setPage(80);
        allp.add(p1);

        Person p2 = new Person();
        p2.setPname("陈恒 2");
        p2.setPassword("78910");
        p2.setPage(90);
        allp.add(p2);
        //返回 ArrayList<Person>对象
        return allp;
        **/

        Map<String, Object> map = new HashMap<String, Object>();
        map.put("pname", "陈恒 2");
        map.put("password", "123456");
        map.put("page", 25);
        //返回一个 Map<String, Object>对象
        //return map;
        //返回一个 List<Map<String, Object>>对象
        List<Map<String, Object>> allp = new ArrayList<Map<String, Object>>();
        allp.add(map);
        Map<String, Object> map1 = new HashMap<String, Object>();
        map1.put("pname", "陈恒 3");
        map1.put("password", "54321");
        map1.put("page", 55);
        allp.add(map1);
        return allp;
    }
}
```

6. 测试应用

右击 ch2_5 应用，选择 Run As|Run on Server 命令发布并运行应用。

2.6　Spring MVC 的基本配置

视频讲解

　　Spring MVC 的定制配置需要配置类实现 WebMvcConfigurer 接口，并在配置类使用@EnableWebMvc 注解来开启对 Spring MVC 的配置支持，这样开发者就可以重写接口方法完成常用的配置。

2.6.1 静态资源配置

应用程序的静态资源（CSS、JS、图片等）需要直接访问，这时需要开发者在配置类重写 public void addResourceHandlers(ResourceHandlerRegistry registry)接口方法来实现。代码如下：

```
package config;
import org.springframework.context.annotation.Bean;
import org.springframework.context.annotation.ComponentScan;
import org.springframework.context.annotation.Configuration;
import org.springframework.web.servlet.config.annotation.EnableWebMvc;
import org.springframework.web.servlet.config.annotation.ResourceHandlerRegistry;
import org.springframework.web.servlet.config.annotation.WebMvcConfigurer;
import org.springframework.web.servlet.view.InternalResourceViewResolver;
@Configuration
@EnableWebMvc
@ComponentScan(basePackages = {"controller","service"})
public class SpringMVCConfig implements WebMvcConfigurer {
    /**
     * 配置视图解析器
     */
    @Bean
    public InternalResourceViewResolver getViewResolver() {
        InternalResourceViewResolver viewResolver = new InternalResourceViewResolver();
        viewResolver.setPrefix("/WEB-INF/jsp/");
        viewResolver.setSuffix(".jsp");
        return viewResolver;
    }
    /**
     * 配置静态资源
     */
    @Override
    public void addResourceHandlers(ResourceHandlerRegistry registry) {
        registry.addResourceHandler("/html/**").addResourceLocations("/html/");
        //addResourceHandler 指的是对外暴露的访问路径
        //addResourceLocations 指的是静态资源存放的位置
    }
}
```

根据上述配置，可以直接访问 Web 应用（如 ch2_3 应用）的/html/目录下的静态资源。假设静态资源如图 2.9 所示，可以通过 http://localhost:8080/ch2_3/html/NewFile.html 直接访问静态资源文件 NewFile.html。

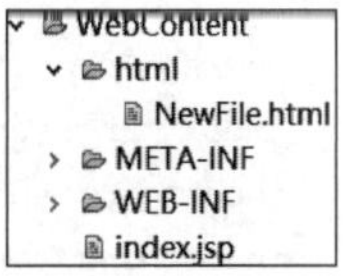

图 2.9 静态资源

2.6.2 拦截器配置

Spring 的拦截器(Interceptor)实现对每一个请求处理前后进行相关的业务处理,类似于 Servlet 的过滤器(Filter)。开发者如需要自定义 Spring 的拦截器,可以通过以下两个步骤完成。

1. 创建自定义拦截器类

自定义拦截器类需要实现 HandlerInterceptor 接口或继承 HandlerInterceptorAdapter 类,代码如下:

```java
package interceptor;
import javax.servlet.http.HttpServletRequest;
import javax.servlet.http.HttpServletResponse;
import org.springframework.web.servlet.HandlerInterceptor;
import org.springframework.web.servlet.ModelAndView;
public class MyInteceptor implements HandlerInterceptor{
    /**
     * 重写 preHandle 方法在请求发生前执行
     */
    @Override
    public boolean preHandle ( HttpServletRequest request, HttpServletResponse response, Object handler)
            throws Exception {
        System.out.println("preHandle 方法在请求发生前执行");
        return true;
    }
    /**
     * 重写 postHandle 方法在请求完成后执行
     */
    @Override
    public void postHandle(HttpServletRequest request, HttpServletResponse response, Object handler,ModelAndView modelAndView) throws Exception {
        System.out.println("postHandle 方法在请求完成后执行");
    }
}
```

2. 配置拦截器

在配置类中,需要首先配置拦截器 Bean,然后重写 addInterceptors 方法注册拦截器,具体代码如下:

```java
package config;
import org.springframework.context.annotation.Bean;
import org.springframework.context.annotation.ComponentScan;
import org.springframework.context.annotation.Configuration;
import org.springframework.web.servlet.config.annotation.EnableWebMvc;
import org.springframework.web.servlet.config.annotation.InterceptorRegistry;
```

```java
import org.springframework.web.servlet.config.annotation.ResourceHandlerRegistry;
import org.springframework.web.servlet.config.annotation.WebMvcConfigurer;
import org.springframework.web.servlet.view.InternalResourceViewResolver;
import interceptor.MyInteceptor;
@Configuration
@EnableWebMvc
@ComponentScan(basePackages = {"controller","service"})
public class SpringMVCConfig implements WebMvcConfigurer {
    /**
     * 配置视图解析器
     */
    @Bean
    public InternalResourceViewResolver getViewResolver() {
        InternalResourceViewResolver viewResolver = new InternalResourceViewResolver();
        viewResolver.setPrefix("/WEB-INF/jsp/");
        viewResolver.setSuffix(".jsp");
        return viewResolver;
    }
    /**
     * 配置静态资源
     */
    @Override
    public void addResourceHandlers(ResourceHandlerRegistry registry) {
        registry.addResourceHandler("/html/**").addResourceLocations("/html/");
        //addResourceHandler 指的是对外暴露的访问路径
        //addResourceLocations 指的是静态资源存放的位置
    }
    /**
     * 配置拦截器 Bean
     */
    @Bean
    public MyInteceptor myInteceptor() {
        return new MyInteceptor();
    }
    /**
     * 重写 addInterceptors 方法注册拦截器
     */
    @Override
    public void addInterceptors(InterceptorRegistry registry) {
        registry.addInterceptor(myInteceptor());
    }
}
```

2.6.3 文件上传配置

文件上传是一个应用中经常使用的功能,Spring MVC 通过配置一个 MultipartResolver 来上传文件。在 Spring MVC 的控制器中,可以通过 MultipartFile myfile 来接收单个文件上传,通过 List<MultipartFile> myfiles 来接收多个文件上传。

由于 Spring MVC 框架的文件上传是基于 commons-fileupload 组件的文件上传，因此，需要将 commons-fileupload 组件相关的 jar(commons-fileupload. jar 和 commons-io. jar)复制到 Spring MVC 应用的 WEB-INF/lib 目录下。下面讲解一下如何下载相关 jar 包。

Commons 是 Apache 开放源代码组织中的一个 Java 子项目，该项目包括文件上传、命令行处理、数据库连接池、XML 配置文件处理等模块。fileupload 就是其中用来处理基于表单的文件上传的子项目，commons-fileupload 组件性能优良，并支持任意大小文件的上传。

commons-fileupload 组件可以从 http://commons. apache. org/proper/commons-fileupload 下载，本书采用的版本是 1. 4。下载它的 Binaries 压缩包(commons-fileupload-1. 4-bin. zip)，解压后有个 JAR 文件：commons-fileupload-1. 4. jar，该文件是 commons-fileupload 组件的类库。

commons-fileupload 组件依赖于 Apache 的另外一个项目：commons-io，该组件可以从 http://commons. apache. org/proper/commons-io/下载，本书采用的版本是 2. 6。下载它的 Binaries 压缩包(commons-io-2. 6-bin. zip)，解压缩后的目录中有 5 个 JAR 文件，其中有一个 commons-io-2. 6. jar 文件，该文件是 commons-io 的类库。

下面通过一个实例讲解如何上传多个文件。

【例 2-6】 上传多个文件。

具体实现步骤如下。

1. 创建 Web 应用并导入相关的 JAR 包

创建 Web 应用 ch2_6，导入如图 2.10 所示的 JAR 包。

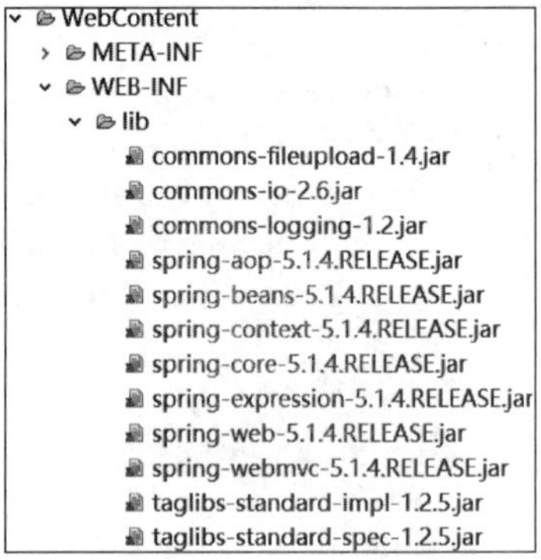

图 2.10　ch2_6 应用的 JAR 包

2. 创建多文件选择页面

在 ch2_6 应用的 WebContent 目录下，创建 JSP 页面 multiFiles. jsp。在该页面中使用表单(别忘了 enctype 属性值为 multipart/form-data)上传多个文件，具体代码如下：

```
<%@ page language = "java" contentType = "text/html; charset = UTF - 8" pageEncoding = "UTF - 8" %>
<! DOCTYPE html >
< html >
< head >
< meta charset = "UTF - 8">
< title > Insert title here </title >
</head >
< body >
< form action = " $ {pageContext. request. contextPath }/multifile" method = "post" enctype =
"multipart/form - data">
    选择文件 1:< input type = "file" name = "myfile">  < br >
    文件描述 1:< input type = "text" name = "description">< br >
    选择文件 2:< input type = "file" name = "myfile">  < br >
    文件描述 2:< input type = "text" name = "description">< br >
    选择文件 3:< input type = "file" name = "myfile">  < br >
    文件描述 3:< input type = "text" name = "description">< br >
```

```
    < input type = "submit" value = "提交">
</form>
</body>
</html>
```

3. 创建 POJO 类

在 ch2_6 应用的 src 目录下,创建 pojo 包,在 pojo 包中创建实体类 MultiFileDomain。上传多文件时,需要 POJO 类 MultiFileDomain 封装文件信息,MultiFileDomain 类的具体代码如下:

```
package pojo;
import java.util.List;
import org.springframework.web.multipart.MultipartFile;
public class MultiFileDomain {
    private List<String> description;
    private List<MultipartFile> myfile;
    //省略 setter 和 getter 方法
}
```

4. 创建控制器类

在 ch2_6 应用的 src 目录下,创建 controller 包,在 controller 包中创建控制器类 MutiFilesController,具体代码如下:

```
package controller;
import java.io.File;
import java.util.List;
import javax.servlet.http.HttpServletRequest;
import org.apache.commons.logging.Log;
import org.apache.commons.logging.LogFactory;
import org.springframework.stereotype.Controller;
import org.springframework.web.bind.annotation.ModelAttribute;
import org.springframework.web.bind.annotation.RequestMapping;
import org.springframework.web.multipart.MultipartFile;
import pojo.MultiFileDomain;
@Controller
public class MutiFilesController {
    //得到一个用来记录日志的对象,这样打印信息时,能够标记打印的是哪个类的信息
    private static final Log logger = LogFactory.getLog(MutiFilesController.class);
    /**
     * 多文件上传
     */
    @RequestMapping("/multifile")
    public String multiFileUpload(@ModelAttribute MultiFileDomain multiFileDomain, HttpServletRequest request){
        String realpath = request.getServletContext().getRealPath("uploadfiles");
        //上传到 eclipse-workspace/.metadata/.plugins/org.eclipse.wst.server.core/tmp0/wtpwebapps/ch2_6/uploadfiles
        File targetDir = new File(realpath);
```

```java
            if(!targetDir.exists()){
                targetDir.mkdirs();
            }
            List<MultipartFile> files = multiFileDomain.getMyfile();
            for (int i = 0; i < files.size(); i++) {
                MultipartFile file = files.get(i);
                String fileName = file.getOriginalFilename();
                File targetFile = new File(realpath,fileName);
                //上传
                try {
                    file.transferTo(targetFile);
                } catch (Exception e) {
                    e.printStackTrace();
                }
            }
            logger.info("成功");
            return "showMulti";
        }
    }
```

5. 创建 Web 与 Spring MVC 配置类

在 ch2_6 应用的 src 目录下，创建名为 config 的包，在该包中创建 Web 配置类 WebConfig 和 Spring MVC 配置类 SpringMVCConfig。

WebConfig 的代码与 ch2_3 应用的代码相同，不再赘述。

在 SpringMVCConfig 配置类中，需要配置 MultipartResolver 进行文件上传，具体代码如下：

```java
package config;
import org.springframework.context.annotation.Bean;
import org.springframework.context.annotation.ComponentScan;
import org.springframework.context.annotation.Configuration;
import org.springframework.web.multipart.MultipartResolver;
import org.springframework.web.multipart.commons.CommonsMultipartResolver;
import org.springframework.web.servlet.config.annotation.EnableWebMvc;
import org.springframework.web.servlet.config.annotation.ResourceHandlerRegistry;
import org.springframework.web.servlet.config.annotation.WebMvcConfigurer;
import org.springframework.web.servlet.view.InternalResourceViewResolver;
@Configuration
@EnableWebMvc
@ComponentScan(basePackages = {"controller","service"})
public class SpringMVCConfig implements WebMvcConfigurer {
    /**
     * 配置视图解析器
     */
    @Bean
    public InternalResourceViewResolver getViewResolver() {
        InternalResourceViewResolver viewResolver = new InternalResourceViewResolver();
        viewResolver.setPrefix("/WEB-INF/jsp/");
        viewResolver.setSuffix(".jsp");
```

```java
        return viewResolver;
    }
    /**
     * 配置静态资源
     */
    @Override
    public void addResourceHandlers(ResourceHandlerRegistry registry) {
        registry.addResourceHandler("/html/**").addResourceLocations("/html/");
        //addResourceHandler 指的是对外暴露的访问路径
        //addResourceLocations 指的是静态资源存放的位置
    }
    /**
     * MultipartResolver 配置
     */
    @Bean
    public MultipartResolver multipartResolver() {
        CommonsMultipartResolver multipartResolver = new CommonsMultipartResolver();
        //设置上传文件的最大值,单位为字节
        multipartResolver.setMaxUploadSize(5400000);
        //设置请求的编码格式,默认为 iso-8859-1
        multipartResolver.setDefaultEncoding("UTF-8");
        return multipartResolver;
    }
}
```

6. 创建成功显示页面

在 ch2_6 应用的 WEB-INF/jsp 目录下,创建多文件上传成功显示页面 showMulti.jsp,具体代码如下:

```jsp
<%@ page language="java" contentType="text/html; charset=UTF-8" pageEncoding="UTF-8"%>
<%@ taglib uri="http://java.sun.com/jsp/jstl/core" prefix="c" %>
<!DOCTYPE html>
<html>
<head>
<meta charset="UTF-8">
<title>Insert title here</title>
</head>
<body>
    <table>
        <tr>
            <td>详情</td><td>文件名</td>
        </tr>
        <!-- 同时取两个数组的元素 -->
        <c:forEach items="${multiFileDomain.description}" var="description" varStatus="loop">
            <tr>
                <td>${description}</td>
                <td>${multiFileDomain.myfile[loop.count-1].originalFilename}</td>
            </tr>
        </c:forEach>
```

```
            <!-- fileDomain.getMyfile().getOriginalFilename() -->
        </table>
</body>
</html>
```

7. 发布并运行应用

发布 ch2_6 应用到 Tomcat 服务器,并启动 Tomcat 服务器。然后,通过地址 http://localhost:8080/ch2_6/multiFiles.jsp 运行多文件选择页面,运行结果如图 2.11 所示。

在图 2.11 中选择文件,并输入文件描述,然后单击"提交"按钮上传多个文件,成功显示如图 2.12 所示的结果。

图 2.11　多文件选择页面

图 2.12　多文件成功上传结果

2.7　本章小结

本章简单介绍了 Spring MVC 框架基础,包括 Spring MVC 的工作流程、控制器、表单标签与数据绑定、JSON 数据交互以及 Spring MVC 的基本配置等内容。第 1 章和本章只是简单介绍了 Spring 和 Spring MVC 的基础知识,关于它们的更多知识,请参考作者的另一本教程《Java EE 框架整合开发入门到实战——Spring＋Spring MVC＋MyBatis(微课版)》。

习题 2

1. 在开发 Spring MVC 应用时,如何配置 DispatcherServlet? 又如何配置 Spring MVC?

2. 简述 Spring MVC 的工作流程。

3. 举例说明数据绑定的优点。

4. Spring MVC 有哪些表单标签? 其中,可以绑定集合数据的标签有哪些?

5. @ModelAttribute 可实现哪些功能?

6. 在 Spring MVC 中,JSON 类型的数据如何交互? 请按照返回实体对象、ArrayList 集合、Map＜String,Object＞集合以及 List＜Map＜String,Object＞＞集合举例说明。

第3章 Spring Boot 入门

学习目的与要求

本章首先介绍什么是 Spring Boot，然后介绍 Spring Boot 应用的开发环境，最后介绍如何快速构建一个 Spring Boot 应用。通过本章的学习，掌握如何构建 Spring Boot 应用的开发环境以及 Spring Boot 应用。

本章主要内容

- Spring Boot 概述。
- Spring Boot 应用的开发环境。
- Maven 构建 Spring Boot 应用。
- 快速构建 Spring Boot 应用。

从前两章的学习可知，Spring 框架非常优秀，但问题在于"配置过多"，造成开发效率低、部署流程复杂以及集成难度大等问题。为解决上述问题，Spring Boot 应运而生。作者在编写本书时，Spring Boot 的最新正式版是 2.1.4.RELEASE，Spring Boot 2.2.0M1 里程碑版本已经发布。为了提高程序的稳定性，本书将以 Spring Boot 2.1.4.RELEASE 版本编写示例代码。读者测试本书示例代码时，建议使用 2.1.4.RELEASE 或更高版本。

3.1 Spring Boot 概述

视频讲解

3.1.1 什么是 Spring Boot

Spring Boot 是由 Pivotal 团队提供的全新框架，其设计目的是用来简化新 Spring 应用的初始搭建以及开发过程。使用 Spring Boot 框架可以做到专注于 Spring 应用的开发，无须过多关注样板化的配置。

在 Spring Boot 框架中，使用"约定优于配置（Convention Over Configuration，COC）"的理念。针对企业应用开发，提供了符合各种场景的 spring-boot-starter 自动配置依赖模块，这些模块都是基于"开箱即用"的原则，进而使企业应用开发更加快捷和高效。可以说，Spring Boot 是开发者和 Spring 框架的中间层，目的是帮助开发者管理应用的配置，提供应用开发中常见配置的默认处理（即约定优于配置），简化 Spring 应用的开发和运维，降低开发人员对框架的关注度，使开发人员把更多精力放在业务逻辑代码上。通过"约定优于配置"的原则，Spring Boot 致力于在蓬勃发展的快速应用开发领域成为领导者。

3.1.2　Spring Boot 的优点

Spring Boot 之所以能够应运而生，是因为它具有以下优点：
（1）使编码变得简单：推荐使用注解。
（2）使配置变得快捷：具有自动配置、快速构建项目、快速集成第三方技术的能力。
（3）使部署变得简便：内嵌 Tomcat、Jetty 等 Web 容器。
（4）使监控变得容易：自带项目监控。

3.1.3　Spring Boot 的主要特性

1. 约定优于配置

Spring Boot 遵循"约定优于配置"的原则，只需很少的配置，大多数情况直接使用默认配置即可。

2. 独立运行的 Spring 应用

Spring Boot 可以以 jar 包的形式独立运行。使用 java -jar 命令或者在项目的主程序中执行 main 方法运行 Spring Boot 应用（项目）。

3. 内嵌 Web 容器

内嵌 Servlet 容器，Spring Boot 可以选择内嵌 Tomcat、Jetty 等 Web 容器，无须以 war 包形式部署应用。

4. 提供 starter 简化 Maven 配置

Spring Boot 提供了一系列的 starter pom 简化 Maven 的依赖加载，基本上可以做到自动化配置，高度封装，开箱即用。

5. 自动配置 Spring

Spring Boot 根据项目依赖（在类路径中的 jar 包、类）自动配置 Spring 框架，极大地减少了项目的配置。

6. 提供准生产的应用监控

Spring Boot 提供基于 HTTP、SSH、TELNET 对运行的项目进行跟踪监控。

7. 无代码生成和 XML 配置

Spring Boot 不是借助于代码生成来实现的，而是通过条件注解来实现的。提倡使用 Java 配置和注解配置相结合的配置方式，方便快捷。

3.2 第一个 Spring Boot 应用

视频讲解

3.2.1 Maven 简介

Apache Maven 是一个软件项目管理工具。基于项目对象模型（Project Object Model，POM）的理念，通过一段核心描述信息来管理项目构建、报告和文档信息。在 Java 项目中，Maven 主要完成两件工作：①统一开发规范与工具；②统一管理 jar 包。

Maven 统一管理项目开发所需要的 jar 包，但这些 jar 包将不再包含在项目内（即不在 lib 目录下），而是存放于仓库中。仓库主要包括以下内容。

1. 中央仓库

中央仓库存放开发过程中的所有 jar 包，例如 JUnit，这些 jar 包都可以通过互联网从中央仓库下载，仓库地址：http://mvnrepository.com。

2. 本地仓库

本地仓库即本地计算机中的仓库。官方下载 Maven 的本地仓库，配置在"%MAVEN_HOME%\conf\settings.xml"文件中，找到 localRepository 即可；Eclipse 中自带 Maven 的默认本地仓库地址在"{user.home}/.m2/repository/settings.xml"文件中，同样找到 localRepository 即可。

Maven 项目首先会从本地仓库中获取所需要的 jar 包，当无法获取指定 jar 包时，本地仓库会从远程仓库（中央仓库）下载 jar 包，并放入本地仓库以备将来使用。

3.2.2 Maven 的 pom.xml

Maven 是基于项目对象模型的理念管理项目的，所以 Maven 的项目都有一个 pom.xml 配置文件来管理项目的依赖以及项目的编译等功能。

在 Maven Web 项目中，重点关注以下元素。

1. properties 元素

在<properties></properties>之间可以定义变量，以便在<dependency></dependency>中引用，代码如下：

```
<properties>
    <!-- spring 版本号 -->
    <spring.version>5.1.5.RELEASE</spring.version>
</properties>
<dependencies>
    <dependency>
        <groupId>org.springframework</groupId>
        <artifactId>spring-core</artifactId>
        <version>${spring.version}</version>
    </dependency>
</dependencies>
```

2. dependencies 元素

<dependencies></dependencies>，此元素包含多个项目依赖需要使用的<dependency></dependency>元素。

3. dependency 元素

<dependency></dependency>元素内部通过<groupId></groupId>、<artifactId></artifactId>、<version></version> 3个子元素确定唯一的依赖，也可以称为3个坐标。代码如下：

```
<dependency>
    <!-- groupId 组织的唯一标识 -->
    <groupId>org.springframework</groupId>
    <!-- artifactId 项目的唯一标识 -->
    <artifactId>spring-core</artifactId>
    <!-- version 项目的版本号 -->
    <version>${spring.version}</version>
</dependency>
```

3.2.3　在 Eclipse 中创建 Maven Web 项目

本节在基于 Eclipse 平台的 Java Web 应用开发环境（见 1.2 节）的基础上，创建 Maven Web 项目。

1. 新建 Maven Web 项目

在 Eclipse 中新建 Maven Web 项目，具体实现步骤如下：

（1）选择菜单 File|New|Maven Project，打开如图 3.1 所示的 Select project name and location 对话框。

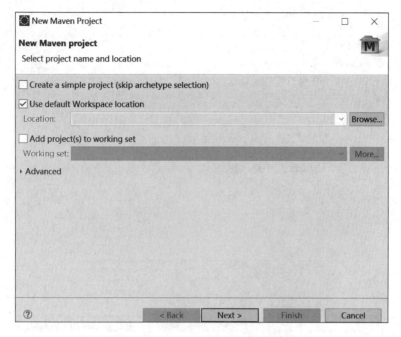

图 3.1 Select project name and location

(2) 单击图 3.1 中的 Next 按钮，打开如图 3.2 所示的 Select an Archetype 对话框，在该对话框中，选择 Archetype 为 webapp。

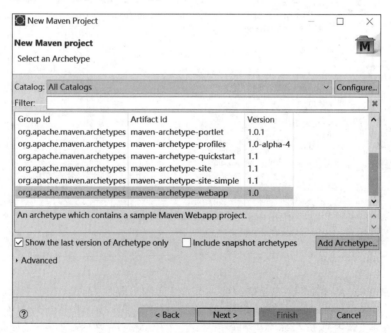

图 3.2 Select an Archetype

(3) 单击图 3.2 中的 Next 按钮，打开如图 3.3 所示的 Specify Archetype parameters 对话框，在该对话框中，输入一些必要信息，单击 Finish 按钮。

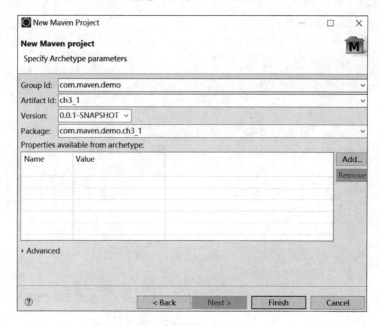

图 3.3　Enter a group id for the artifact

(4) 创建的 Maven Web 项目 ch3_1 的目录结构如图 3.4 所示。

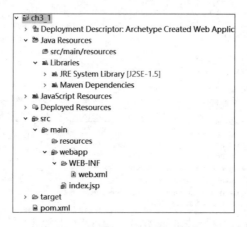

图 3.4　项目 ch3_1 的目录结构

2. 配置 Maven Web 项目

新建的 Maven Web 项目，需要修改一些配置，具体步骤如下：

(1) 右击项目 ch3_1，选择 Build Path | Configure Build Path，打开如图 3.5 所示的 Properties for ch3_1 对话框。

(2) 在图 3.5 中，选择 Libraries 标签，选中 JRE System Library，单击 Edit 按钮，打开如图 3.6 所示的 Select JRE for the project build path 对话框。

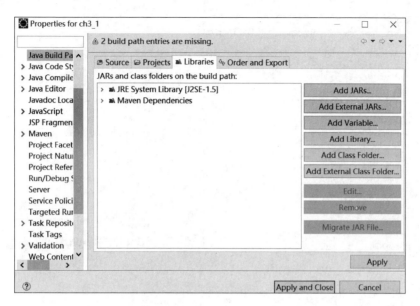

图 3.5　Properties for ch3_1

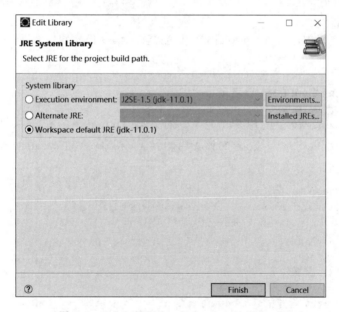

图 3.6　Select JRE for the project build path

（3）在图 3.6 中，选择 Workspace 默认的 JRE，单击 Finish 按钮，最后，单击 Apply and Close 即可。完成后，项目结构如图 3.7 所示。

图 3.7 中 src/main/java 目录包含项目的 Java 源代码；src/main/resources 目录包含项目所需的资源（如配置文件）；src/test/java 目录包含用于测试的 Java 代码；src/main/webapp 目录包含 Java Web 应用程序；目录由 Maven 创建。target 包含所有编译的类、JAR 文件等。当执行 mvn clean 命令时，Maven 将清除此目录。

（4）右击项目名 ch3_1，选择 Run As|Run on Server，运行 ch3_1 项目，运行结果如图 3.8 所示。

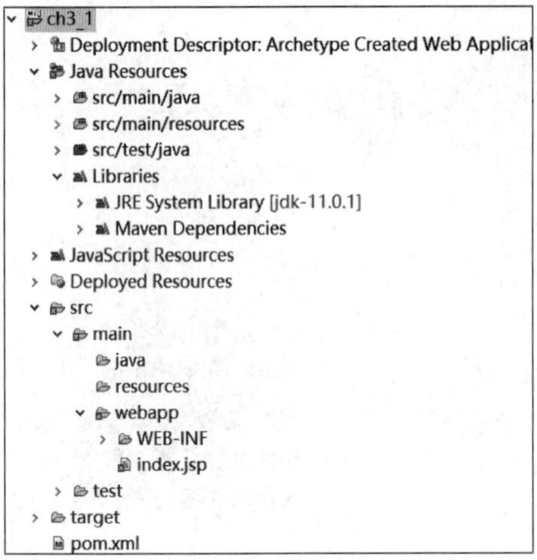

```
> ch3_1
  > Deployment Descriptor: Archetype Created Web Applicat
  > Java Resources
    > src/main/java
    > src/main/resources
    > src/test/java
    > Libraries
      > JRE System Library [jdk-11.0.1]
      > Maven Dependencies
  > JavaScript Resources
  > Deployed Resources
  > src
    > main
      > java
      > resources
      > webapp
        > WEB-INF
          index.jsp
    > test
  > target
    pom.xml
```

图 3.7　项目结构图

图 3.8　运行 ch3_1 项目

3.2.4　Maven 手工构建第一个 Spring Boot 应用

本节在 3.2.3 节 Maven Web 项目 ch3_1 的基础上，手工构建一个 Spring Boot 应用，具体实现步骤如下。

1. 配置 Spring Boot 的核心启动器

构建基于 Spring Boot 的应用。首先，在 pom.xml 文件的<url…/>元素之后添加<parent…/>元素配置 Spring Boot 的核心启动器 spring-boot-starter-parent，代码如下：

```
<parent>
    <groupId>org.springframework.boot</groupId>
    <artifactId>spring-boot-starter-parent</artifactId>
    <version>2.1.4.RELEASE</version>
</parent>
```

2. 添加 starter 模块

然后，在 pom.xml 文件的<dependencies…/>元素中增加一个<dependency…/>元素添加需要的 starter 模块，此处只添加了 spring-boot-starter-web 模块。Spring Boot 包含了很多 starter 模块（见 4.1.6 节），每一个 starter 模块就是一系列的依赖组合。如 spring-boot-starter-web 模块包含 Spring Boot 预定义的 Web 开发的常用依赖包（包括 Tomcat 和 spring-webmvc）。由于指定了 spring-boot-starter-parent 的版本，所以此处的 spring-boot-starter-web 模块不需要指定版本，Spring Boot 将自动选择匹配的版本进行加载。代码如下：

```
<dependency>
    <groupId>org.springframework.boot</groupId>
    <artifactId>spring-boot-starter-web</artifactId>
</dependency>
```

Web 项目 ch3_1 的 pom.xml 文件修改后的内容如下：

```xml
<project xmlns="http://maven.apache.org/POM/4.0.0" xmlns:xsi="http://www.w3.org/2001/XMLSchema-instance"
  xsi:schemaLocation="http://maven.apache.org/POM/4.0.0 http://maven.apache.org/maven-v4_0_0.xsd">
  <modelVersion>4.0.0</modelVersion>
  <groupId>com.maven.demo</groupId>
  <artifactId>ch3_1</artifactId>
  <packaging>war</packaging>
  <version>0.0.1-SNAPSHOT</version>
  <name>ch3_1 Maven Webapp</name>
  <url>http://maven.apache.org</url>
  <parent>
        <groupId>org.springframework.boot</groupId>
        <artifactId>spring-boot-starter-parent</artifactId>
        <version>2.1.4.RELEASE</version>
  </parent>
  <dependencies>
     <dependency>
        <groupId>org.springframework.boot</groupId>
        <artifactId>spring-boot-starter-web</artifactId>
     </dependency>
     <dependency>
        <groupId>junit</groupId>
        <artifactId>junit</artifactId>
        <scope>test</scope>
     </dependency>
  </dependencies>
  <build>
     <finalName>ch3_1</finalName>
  </build>
</project>
```

pom.xml 文件修改保存后，Maven 将自动在互联网环境下，下载所需的所有 jar 文件。

3. 编写测试代码

在 src/main/java 目录下，创建 com.test 包，并在该包中创建 TestController 类，具体代码如下：

```java
package com.test;
import org.springframework.web.bind.annotation.RequestMapping;
import org.springframework.web.bind.annotation.RestController;
@RestController
public class TestController {
    @RequestMapping("/hello")
    public String hello() {
        return "您好,Spring Boot!";
    }
}
```

上述代码中使用的@RestController 注解是一个组合注解,相当于 Spring MVC 中的@Controller 和@ResponseBody 注解的组合,具体应用如下:

(1) 如果只是使用@RestController 注解 Controller,则 Controller 中的方法无法返回 JSP 或者 html 页面,返回的内容就是 return 的内容。

(2) 如果需要返回指定页面,则需要用@Controller 注解。如果需要返回 JSON,XML 或自定义 mediaType 内容到页面,则需要在对应的方法上加上@ResponseBody 注解。

4. 创建应用程序的 App 类(启动类)

在 com.test 包中创建 Ch3_1Application 类,具体代码如下:

```
package com.test;
import org.springframework.boot.SpringApplication;
import org.springframework.boot.autoconfigure.SpringBootApplication;
@SpringBootApplication
public class Ch3_1Application {
    public static void main(String[] args) {
        SpringApplication.run(Ch3_1Application.class, args);
    }
}
```

上述代码中使用@SpringBootApplication 注解指定该程序是一个 Spring Boot 应用,该注解也是一个组合注解,相当于@Configuration、@EnableAutoConfiguration 和@ComponentScan 注解的组合,具体细节在第 4 章讲解。SpringApplication 类调用 run 方法启动 Spring Boot 应用。

5. 运行 main 方法启动 Spring Boot 应用

运行 Ch3_1Application 类的 main 方法后,控制台信息如图 3.9 所示。

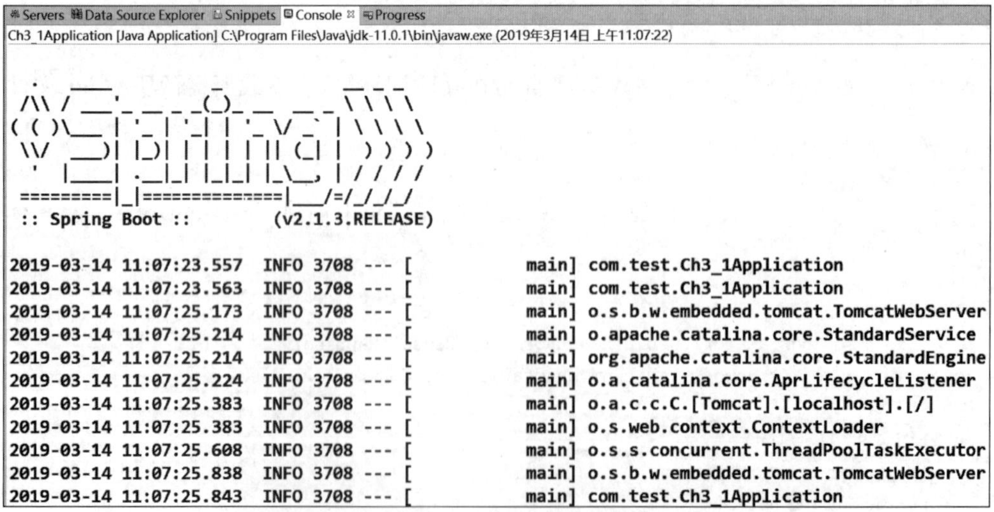

图 3.9 启动 Spring Boot 应用后的控制台信息

从上面的控制台信息可以看到 Tomcat 的启动过程、Spring MVC 的加载过程。注意 Spring Boot 内嵌 Tomcat 容器,因此 Spring Boot 应用不需要开发者配置与启动 Tomcat。

6. 测试 Spring Boot 应用

启动 Spring Boot 应用后,默认访问地址为 http://localhost:8080/,将项目路径直接设为根路径,这是 Spring Boot 的默认设置。因此,可以通过 http://localhost:8080/hello 测试应用(hello 与测试类 TestController 中的 @RequestMapping("/hello") 对应),测试效果如图 3.10 所示。

图 3.10 访问 Spring Boot 应用

3.3 Spring Boot 快速构建

视频讲解

在 3.2.4 节使用 Maven 方便快捷地手工构建了一个 Spring Boot 应用,但还可以使用更便捷的方法构建 Spring Boot 应用。本节将讲解两个方便快捷的构建方法:http://start.spring.io 和 Spring Tool Suite(STS)。

3.3.1 http://start.spring.io

使用 http://start.spring.io 快速构建 Spring Boot 应用的具体步骤如下。

1. 打开 spring.io

在浏览器地址栏中输入"http://start.spring.io",打开如图 3.11 所示的界面。

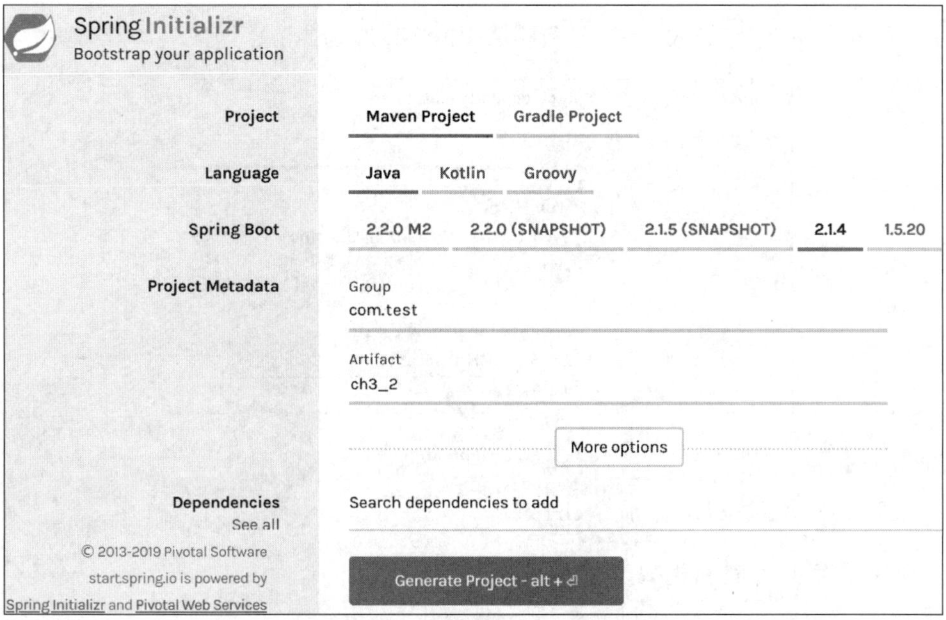

图 3.11 打开 spring.io

2. 填写项目信息

在图 3.11 中，Project 默认选择 Maven Project，Language 默认选择 Java，Spring Boot 默认选择最新正式版。Project Metadata 的 Group 输入"com.test"，Artifact 输入"ch3_2"，单击 More options 按钮，Java Version 选择 11。Project Metadata 的信息如图 3.12 所示。

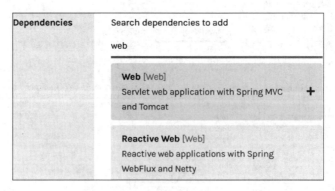

图 3.12　Project Metadata

在 Dependencies 处，输入"web"进行搜索，如图 3.13 所示。

图 3.13　Search dependencies to add

单击图 3.13 中的"＋"号添加 Web 依赖。

3. 创建应用并下载源代码

在图 3.11 中单击 Generate Project 按钮，创建 Spring Boot 应用，并下载源代码。此处

下载的源代码是一个简单的基于 Maven 的应用，解压后的目录如图 3.14 所示。

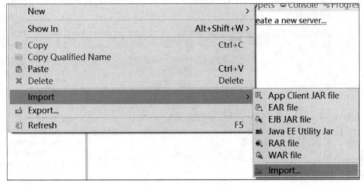

图 3.14　项目目录

4. 导入 Maven 项目到集成开发工具中

可将 ch3_2 这个 Maven 项目导入读者常用的集成开发工具中，如 Eclipse。Eclipse 导入 Maven 项目过程如图 3.15 所示。

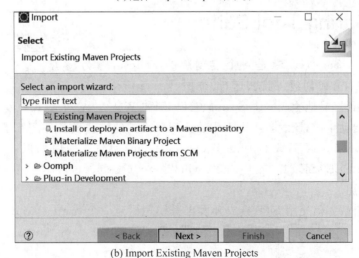

(a) 选择Eclipse的Import菜单项

(b) Import Existing Maven Projects

图 3.15　Eclipse 导入 Maven 项目过程

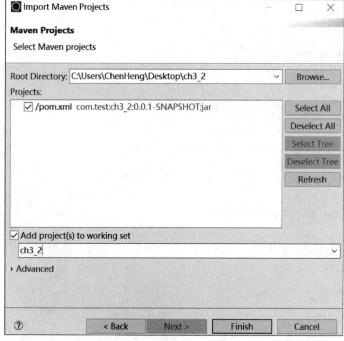

(c) Select Maven projects

图 3.15 （续）

成功导入 ch3_2 这个 Maven 项目后，在 Eclipse 中显示如图 3.16 所示的项目结构。

这时，就可以在 ch3_2 项目中编写自己的 Spring Boot 应用程序了，如 Web 应用程序。

3.3.2 Spring Tool Suite

图 3.16 ch3_2 项目结构

Spring Tool Suite(STS)是一个定制版的 Eclipse，专为 Spring 开发定制，方便创建、调试、运行、维护 Spring 应用。通过该工具，可以很轻易地生成一个 Spring 工程，例如 Web 工程，最令人兴奋的是工程中的配置文件都将自动生成，开发者再也不用关注配置文件的格式及各种配置了。可通过官网 https://spring.io/tools 下载 Spring Tools for Eclipse，本书采用的版本是 spring-tool-suite-4-4.1.1.RELEASE-e4.10.0-win32.win32.x86_64.zip。该版本与 Eclipse 一样，免安装，解压即可使用。

下面详细讲解如何使用 STS 集成开发工具快速构建一个 Spring Boot 应用，具体实现步骤如下。

1. 新建 Spring Starter Project

选择菜单 File|New|Spring Starter Project，打开如图 3.17 所示的 New Spring Starter Project 对话框。

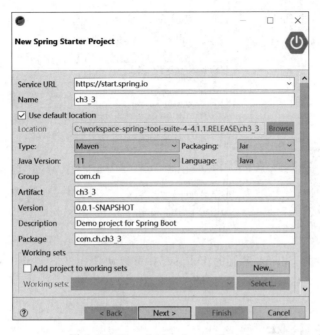

图 3.17　New Spring Starter Project

2. 选择项目依赖

在图 3.17 中输入项目信息后,单击 Next 按钮,打开如图 3.18 所示的 New Spring Starter Project Dependencies 对话框,并在图中选择项目依赖,如 Web。

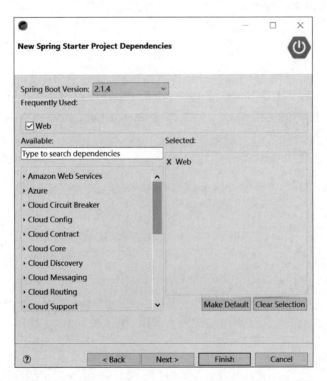

图 3.18　New Spring Starter Project Dependencies

单击图 3.18 中的 Finish 按钮,即可完成 Web 应用的创建。ch3_3 的项目结构如图 3.19 所示。此时,就可以在项目 ch3_3 中编写自己的 Web 应用程序了。

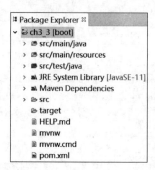

图 3.19　ch3_3 的项目结构

STS 是一款专为 Spring 定制的 Eclipse,并可以快速构建 Spring Boot 应用。因此,本书后续章节都采用 STS IDE 编写示例代码。

3.4　本章小结

本章首先简单介绍了 Spring Boot 应运而生的缘由,然后讲述了如何使用 Maven 手工构建一个 Spring Boot 应用,最后演示了如何使用 http://start.spring.io 或 Spring Tool Suite(STS)快速构建 Spring Boot 应用。开发者如何构建 Spring Boot 应用,可根据实际工程需要选择合适的 IDE。

习题 3

1. Spring、Spring MVC、Spring Boot 三者有什么联系?为什么还要学习 Spring Boot?
2. 在 Eclipse 中如何使用 Maven 手工构建 Spring Boot 的一个 Web 应用?

Spring Boot 核心

学习目的与要求

本章将详细介绍 Spring Boot 的核心注解、基本配置、自动配置原理以及条件注解。通过本章的学习,掌握 Spring Boot 的核心注解与基本配置,理解 Spring Boot 的自动配置原理与条件注解。

本章主要内容

- Spring Boot 的基本配置。
- 读取应用配置。
- Spring Boot 的自动配置原理。
- Spring Boot 的条件注解。

在 Spring Boot 产生之前,Spring 项目会存在多个配置文件,例如 web.xml、application.xml,应用程序自身也需要多个配置文件,同时需要编写程序读取这些配置文件。现在 Spring Boot 简化了 Spring 项目配置的管理和读取,仅需要一个 application.properties 文件,并提供了多种读取配置文件的方式。本章将学习 Spring Boot 的基本配置与运行原理。

4.1 Spring Boot 的基本配置

4.1.1 启动类和核心注解@SpringBootApplication

Spring Boot 应用通常都有一个名为 *Application 的程序入口类,该入口类需要使用 Spring Boot 的核心注解@SpringBootApplication 标注为应用的启动类。另外,该入口类有一个标准的Java应用程序的main方法,在main方法中通过"SpringApplication.run(*Application.class,args);"启动Spring Boot应用。启动类示例代码如下:

```
package com.ch.ch4_1;
import org.springframework.boot.SpringApplication;
import org.springframework.boot.autoconfigure.SpringBootApplication;
@SpringBootApplication
public class Ch41Application {
    public static void main(String[] args) {
        SpringApplication.run(Ch41Application.class, args);
    }
}
```

Spring Boot 的核心注解@SpringBootApplication是一个组合注解,主要组合了@SpringBootConfiguration、@EnableAutoConfiguration和@ComponentScan注解。源代码可以从spring-boot-autoconfigure-2.1.4.RELEASE.jar依赖包中查看org/springframework/boot/autoconfigure/SpringBootApplication.java。

1. @SpringBootConfiguration 注解

@SpringBootConfiguration是Spring Boot应用的配置注解,该注解也是一个组合注解,源代码可以从spring-boot-2.1.4.RELEASE.jar依赖包中查看org/springframework/boot/SpringBootConfiguration.java。在Spring Boot应用中推荐使用@SpringBootConfiguration注解替代@Configuration注解。

2. @EnableAutoConfiguration 注解

@EnableAutoConfiguration注解可以让Spring Boot根据当前应用项目所依赖的jar自动配置项目的相关配置。例如,在Spring Boot项目的pom.xml文件中添加了spring-boot-starter-web依赖,Spring Boot项目会自动添加Tomcat和Spring MVC的依赖,同时对Tomcat和Spring MVC进行自动配置。打开pom.xml文件,选择Dependency Hierarchy页面查看spring-boot-starter-web的自动配置,如图4.1所示。

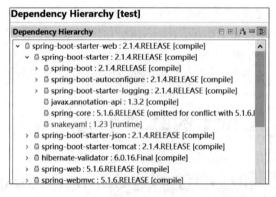

图4.1 spring-boot-starter-web的自动配置

3. @ComponentScan 注解

该注解的功能是让 Spring Boot 自动扫描@SpringBootApplication 所在类的同级包以及它的子包中的配置，所以建议将@SpringBootApplication 注解的入口类放置在项目包下（Group Id＋Artifact Id 组合的包名），这样可以保证 Spring Boot 自动扫描项目所有包中的配置。

4.1.2 关闭某个特定的自动配置

从 4.1.1 小节可知，使用@EnableAutoConfiguration 注解可以让 Spring Boot 根据当前应用项目所依赖的 jar 自动配置项目的相关配置，如图 4.2 所示。

图 4.2　Spring Boot 的自动配置项

如果开发者不需要 Spring Boot 的某一项自动配置，该如何实现呢？通过查看@SpringBootApplication 的源代码可知，应该使用@SpringBootApplication 注解的 exclude 参数关闭特定的自动配置。以关闭 neo4j 自动配置为例，代码如下：

```
@SpringBootApplication(exclude = {Neo4jDataAutoConfiguration.class})
```

4.1.3 定制 Banner

Spring Boot 项目启动时，在控制台可以看到如图 4.3 所示的默认启动图案。

如果开发者希望指定自己的启动信息，又该如何配置呢？首先，在 src/main/resources 目录下新建 banner.txt 文件，并在文件中添加任意字符串内容，如"♯ Hello, Spring

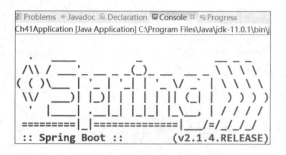

图 4.3 Spring Boot 项目的默认启动图案

Boot！"。然后，重新启动 Spring Boot 项目，将发现控制台启动信息已经发生改变。如果开发者想把启动字符串信息换成字符串图案，可以通过如下操作实现。

1. 生成字符串图案

首先，打开网页 http://patorjk.com/software/taag，输入自定义字符串，单击网页下方的 Select & Copy 按钮，如图 4.4 所示。然后，将自定义 banner 字符串图案复制到 src/main/resources 目录下的 banner.txt 文件中。

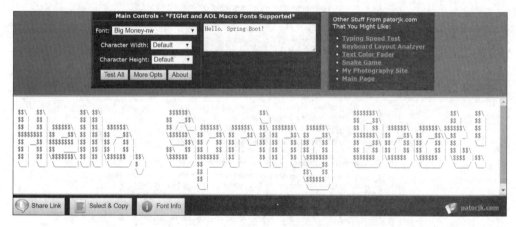

图 4.4 自定义 banner 图案

2. 重新启动 Spring Boot 项目

重新启动 Spring Boot 项目，效果如图 4.5 所示。

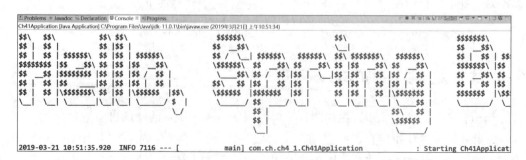

图 4.5 banner 的启动效果

4.1.4 关闭 banner

开发者如果需要关闭 banner，可以在 src/main/resources 目录下的 application.properties 文件中添加如下配置：

```
spring.main.banner-mode = off
```

4.1.5 Spring Boot 的全局配置文件

Spring Boot 的全局配置文件（application.properties 或 application.yml）可以放置在 Spring Boot 项目的 src/main/resources 目录下或者类路径的/config 目录下。

1. 设置端口号

全局配置文件主要用于修改项目的默认配置，如修改内嵌的 Tomcat 的默认端口。例如，在 Spring Boot 项目 ch4_1 的 src/main/resources 目录下找到名为 application.properties 的全局配置文件，添加如下配置内容：

```
server.port = 8888
```

可以将内嵌的 Tomcat 的默认端口改为 8888。启动项目时，端口修改为 8888，如图 4.6 所示。

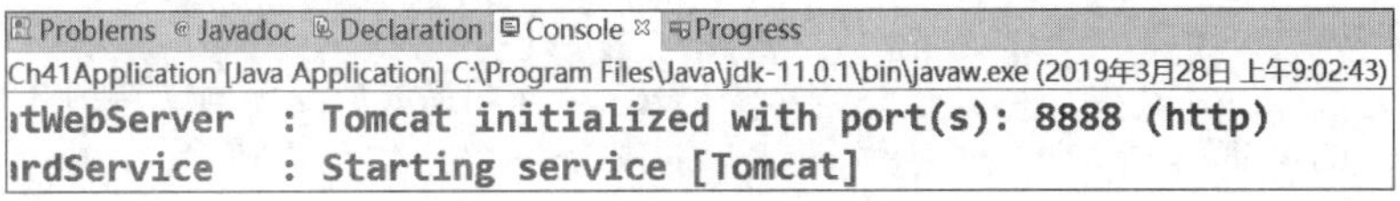

图 4.6 修改 Tomcat 的默认端口

2. 设置 Web 应用的上下文路径

如果开发者想设置一个 Web 应用程序的上下文路径，可以在 application.properties 文件中配置如下内容：

```
server.servlet.context-path = /XXX
```

这时应该通过"http://localhost:8080/XXX/testStarters"访问如下控制器类中的请求处理方法：

```
@RequestMapping("/testStarters")
public String index() {
}
```

3. 配置文档

在 Spring Boot 的全局配置文件中，可以配置与修改多个参数，读者想了解参数的详细说明和描述，可以查看官方文档说明：https://docs.spring.io/spring-boot/docs/2.1.4.

RELEASE/reference/htmlsingle/#common-application-properties。

4.1.6　Spring Boot 的 Starters

Spring Boot 提供了很多简化企业级开发的"开箱即用"的 Starters。Spring Boot 项目只要使用了所需要的 Starters，Spring Boot 就可以自动关联项目开发所需要的相关依赖。例如，在 ch4_1 的 pom.xml 文件中，添加如下依赖配置：

```
<dependency>
    <groupId>org.springframework.boot</groupId>
    <artifactId>spring-boot-starter-web</artifactId>
</dependency>
```

Spring Boot 将自动关联 Web 开发的相关依赖，如 tomcat、spring-webmvc 等，进而对 Web 开发提供支持，并对相关技术的配置实现自动配置。

通过访问"https://docs.spring.io/spring-boot/docs/2.1.4.RELEASE/reference/htmlsingle/#using-boot-starter"官网，可以查看 Spring Boot 官方提供的 Starters，如表 4.1 所示。

表 4.1　Spring Boot 的 Starters

名　　称	描　　述
spring-boot-starter	核心 starter，包括自动配置、日志记录和 YAML 文件的支持
spring-boot-starter-actuator	支持准生产特性的 starter，用来监控和管理应用
spring-boot-starter-activemq	为 JMS 使用 Apache ActiveMQ 进行消息传递的 starter。ActiveMQ 是 Apache 出品的最流行、能力强的开源消息总线
spring-boot-starter-amqp	使用 spring-rabbit 支持 AMQP 协议（Advanced Message Queuing Protocol）的 starter
spring-boot-starter-aop	支持面向切面编程（AOP）的 starter，包括 spring-aop 和 AspectJ
spring-boot-starter-artemis	使用 Apache Artemis 支持 JMS 消息传递的 starter
spring-boot-starter-batch	支持 Spring Batch 的 starter，包括 HSQLDB（HyperSQL DataBase）数据库
spring-boot-starter-cache	支持 Spring Cache 的 starter
spring-boot-starter-cloud-connectors	支持 Spring Cloud Connectors 的 starter，简化了 Cloud Foundry、Heroku 等云平台中的服务连接
spring-boot-starter-data-cassandra	使用 Spring Data Cassandra 支持 Cassandra 分布式数据库的 starter
spring-boot-starter-data-couchbase	使用 Spring Data Couchbase 支持 Couchbase 文件存储数据库的 starter
spring-boot-starter-data-elasticsearch	使用 Spring Data Elasticsearch 支持 Elasticsearch 搜索和分析引擎的 starter
spring-boot-starter-data-jdbc	支持 Spring Data JDBC 的 starter
spring-boot-starter-data-jpa	支持 JPA（Java Persistence API）的 starter，包括 spring-data-jpa、spring-orm 和 Hibernate

续表

名称	描述
spring-boot-starter-data-ldap	支持 Spring Data LDAP（Lightweight Directory Access Protocol，轻量级目录访问协议）的 starter
spring-boot-starter-data-mongodb	使用 Spring Data MongoDB 支持 MongoDB 的 starter
spring-boot-starter-data-neo4j	使用 Spring Data Neo4j 支持 Neo4j 图数据库的 starter
spring-boot-starter-data-redis	使用 Spring Data Redis 支持 Redis 键值存储数据库的 starter，包括 Lettuce 客户端
spring-boot-starter-data-rest	使用 Spring Data REST 支持通过 REST 公开 Spring Data 数据仓库的 starter
spring-boot-starter-data-solr	使用 Spring Data Solr 支持 Apache Solr 搜索平台的 starter
spring-boot-starter-freemarker	支持 FreeMarker 模板引擎构建 MVC Web 应用的 starter
spring-boot-starter-groovy-templates	支持 Groovy 模板引擎构建 MVC Web 应用的 starter
spring-boot-starter-hateoas	使用 Spring MVC、Spring HATEOAS 构建基于超媒体的 RESTful Web 应用程序的 starter
spring-boot-starter-integration	支持通用的 spring-integration 模块的 starter
spring-boot-starter-jdbc	支持 JDBC 的 starter，包括 HikariCP 连接池
spring-boot-starter-jersey	使用 JAX-RS 和 Jersey 支持 RESTful Web 应用程序的 starter，替代 spring-boot-starter-web
spring-boot-starter-jetty	使用 Jetty 作为嵌入式 servlet 容器替代 Tomcat 的 starter
spring-boot-starter-jooq	使用 jOOQ 支持访问 SQL 数据库的 starter，替代 spring-boot-starter-jpa 或 spring-boot-starter-jdbc
spring-boot-starter-json	读写 json 的 starter
spring-boot-starter-jta-atomikos	使用 Atomikos 支持 JTA 分布式事务处理的 starter
spring-boot-starter-jta-bitronix	使用 Bitronix 支持 JTA 分布式事务处理的 starter
spring-boot-starter-log4j2	支持使用 Log4j2 日志框架的 starter，替代 spring-boot-starter-logging
spring-boot-starter-logging	支持 Spring Boot 的默认日志框架 Logback 的 starter
spring-boot-starter-mail	支持 javax.mail 的 starter
spring-boot-starter-mustache	支持 Mustache 模板引擎构建 Web 应用的 starter
spring-boot-starter-oauth2-client	使用 Spring Security 的 OAuth2/OpenID Connect 支持客户端功能的 starter
spring-boot-starter-oauth2-resource-server	使用 Spring Security 的 OAuth2 支持资源服务器功能的 starter
spring-boot-starter-quartz	支持 Quartz 调度器的 starter
spring-boot-starter-reactor-netty	使用 Reactor Netty 作为嵌入式响应 HTTP 服务器的 starter
spring-boot-starter-security	支持 spring-security 的 starter
spring-boot-starter-test	支持常规的测试依赖的 starter，包括 Junit、Hamcrest、Mockito 以及 spring-test 模块
spring-boot-starter-thymeleaf	支持 Thymeleaf 模板引擎构建 MVC Web 应用的 starter
spring-boot-starter-tomcat	支持 Spring Boot 的默认 Servlet 容器 Tomcat 的 starter
spring-boot-starter-undertow	使用 Undertow 作为嵌入式 servlet 容器替代 Tomcat 的 starter
spring-boot-starter-validation	支持 Java Bean 验证的 starter，包括 Hibernate 验证器

续表

名称	描述
spring-boot-starter-web	支持Web应用开发的starter,包括Tomcat(默认嵌入式容器)和spring-webmvc
spring-boot-starter-web-services	支持Spring Web Services的starter
spring-boot-starter-webflux	支持WebFlux(一个非阻塞异步框架)开发的starter
spring-boot-starter-websocket	支持WebSocket开发的starter

除了Spring Boot官方提供的Starters外,还可以通过访问"https://github.com/spring-projects/spring-boot/blob/master/spring-boot-project/spring-boot-starters/README.adoc"网站,查看第三方为Spring Boot贡献的Starters。编写本书时,共有如表4.2所示的第三方Starters。

表4.2 第三方Starters

名称	地址
Apache Camel	https://github.com/apache/camel/tree/master/components/camel-spring-boot
Apache CXF	https://github.com/apache/cxf
Apache Qpid	https://github.com/amqphub/amqp-10-jms-spring-boot
Apache Wicket	https://github.com/MarcGiffing/wicket-spring-boot
ArangoDB	https://github.com/arangodb/spring-boot-starter
Axon Framework	https://github.com/AxonFramework/AxonFramework
Azure	https://github.com/Microsoft/azure-spring-boot-starters
Azure Application Insights	https://github.com/Microsoft/ApplicationInsights-Java/tree/master/azure-application-insights-spring-boot-starter
Bucket4j	https://github.com/MarcGiffing/bucket4j-spring-boot-starter
Camunda BPM	https://github.com/camunda/camunda-bpm-spring-boot-starter
Charon reverse proxy	https://github.com/mkopylec/charon-spring-boot-starter
Cloudant	https://github.com/icha024/cloudant-spring-boot-starter
Couchbase HTTP session	https://github.com/mkopylec/session-couchbase-spring-boot-starter
DataSource decorating	https://github.com/gavlyukovskiy/spring-boot-data-source-decorator
Docker Java and Docker Client	https://github.com/jliu666/docker-api-spring-boot
Dozer	https://github.com/DozerMapper/dozer
ErroREST exception handler	https://github.com/mkopylec/errorest-spring-boot-starter
Flowable	https://github.com/flowable/flowable-engine/tree/master/modules/flowable-spring-boot/flowable-spring-boot-starters
Google's reCAPTCHA	https://github.com/mkopylec/recaptcha-spring-boot-starter
GraphQL and GraphiQL with GraphQL Java	https://github.com/graphql-java/graphql-spring-boot
gRPC	https://github.com/LogNet/grpc-spring-boot-starter
HA JDBC	https://github.com/lievendoclo/hajdbc-spring-boot
Handlebars	https://github.com/allegro/handlebars-spring-boot-starter
HDIV	https://github.com/hdiv/spring-boot-starter-hdiv
Hiatus for Spring Boot	https://github.com/jihor/hiatus-spring-boot

续表

名 称	地 址
Infinispan	https://github.com/infinispan/infinispan-spring-boot
Jade Templates (Jade4J)	https://github.com/domix/jade4j-spring-boot-starter
JavaMelody monitoring	https://github.com/javamelody/javamelody/wiki/SpringBootStarter
JaVers	https://github.com/javers/javers
JODConverter	https://github.com/sbraconnier/jodconverter
JSF integration for various libraries	http://joinfaces.org
Liquigraph	https://github.com/liquigraph/liquigraph
Logback-access	https://github.com/akihyro/logback-access-spring-boot-starter
MyBatis	https://github.com/mybatis/mybatis-spring-boot
Narayana	https://github.com/snowdrop/narayana-spring-boot
Nutz	https://github.com/nutzam/nutzmore
OkHttp	https://github.com/freefair/okhttp-spring-boot
Okta	https://github.com/okta/okta-spring-boot
Orika	https://github.com/akihyro/orika-spring-boot-starter
RabbitMQ (Advanced usage)	https://github.com/societe-generale/rabbitmq-advanced-spring-boot-starter
RESTEasy	https://github.com/resteasy/resteasy-spring-boot
Rollbar	https://github.com/olmero/rollbar-spring-boot-starter
Sentry	https://github.com/getsentry/sentry-java/tree/master/sentry-spring-boot-starter
SOAP Web Services support with Apache CXF	https://github.com/codecentric/cxf-spring-boot-starter
Spring Batch (Advanced usage)	https://github.com/codecentric/spring-boot-starter-batch-web
Spring Shell	https://github.com/fonimus/ssh-shell-spring-boot
Sprout Platform	https://github.com/savantly-net/sprout-platform/tree/master/spring/sprout-spring-boot-starter
SSH Daemon	https://github.com/anand1st/sshd-shell-spring-boot
Stripe API	https://github.com/pankajtandon/stripe-starter
Stripes	https://github.com/juanpablo-santos/stripes-spring-boot
Structurizr	https://github.com/Catalysts/structurizr-extensions
Vaadin	https://github.com/vaadin/spring/tree/master/vaadin-spring-boot-starter
Valiktor	https://github.com/valiktor/valiktor/tree/master/valiktor-spring/valiktor-spring-boot-starter
WireMock and Spring REST Docs	https://github.com/ePages-de/restdocs-wiremock
Wro4j	https://github.com/michael-simons/wro4j-spring-boot-starter

4.2 读取应用配置

视频讲解

Spring Boot 提供了 3 种方式读取项目的 application.properties 配置文件的内容。这 3 种方式分别为 Environment 类、@Value 注解以及@ConfigurationProperties 注解。

4.2.1 Environment

Environment 是一个通用的读取应用程序运行时的环境变量的类,可以通过 key-value 方式读取 application.properties、命令行输入参数、系统属性、操作系统环境变量等。下面通过一个实例来演示如何使用 Environment 类读取 application.properties 配置文件的内容。

【例 4-1】 使用 Environment 类读取 application.properties 配置文件的内容。

具体实现步骤如下。

1. 创建 Spring Boot 项目 ch4_1

使用 STS 快速创建 Spring Boot Web 应用 ch4_1 (参考 3.3.2 节),项目目录结构如图 4.7 所示。

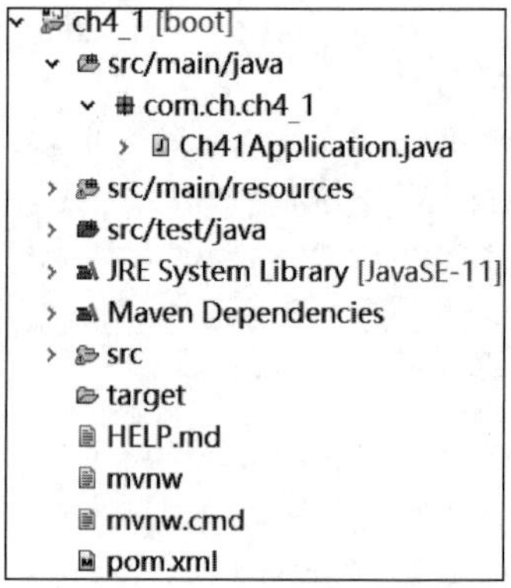

2. 添加配置文件内容

在 src/main/resources 目录下,找到全局配置文件 application.properties,并添加如下内容:

```
test.msg = read config
```

3. 创建控制器类 EnvReaderConfigController

图 4.7 Web 应用 ch4_1 的目录结构

在 src/main/java 目录下,创建名为 com.ch.ch4_1.controller 的包(是 com.ch.ch4_1 包的子包,保障注解全部被扫描),并在该包下创建控制器类 EnvReaderConfigController。在控制器类 EnvReaderConfigController 中,使用@Autowired 注解依赖注入 Environment 类的对象,具体代码如下:

```java
package com.ch.ch4_1.controller;
import org.springframework.beans.factory.annotation.Autowired;
import org.springframework.core.env.Environment;
import org.springframework.web.bind.annotation.RequestMapping;
import org.springframework.web.bind.annotation.RestController;
@RestController
public class EnvReaderConfigController{
    @Autowired
    private Environment env;
    @RequestMapping("/testEnv")
    public String testEnv() {
```

```
        return "方法一:" + env.getProperty("test.msg");
        //test.msg 为配置文件 application.properties 中的 key
    }
}
```

4. 启动 Spring Boot 应用

运行 Ch41Application 类的 main 方法,启动 Spring Boot 应用。

5. 测试应用

启动 Spring Boot 应用后,默认访问地址为 http://localhost:8080/,将项目路径直接设为根路径,这是 Spring Boot 的默认设置。因此,我们可以通过 http://localhost:8080/testEnv 测试应用(testEnv 与控制器类 ReaderConfigController 中的 @RequestMapping("/testEnv")对应),测试效果如图 4.8 所示。

图 4.8 使用 Environment 类读取 application.properties 文件内容

4.2.2 @Value

使用@Value 注解读取配置文件内容示例代码如下:

```
@Value("${test.msg}")         //test.msg 为配置文件 application.properties 中的 key
private String msg;           //通过@Value 注解将配置文件中 key 对应的 value 赋值给变量 msg
```

下面通过实例讲解如何使用@Value 注解读取配置文件内容。

【例 4-2】 使用@Value 注解读取配置文件内容。

具体实现步骤如下。

1. 创建控制器类 ValueReaderConfigController

在 ch4_1 项目的 com.ch.ch4_1.controller 包中,创建名为 ValueReaderConfigController 的控制器类,在该控制器类中使用@Value 注解读取配置文件内容,具体代码如下:

```
package com.ch.ch4_1.controller;
import org.springframework.beans.factory.annotation.Value;
import org.springframework.web.bind.annotation.RequestMapping;
import org.springframework.web.bind.annotation.RestController;
@RestController
public class ValueReaderConfigController {
    @Value("${test.msg}")
    private String msg;
    @RequestMapping("/testValue")
    public String testValue() {
        return "方法二:" + msg ;
    }
}
```

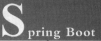

2. 启动并测试应用

首先，运行 Ch41Application 类的 main 方法，启动 Spring Boot 应用。然后，通过 http://localhost:8080/testValue 测试应用。

4.2.3 @ConfigurationProperties

使用@ConfigurationProperties 首先建立配置文件与对象的映射关系，然后在控制器方法中使用@Autowired 注解将对象注入。

下面通过实例讲解如何使用@ConfigurationProperties 读取配置文件内容。

【例 4-3】 使用@ConfigurationProperties 读取配置文件内容。

具体实现步骤如下。

1. 添加配置文件内容

在 ch4_1 项目的 src/main/resources 目录下，找到全局配置文件 application.properties，并添加如下内容：

```
obj.sname = chenheng
obj.sage = 88
```

2. 建立配置文件与对象的映射关系

在 ch4_1 项目的 src/main/java 目录下创建名为 com.ch.ch4_1.model 的包，并在包中创建实体类 StudentProperties，在该类中使用@ConfigurationProperties 注解建立配置文件与对象的映射关系，具体代码如下：

```java
package com.ch.ch4_1.model;
import org.springframework.boot.context.properties.ConfigurationProperties;
import org.springframework.stereotype.Component;
@Component                            //使用 Component 注解，声明一个组件，被控制器依赖注入
@ConfigurationProperties(prefix = "obj")     //obj 为配置文件中 key 的前缀
public class StudentProperties {
    private String sname;
    private int sage;
    //省略 set 和 get 方法
    @Override
    public String toString() {
        return "StudentProperties [sname = " + sname + ", sage = " + sage + "]";
    }
}
```

3. 创建控制器类 ConfigurationPropertiesController

在 ch4_1 项目的 com.ch.ch4_1.controller 包中，创建名为 ConfigurationPropertiesController 的控制器类，在该控制器类中使用@Autowired 注解依赖注入 StudentProperties 对象，具体

代码如下：

```
package com.ch.ch4_1.controller;
import org.springframework.beans.factory.annotation.Autowired;
import org.springframework.web.bind.annotation.RequestMapping;
import org.springframework.web.bind.annotation.RestController;
import com.ch.ch4_1.model.StudentProperties;
@RestController
public class ConfigurationPropertiesController {
    @Autowired
    StudentProperties studentProperties;
    @RequestMapping("/testConfigurationProperties")
    public String testConfigurationProperties() {
        return studentProperties.toString();
    }
}
```

4. 启动并测试应用

首先，运行 Ch41Application 类的 main 方法，启动 Spring Boot 应用。然后，通过 http://localhost:8080/testConfigurationProperties 测试应用。

4.2.4　@PropertySource

开发者希望读取项目的其他配置文件，而不是全局配置文件 application.properties，该如何实现呢？可以使用@PropertySource 注解找到项目的其他配置文件，然后结合 4.2.1 节～4.2.3 节中任意一种方式读取即可。

下面通过实例讲解如何使用@PropertySource ＋ @Value 读取其他配置文件内容。

【例 4-4】 使用@PropertySource ＋ @Value 读取其他配置文件内容。

具体实现步骤如下。

1. 创建配置文件

在 ch4_1 的 src/main/resources 目录下创建配置文件 ok.properties 和 test.properties，并在 ok.properties 文件中添加如下内容：

your.msg = hello.

在 test.properties 文件中添加如下内容：

my.msg = test PropertySource

2. 创建控制器类 PropertySourceValueReaderOhterController

在 ch4_1 项目的 com.ch.ch4_1.controller 包中，创建名为 PropertySourceValueReaderOhterController 的控制器类。在该控制器类中，首先使用@PropertySource 注解找到其他配置文件，然后使用@Value 注解读取配置文件内容，具体代码如下：

```
package com.ch.ch4_1.controller;
import org.springframework.beans.factory.annotation.Value;
import org.springframework.context.annotation.PropertySource;
import org.springframework.web.bind.annotation.RequestMapping;
import org.springframework.web.bind.annotation.RestController;
@RestController
@PropertySource({"test.properties","ok.properties"})
public class PropertySourceValueReaderOhterController {
    @Value("${my.msg}")
    private String mymsg;
    @Value("${your.msg}")
    private String yourmsg;
    @RequestMapping("/testProperty")
    public String testProperty() {
           return "其他配置文件 test.properties:" + mymsg + "<br>"
                + "其他配置文件 ok.properties:" + yourmsg;
    }
}
```

3. 启动并测试应用

首先，运行 Ch41Application 类的 main 方法，启动 Spring Boot 应用。然后，通过 http://localhost:8080/testProperty 测试应用。测试结果如图 4.9 所示。

图 4.9 读取其他配置文件内容

4.3 日志配置

视频讲解

默认情况下，Spring Boot 项目使用 LogBack 实现日志，使用 apache Commons Logging 作为日志接口，因此代码中通常如下使用日志：

```
package com.ch.ch4_1.controller;
import org.apache.commons.logging.Log;
import org.apache.commons.logging.LogFactory;
import org.springframework.web.bind.annotation.RequestMapping;
import org.springframework.web.bind.annotation.RestController;
@RestController
public class LogTestController {
    private Log log = LogFactory.getLog(LogTestController.class);
    @RequestMapping("/testLog")
    public String testLog() {
        log.info("测试日志");
        return "测试日志";
    }
}
```

通过地址 http://localhost:8080/testLog 运行上述控制器类代码,可以在控制台输出如图 4.10 所示的日志。

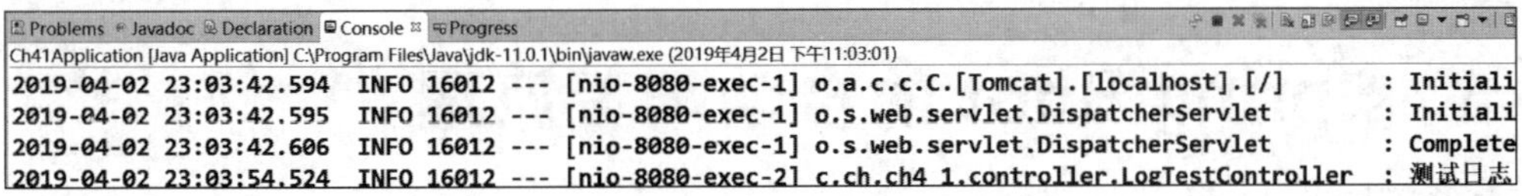

```
Problems  Javadoc  Declaration  Console  Progress
Ch41Application [Java Application] C:\Program Files\Java\jdk-11.0.1\bin\javaw.exe (2019年4月2日 下午11:03:01)
2019-04-02 23:03:42.594  INFO 16012 --- [nio-8080-exec-1] o.a.c.c.C.[Tomcat].[localhost].[/]       : Initiali
2019-04-02 23:03:42.595  INFO 16012 --- [nio-8080-exec-1] o.s.web.servlet.DispatcherServlet        : Initiali
2019-04-02 23:03:42.606  INFO 16012 --- [nio-8080-exec-1] o.s.web.servlet.DispatcherServlet        : Complete
2019-04-02 23:03:54.524  INFO 16012 --- [nio-8080-exec-2] c.ch.ch4_1.controller.LogTestController  : 测试日志
```

图 4.10　默认日志

日志级别有 ERROR、WARN、INFO、DEBUG 和 TRACE。Spring Boot 默认的日志级别为 INFO,日志信息可以打印到控制台。但开发者可以自己设定 Spring Boot 项目的日志输出级别,例如在 application.properties 配置文件中加入以下配置:

```
#设定日志的默认级别为 info
logging.level.root = info
#设定 org 包下的日志级别为 warn
logging.level.org = warn
#设定 com.ch.ch4_1 包下的日志级别为 debug
logging.level.com.ch.ch4_1 = debug
```

Spring Boot 项目默认并没有输出日志到文件,但开发者可以在 application.properties 配置文件中指定日志输出到文件,代码如下:

```
logging.file = my.log
```

日志输出到 my.log 文件,该日志文件位于 Spring Boot 项目运行的当前目录(项目工程目录下)。也可以指定日志文件目录,代码如下:

```
logging.file = C:/log/my.log
```

这样将 C:/log 目录下生成一个名为 my.log 的日志文件。不管日志文件位于何处,当日志文件大小到达 10MB 时,将自动生成一个新日志文件。

Spring Boot 使用内置的 LogBack 支持对控制台日志输出和文件输出进行格式控制,例如开发者可以在 application.properties 配置文件中添加如下配置:

```
logging.pattern.console = %level %date{yyyy-MM-dd HH:mm:ss:SSS} %logger{50}.%M %L:%m%n
logging.pattern.file = %level %date{ISO8601} %logger{50}.%M %L:%m%n
```

logging.pattern.console:指定控制台日志格式。

logging.pattern.file:指定日志文件格式。

%level:指定输出日志级别。

%date:指定日志发生的时间。ISO8601 表示标准日期,相当于 yyyy-MM-dd HH:mm:ss:SSS。

%logger:指定输出 Logger 的名字,包名+类名,{n}限定了输出长度。

%M:指定日志发生时的方法名。

%L:指定日志调用时所在代码行,适用于开发调试,线上运行时不建议使用此参数,因为获取代码行对性能有消耗。

%m：表示日志消息。
%n：表示日志换行。

4.4 Spring Boot 的自动配置原理

视频讲解

从 4.1.1 节可知，Spring Boot 使用核心注解@SpringBootApplication 将一个带有 main 方法的类标注为应用的启动类。@SpringBootApplication 注解最主要的功能之一是为 Spring Boot 开启了一个@EnableAutoConfiguration 注解的自动配置功能。

@EnableAutoConfiguration 注解主要利用了一个类名为 AutoConfigurationImportSelector 的选择器向 Spring 容器自动配置一些组件。@EnableAutoConfiguration 注解的源代码可以从 spring-boot-autoconfigure-2.1.4.RELEASE.jar(org.springframework.boot.autoconfigure)依赖包中查看，核心代码如下：

```
@Import(AutoConfigurationImportSelector.class)
public @interface EnableAutoConfiguration {
String ENABLED_OVERRIDE_PROPERTY = "spring.boot.enableautoconfiguration";
/**
 * Exclude specific auto-configuration classes such that they will never be applied.
 * @return the classes to exclude
 */
Class<?>[] exclude() default {};
/**
 * Exclude specific auto-configuration class names such that they will never be
 * applied.
 * @return the class names to exclude
 * @since 1.3.0
 */
String[] excludeName() default {};
}
```

AutoConfigurationImportSelector(源代码位于 org.springframework.boot.autoconfigure 包)类中有一个名为 selectImports 的方法，该方法规定了向 Spring 容器自动配置的组件。

selectImports 方法的代码如下：

```
@Override
public String[] selectImports(AnnotationMetadata annotationMetadata) {
    //判断@EnableAutoConfiguration 注解有没有开启，默认开启
    if (!isEnabled(annotationMetadata)) {
        return NO_IMPORTS;
    }
    //获取 META-INF/spring-autoconfigure-metadata.properties 的配置数据
    AutoConfigurationMetadata autoConfigurationMetadata = AutoConfigurationMetadataLoader
            .loadMetadata(this.beanClassLoader);
    //获得自动配置
    AutoConfigurationEntry autoConfigurationEntry = getAutoConfigurationEntry(
            autoConfigurationMetadata, annotationMetadata);
    return StringUtils.toStringArray(autoConfigurationEntry.getConfigurations());
}
```

在方法 selectImports 中,调用 getAutoConfigurationEntry 方法获得自动配置。进入该方法,查看到的源代码如下:

```java
protected AutoConfigurationEntry getAutoConfigurationEntry(
        AutoConfigurationMetadata autoConfigurationMetadata,
        AnnotationMetadata annotationMetadata) {
    if (!isEnabled(annotationMetadata)) {
        return EMPTY_ENTRY;
    }
    AnnotationAttributes attributes = getAttributes(annotationMetadata);
    //获取 META-INF/spring.factories 的配置数据
    List<String> configurations = getCandidateConfigurations(annotationMetadata,
            attributes);
    //去重
    configurations = removeDuplicates(configurations);
    //去除一些多余的类
    Set<String> exclusions = getExclusions(annotationMetadata, attributes);
    checkExcludedClasses(configurations, exclusions);
    configurations.removeAll(exclusions);
    //过滤掉一些条件没有满足的配置
    configurations = filter(configurations, autoConfigurationMetadata);
    fireAutoConfigurationImportEvents(configurations, exclusions);
    return new AutoConfigurationEntry(configurations, exclusions);
}
```

在方法 getAutoConfigurationEntry 中,调用 getCandidateConfigurations 方法获取 META-INF/spring.factories 的配置数据。进入该方法,查看到的源代码如下:

```java
protected List<String> getCandidateConfigurations(AnnotationMetadata metadata,
        AnnotationAttributes attributes) {
    List<String> configurations = SpringFactoriesLoader.loadFactoryNames(
            getSpringFactoriesLoaderFactoryClass(), getBeanClassLoader());
    Assert.notEmpty(configurations,
            "No auto configuration classes found in META-INF/spring.factories. If you "
                    + "are using a custom packaging, make sure that file is correct.");
    return configurations;
}
```

在方法 getCandidateConfigurations 中,调用了 loadFactoryNames 方法。进入该方法,查看到的源代码如下:

```java
public static List<String> loadFactoryNames(Class<?> factoryClass, @Nullable ClassLoader classLoader) {
    String factoryClassName = factoryClass.getName();
    return loadSpringFactories(classLoader).getOrDefault(factoryClassName, Collections.emptyList());
}
```

在方法 loadFactoryNames 中,调用了 loadSpringFactories 方法。进入该方法,查看到的源代码如下:

```java
private static Map<String, List<String>> loadSpringFactories(@Nullable ClassLoader classLoader) {
    MultiValueMap<String, String> result = cache.get(classLoader);
    if (result != null) {
        return result;
    }
    try {
        Enumeration<URL> urls = (classLoader != null ?
                classLoader.getResources(FACTORIES_RESOURCE_LOCATION) :
                ClassLoader.getSystemResources(FACTORIES_RESOURCE_LOCATION));
        result = new LinkedMultiValueMap<>();
        while (urls.hasMoreElements()) {
            URL url = urls.nextElement();
            UrlResource resource = new UrlResource(url);
            Properties properties = PropertiesLoaderUtils.loadProperties(resource);
            for (Map.Entry<?, ?> entry : properties.entrySet()) {
                String factoryClassName = ((String) entry.getKey()).trim();
        for (String factoryName : StringUtils.commaDelimitedListToStringArray((String) entry.getValue())) {
                    result.add(factoryClassName, factoryName.trim());
                }
            }
        }
        cache.put(classLoader, result);
        return result;
    }
    catch (IOException ex) {
        throw new IllegalArgumentException("Unable to load factories from location [" +
                FACTORIES_RESOURCE_LOCATION + "]", ex);
    }
}
```

在方法 loadSpringFactories 中,可以看到加载一个常量:FACTORIES_RESOURCE_LOCATION,该常量的源代码如下:

```java
/**
 * The location to look for factories.
 * <p>Can be present in multiple JAR files.
 */
public static final String FACTORIES_RESOURCE_LOCATION = "META-INF/spring.factories";
```

从上述源代码中可以看出,最终 Spring Boot 是通过加载所有(in multiple JAR files) META-INF/spring.factories 配置文件进行自动配置的。所以,@SpringBootApplication 注解通过使用@EnableAutoConfiguration 注解自动配置的原理是:从 classpath 中搜索所有 META-INF/spring.factories 配置文件,并将其中 org.springframework.boot.autoconfigure.EnableAutoConfiguration 对应的配置项通过 Java 反射机制进行实例化,然后汇总并加载到 Spring 的 IoC 容器。

在 Spring Boot 项目的 Maven Dependencies 的 spring-boot-autoconfigure-2.1.4.RELEASE.jar 目录下,找到 META-INF/spring.factories 配置文件,如图 4.11 所示。

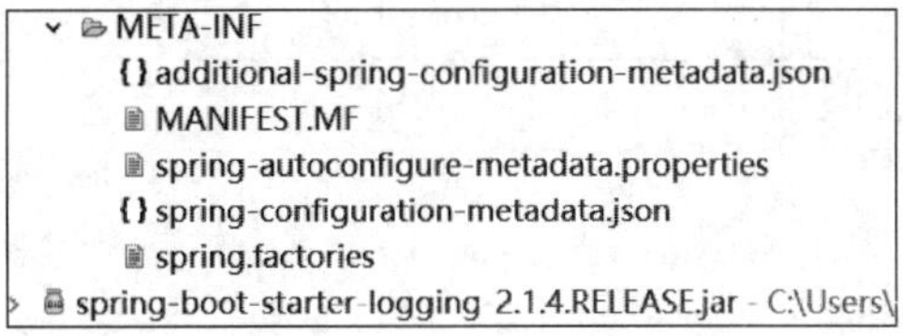

第4章 Spring Boot 核心

图 4.11 spring.factories 配置文件位置

视频讲解

4.5 Spring Boot 的条件注解

打开图 4.11 spring.factories 配置文件中任意一个 AutoConfiguration，一般都可以找到条件注解。例如，打开 org.springframework.boot.autoconfigure.aop.AopAutoConfiguration 的源代码，可以看到@ConditionalOnClass 和@ConditionalOnProperty 等条件注解。

org.springframework.boot.autoconfigure.aop.AopAutoConfiguration 的源代码如下：

```
@Configuration
@ConditionalOnClass({ EnableAspectJAutoProxy.class, Aspect.class, Advice.class,
        AnnotatedElement.class })
@ConditionalOnProperty(prefix = "spring.aop", name = "auto", havingValue = "true",
matchIfMissing = true)
public class AopAutoConfiguration {
    @Configuration
    @EnableAspectJAutoProxy(proxyTargetClass = false)
    @ConditionalOnProperty(prefix = "spring.aop", name = "proxy-target-class",
havingValue = "false", matchIfMissing = false)
    public static class JdkDynamicAutoProxyConfiguration {
    }
    @Configuration
    @EnableAspectJAutoProxy(proxyTargetClass = true)
    @ConditionalOnProperty(prefix = "spring.aop", name = "proxy-target-class",
havingValue = "true", matchIfMissing = true)
    public static class CglibAutoProxyConfiguration {
    }
}
```

通过上述源代码可以看出，Spring Boot 的自动配置是使用 Spring 的@Conditional 注解实现的。因此，本节将介绍相关的条件注解，并讲述如何自定义 Spring 的条件注解。

4.5.1 条件注解

所谓 Spring 的条件注解，就是应用程序的配置类的配置项，在满足某些特定条件后才会被自动启用。Spring Boot 的条件注解位于 spring-boot-autoconfigure-2.1.4.RELEASE.jar 的 org.springframework.boot.autoconfigure.condition 包下，具体如表 4.3 所示。

表 4.3　Spring Boot 的条件注解

注　解　名	条件实现类	条　　件
@ConditionalOnBean	OnBeanCondition	Spring 容器中存在指定的实例 Bean
@ConditionalOnClass	OnClassCondition	类加载器（类路径）中存在对应的类
@ConditionalOnCloudPlatform	OnCloudPlatformCondition	是否在云平台
@ConditionalOnExpression	OnExpressionCondition	判断 SpEL 表达式是否成立
@ConditionalOnJava	OnJavaCondition	指定 Java 版本是否符合要求
@ConditionalOnJndi	OnJndiCondition	在 JNDI（Java 命名和目录接口）存在的条件下查找指定的位置
@ConditionalOnMissingBean	OnBeanCondition	Spring 容器中不存在指定的实例 Bean
@ConditionalOnMissingClass	OnClassCondition	类加载器（类路径）中不存在对应的类
@ConditionalOnNotWebApplication	OnWebApplicationCondition	当前应用程序不是 Web 程序
@ConditionalOnProperty	OnPropertyCondition	应用环境中属性是否存在指定的值
@ConditionalOnResource	OnResourceCondition	是否存在指定的资源文件
@ConditionalOnSingleCandidate	OnBeanCondition	Spring 容器中是否存在且只存在一个对应的实例 Bean
@ConditionalOnWebApplication	OnWebApplicationCondition	当前应用程序是 Web 程序

表 4.3 中的条件注解都是组合了 @Conditional 元注解，只是针对不同的条件去实现。

【例 4-5】　通过 @ConditionalOnWebApplication 注解的源码分析，讲解条件注解的实现方法。

@ConditionalOnWebApplication 是一个标记注解，源代码如下：

```
package org.springframework.boot.autoconfigure.condition;
import java.lang.annotation.Documented;
import java.lang.annotation.ElementType;
import java.lang.annotation.Retention;
import java.lang.annotation.RetentionPolicy;
import java.lang.annotation.Target;
import org.springframework.context.annotation.Conditional;
@Target({ ElementType.TYPE, ElementType.METHOD })
@Retention(RetentionPolicy.RUNTIME)
@Documented
@Conditional(OnWebApplicationCondition.class)
public @interface ConditionalOnWebApplication {
}
```

下面通过代码注释的形式，分析了 @ConditionalOnWebApplication 注解的实现类 OnWebApplicationCondition 的源代码，具体分析如下：

```
class OnWebApplicationCondition extends FilteringSpringBootCondition {
    …
    /** ConditionOutcome 记录了匹配结果 */
    @Override
    public ConditionOutcome getMatchOutcome(ConditionContext context,
```

```java
        AnnotatedTypeMetadata metadata) {
    //1. 检查是否被@ConditionalOnWebApplication注解
    boolean required = metadata
            .isAnnotated(ConditionalOnWebApplication.class.getName());
    //2. 判断是否是 WebApplication
    ConditionOutcome outcome = isWebApplication(context, metadata, required);
    /** 3. 如果有@ConditionalOnWebApplication 注解,但不是 WebApplication 环境,则返回
     不匹配 */
    if (required && !outcome.isMatch()) {
        return ConditionOutcome.noMatch(outcome.getConditionMessage());
    }
    /** 4. 如果没有被@ConditionalOnWebApplication 注解,但是 WebApplication 环境,则返
    回不匹配 */
    if (!required && outcome.isMatch()) {
        return ConditionOutcome.noMatch(outcome.getConditionMessage());
    }
    /** 5. 如果被@ConditionalOnWebApplication 注解,并且是 WebApplication 环境,则返回
    匹配 */
    return ConditionOutcome.match(outcome.getConditionMessage());
}
/** 判断是否是 Web 环境 */
private ConditionOutcome isWebApplication(ConditionContext context,
        AnnotatedTypeMetadata metadata, boolean required) {
    switch (deduceType(metadata)) {
    //1. 基于 servlet 的 Web 应用程序
    case SERVLET:
        return isServletWebApplication(context);
    //2. 基于 reactive(响应)的 Web 应用程序
    case REACTIVE:
        return isReactiveWebApplication(context);
    //3. 任意的 Web 应用程序
    default:
        return isAnyWebApplication(context, required);
    }
}
/** 任意的 Web 应用程序 */
private ConditionOutcome isAnyWebApplication(ConditionContext context,
        boolean required) {
    ConditionMessage.Builder message = ConditionMessage.forCondition(
            ConditionalOnWebApplication.class, required ? "(required)" : "");
    ConditionOutcome servletOutcome = isServletWebApplication(context);
    if (servletOutcome.isMatch() && required) {
        return new ConditionOutcome(servletOutcome.isMatch(),
                message.because(servletOutcome.getMessage()));
    }
    ConditionOutcome reactiveOutcome = isReactiveWebApplication(context);
    if (reactiveOutcome.isMatch() && required) {
        return new ConditionOutcome(reactiveOutcome.isMatch(),
                message.because(reactiveOutcome.getMessage()));
    }
    return new ConditionOutcome(servletOutcome.isMatch() || reactiveOutcome.isMatch(),
```

```java
            message.because(servletOutcome.getMessage()).append("and")
                    .append(reactiveOutcome.getMessage())));
}
/** 基于 servlet 的 Web 应用程序判断 */
private ConditionOutcome isServletWebApplication(ConditionContext context) {
    ConditionMessage.Builder message = ConditionMessage.forCondition("");
    //1. 判断 GenericWebApplicationContext 是否在类路径中,如果不存在,则返回不匹配
    if (!ClassNameFilter.isPresent(SERVLET_WEB_APPLICATION_CLASS,
            context.getClassLoader())) {
        return ConditionOutcome.noMatch(
                message.didNotFind("servlet web application classes").atAll());
    }
    //2. 容器里是否有名为 session 的 scope,如果存在,则返回匹配
    if (context.getBeanFactory() != null) {
        String[] scopes = context.getBeanFactory().getRegisteredScopeNames();
        if (ObjectUtils.containsElement(scopes, "session")) {
            return ConditionOutcome.match(message.foundExactly("'session' scope"));
        }
    }
    //3. Environment 是否为 ConfigurableWebEnvironment,如果是,则返回匹配
    if (context.getEnvironment() instanceof ConfigurableWebEnvironment) {
        return ConditionOutcome
                .match(message.foundExactly("ConfigurableWebEnvironment"));
    }
    //4. 当前 ResourceLoader 是否为 WebApplicationContext,如果是,则返回匹配
    if (context.getResourceLoader() instanceof WebApplicationContext) {
        return ConditionOutcome.match(message.foundExactly("WebApplicationContext"));
    }
    //5. 其他情况,返回不匹配
    return ConditionOutcome.noMatch(message.because("not a servlet web application"));
}
/** 基于 reactive(响应)的 Web 应用程序 */
private ConditionOutcome isReactiveWebApplication(ConditionContext context) {
    ConditionMessage.Builder message = ConditionMessage.forCondition("");
    //1. 判断 HandlerResult 是否在类路径中,如果不存在,则返回不匹配
    if (!ClassNameFilter.isPresent(REACTIVE_WEB_APPLICATION_CLASS,
            context.getClassLoader())) {
        return ConditionOutcome.noMatch(
                message.didNotFind("reactive web application classes").atAll());
    }
    //2. Environment 是否为 ConfigurableReactiveWebEnvironment,如果是,则返回匹配
    if (context.getEnvironment() instanceof ConfigurableReactiveWebEnvironment) {
        return ConditionOutcome
                .match(message.foundExactly("ConfigurableReactiveWebEnvironment"));
    }
    //3. 当前 ResourceLoader 是否为 ReactiveWebApplicationContext,如果是,则返回匹配
    if (context.getResourceLoader() instanceof ReactiveWebApplicationContext) {
        return ConditionOutcome
                .match(message.foundExactly("ReactiveWebApplicationContext"));
    }
    //4. 其他情况,返回不匹配
```

```
            return ConditionOutcome
                    .noMatch(message.because("not a reactive web application"));
        }
        …
}
```

从上述源代码可以看出,实现类 OnWebApplicationCondition 的 getMatchOutcome 方法返回条件匹配结果。其中,最重要的一步是判断是否是 Web 环境(ConditionOutcome outcome = isWebApplication(context, metadata, required);)。

4.5.2 实例分析

在 4.5.1 节了解了 Spring Boot 的条件注解后,下面通过实例来分析 Spring Boot 内置的一个简单的自动配置:HTTP 编码配置。

【例 4-6】 分析 HTTP 编码配置。

在 Spring MVC 常规项目中,可以在 web.xml 中配置一个 filter 设置 HTTP 编码,例如:

```xml
<filter>
    <filter-name>characterEncodingFilter</filter-name>
    <filter-class>org.springframework.web.filter.CharacterEncodingFilter</filter-class>
    <init-param>
        <param-name>encoding</param-name>
        <param-value>UTF-8</param-value>
    </init-param>
    <init-param>
        <param-name>forceEncoding</param-name>
        <param-value>true</param-value>
    </init-param>
</filter>
<filter-mapping>
    <filter-name>characterEncodingFilter</filter-name>
    <url-pattern>/*</url-pattern>
</filter-mapping>
```

Spring Boot 自动配置 HTTP 编码需要满足以下两个条件:
- 配置 encoding 和 forceEncoding 两个参数;
- 配置 CharacterEncodingFilter 的 Bean。

1. 配置 HTTP 编码参数

通过查看源代码可知 org.springframework.boot.autoconfigure.http.HttpProperties 类配置了 HTTP 编码参数,具体代码如下:

```
package org.springframework.boot.autoconfigure.http;
import java.nio.charset.Charset;
import java.nio.charset.StandardCharsets;
import java.util.Locale;
```

```java
import java.util.Map;
import org.springframework.boot.context.properties.ConfigurationProperties;
/**
 * HTTP 属性
 */
@ConfigurationProperties(prefix = "spring.http")      //在 application.properties 中配置
                                                      //前缀是 spring.http
public class HttpProperties {
    private boolean logRequestDetails;
    //HTTP 编码属性
    private final Encoding encoding = new Encoding();
    public boolean isLogRequestDetails() {
        return this.logRequestDetails;
    }
    public void setLogRequestDetails(boolean logRequestDetails) {
        this.logRequestDetails = logRequestDetails;
    }
    public Encoding getEncoding() {
        return this.encoding;
    }
    /**
     *    HTTP 编码的配置属性
     */
    public static class Encoding {
        public static final Charset DEFAULT_CHARSET = StandardCharsets.UTF_8;
        private Charset charset = DEFAULT_CHARSET;        //默认编码为 UTF-8
        private Boolean force;                            //设置 forceEncoding,默认为 true
        private Boolean forceRequest;
        private Boolean forceResponse;
        private Map<Locale, Charset> mapping;
        public Charset getCharset() {
            return this.charset;
        }
        public void setCharset(Charset charset) {
            this.charset = charset;
        }
        public boolean isForce() {
            return Boolean.TRUE.equals(this.force);
        }
        public void setForce(boolean force) {
            this.force = force;
        }
        public boolean isForceRequest() {
            return Boolean.TRUE.equals(this.forceRequest);
        }
        public void setForceRequest(boolean forceRequest) {
            this.forceRequest = forceRequest;
        }
        public boolean isForceResponse() {
            return Boolean.TRUE.equals(this.forceResponse);
        }
```

```java
    public void setForceResponse(boolean forceResponse) {
        this.forceResponse = forceResponse;
    }
    public Map<Locale, Charset> getMapping() {
        return this.mapping;
    }
    public void setMapping(Map<Locale, Charset> mapping) {
        this.mapping = mapping;
    }
    public boolean shouldForce(Type type) {
        Boolean force = (type != Type.REQUEST) ? this.forceResponse
                : this.forceRequest;
        if (force == null) {
            force = this.force;
        }
        if (force == null) {
            force = (type == Type.REQUEST);
        }
        return force;
    }
    public enum Type {
        REQUEST, RESPONSE
    }
  }
}
```

通过分析 HTTP 属性类 HttpProperties 的源码可知：

① 在 application.properties 中配置前缀是 spring.http(@ConfigurationProperties(prefix = "spring.http"))。

② 默认编码方式为 UTF-8(private Charset charset = DEFAULT_CHARSET;)，若修改可使用 spring.http.encoding.charset=编码。

③ 设置 forceEncoding，默认为 true(private Boolean force;)，若修改可使用 spring.http.encoding.force=false。

2. 配置 CharacterEncodingFilter 的 Bean

通过查看源代码可知 org.springframework.boot.autoconfigure.web.servlet.HttpEncodingAutoConfiguration 类，根据条件注解配置了 CharacterEncodingFilter 的 Bean。代码分析见代码中的注释部分，具体分析如下：

```java
package org.springframework.boot.autoconfigure.web.servlet;
import org.springframework.boot.autoconfigure.EnableAutoConfiguration;
import org.springframework.boot.autoconfigure.condition.ConditionalOnClass;
import org.springframework.boot.autoconfigure.condition.ConditionalOnMissingBean;
import org.springframework.boot.autoconfigure.condition.ConditionalOnProperty;
import org.springframework.boot.autoconfigure.condition.ConditionalOnWebApplication;
import org.springframework.boot.autoconfigure.http.HttpProperties;
import org.springframework.boot.autoconfigure.http.HttpProperties.Encoding.Type;
import org.springframework.boot.context.properties.EnableConfigurationProperties;
```

```java
import org.springframework.boot.web.server.WebServerFactoryCustomizer;
import org.springframework.boot.web.servlet.filter.OrderedCharacterEncodingFilter;
import org.springframework.boot.web.servlet.server.ConfigurableServletWebServerFactory;
import org.springframework.context.annotation.Bean;
import org.springframework.context.annotation.Configuration;
import org.springframework.core.Ordered;
import org.springframework.web.filter.CharacterEncodingFilter;
@Configuration                                          //配置类
@EnableConfigurationProperties(HttpProperties.class)    //开启属性注入,使用@Autowired注入
@ConditionalOnWebApplication(type = ConditionalOnWebApplication.Type.SERVLET)/** 基于
Servlet 的 Web 应用程序 */
@ConditionalOnClass(CharacterEncodingFilter.class) //当 CharacterEncodingFilter 在类路径下
@ConditionalOnProperty(prefix = "spring.http.encoding", value = "enabled", matchIfMissing =
true)/** 当设置 spring.http.encoding = enabled 时,如果没有该设置默认为 true,即条件符合 */
public class HttpEncodingAutoConfiguration {
    private final HttpProperties.Encoding properties;
    public HttpEncodingAutoConfiguration(HttpProperties properties) {
        this.properties = properties.getEncoding();
    }
    @Bean                       //使用 Java 配置的方式配置 CharacterEncodingFilter 的 Bean
    @ConditionalOnMissingBean   /** 当 Spring 容器中没有 CharacterEncodingFilter 的 Bean
时,新建该 Bean */
    public CharacterEncodingFilter characterEncodingFilter() {
        CharacterEncodingFilter filter = new OrderedCharacterEncodingFilter();
        filter.setEncoding(this.properties.getCharset().name());
        filter.setForceRequestEncoding(this.properties.shouldForce(Type.REQUEST));
        filter.setForceResponseEncoding(this.properties.shouldForce(Type.RESPONSE));
        return filter;
    }
    @Bean
    public LocaleCharsetMappingsCustomizer localeCharsetMappingsCustomizer() {
        return new LocaleCharsetMappingsCustomizer(this.properties);
    }
    private static class LocaleCharsetMappingsCustomizer implements
            WebServerFactoryCustomizer<ConfigurableServletWebServerFactory>, Ordered {
        private final HttpProperties.Encoding properties;
        LocaleCharsetMappingsCustomizer(HttpProperties.Encoding properties) {
            this.properties = properties;
        }
        @Override
        public void customize(ConfigurableServletWebServerFactory factory) {
            if (this.properties.getMapping() != null) {
                factory.setLocaleCharsetMappings(this.properties.getMapping());
            }
        }
        @Override
        public int getOrder() {
            return 0;
        }
    }
}
```

4.5.3 自定义条件

Spring 的@Conditional 注解根据满足某特定条件创建一个特定的 Bean。例如,当某 jar 包在类路径下时,自动配置一个或多个 Bean。这就是根据特定条件控制 Bean 的创建行为,这样就可以利用这个特性进行一些自动配置。那么,开发者如何自己构造条件呢? 在 Spring 框架中,可以通过实现 Condition 接口,并重写 matches 方法来构造条件。下面通过实例讲解条件的构造过程。

【例 4-7】 如果类路径 classpath(src/main/resources)下存在文件 test.properties,则输出"test.properties 文件存在。"; 否则输出"test.properties 文件不存在!"。

具体实现步骤如下。

1. 构造条件

在 Spring Boot 应用 ch4_1 的 src/main/java 目录下创建 com.ch.ch4_1.conditional 包,并在该包中分别创建条件实现类 MyCondition(存在文件 test.properties)和 YourCondition(不存在文件 test.properties)。

MyCondition 的代码如下:

```java
package com.ch.ch4_1.conditional;
import org.springframework.context.annotation.Condition;
import org.springframework.context.annotation.ConditionContext;
import org.springframework.core.type.AnnotatedTypeMetadata;
public class MyCondition implements Condition{
    @Override
    public boolean matches(ConditionContext context, AnnotatedTypeMetadata metadata) {
        return context.getResourceLoader().getResource("classpath:test.properties").exists();
    }
}
```

YourCondition 的代码如下:

```java
package com.ch.ch4_1.conditional;
import org.springframework.context.annotation.Condition;
import org.springframework.context.annotation.ConditionContext;
import org.springframework.core.type.AnnotatedTypeMetadata;
public class YourCondition implements Condition{
    @Override
    public boolean matches(ConditionContext context, AnnotatedTypeMetadata metadata) {
        return !context.getResourceLoader().getResource("classpath:test.properties").exists();
    }
}
```

2. 创建不同条件下 Bean 的类

在 com.ch.ch4_1.conditional 包中,创建 MessagePrint 接口,并分别创建该接口的实

现类 MyMessagePrint 和 YourMessagePrint。

MessagePrint 的代码如下：

```
package com.ch.ch4_1.conditional;
public interface MessagePrint {
    public String showMessage();
}
```

MyMessagePrint 的代码如下：

```
package com.ch.ch4_1.conditional;
public class MyMessagePrint implements MessagePrint{
    @Override
    public String showMessage() {
        return "test.properties 文件存在。";
    }
}
```

YourMessagePrint 的代码如下：

```
package com.ch.ch4_1.conditional;
public class YourMessagePrint implements MessagePrint{
    @Override
    public String showMessage() {
        return "test.properties 文件不存在！";
    }
}
```

3．创建配置类

在 com.ch.ch4_1.conditional 包中，创建配置类 ConditionConfig，并在该配置类中使用 @Bean 和 @Conditional 实例化符合条件的 Bean。

ConditionConfig 的代码如下：

```
package com.ch.ch4_1.conditional;
import org.springframework.context.annotation.Bean;
import org.springframework.context.annotation.Conditional;
import org.springframework.context.annotation.Configuration;
@Configuration
public class ConditionConfig {
    @Bean
    @Conditional(MyCondition.class)
    public MessagePrint myMessage() {
        return new MyMessagePrint();
    }
    @Bean
    @Conditional(YourCondition.class)
    public MessagePrint yourMessage() {
        return new YourMessagePrint();
    }
}
```

4. 创建测试类

在 com.ch.ch4_1.conditional 包中，创建测试类 TestMain，具体代码如下：

```
package com.ch.ch4_1.conditional;
import org.springframework.context.annotation.AnnotationConfigApplicationContext;
public class TestMain {
    private static AnnotationConfigApplicationContext context;
    public static void main(String[] args) {
        context = new AnnotationConfigApplicationContext(ConditionConfig.class);
        MessagePrint mp = context.getBean(MessagePrint.class);
        System.out.println(mp.showMessage());
    }
}
```

5. 运行

当 Spring Boot 应用 ch4_1 的 src/main/resources 目录下存在 test.properties 文件时，运行测试类，结果如图 4.12 所示；当 Spring Boot 应用 ch4_1 的 src/main/resources 目录下不存在 test.properties 文件时，运行测试类，结果如图 4.13 所示。

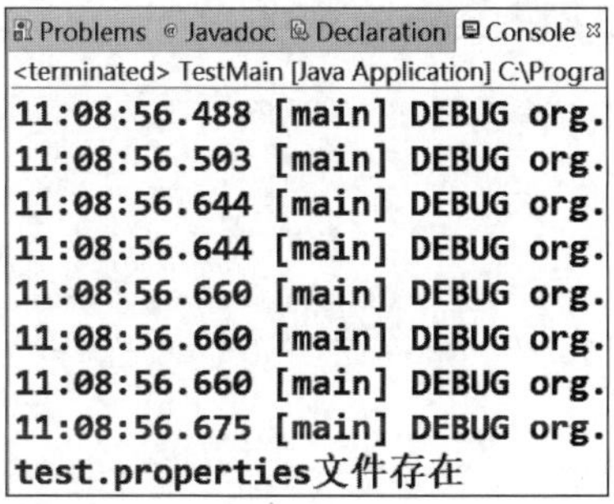

图 4.12　存在 test.properties 文件

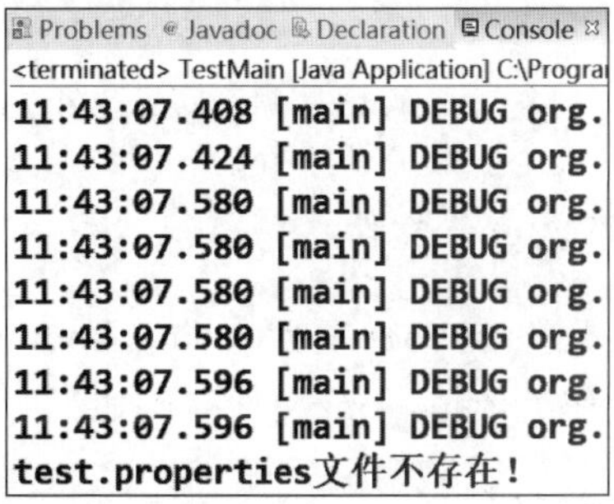

图 4.13　不存在 test.properties 文件

4.5.4　自定义 Starters

从 4.1.6 节可知，第三方为 Spring Boot 贡献了许多 Starters。那么，我们作为开发者是否也可以贡献自己的 Starters？学习 Spring Boot 的自动配置机制后，答案是肯定的。下面通过实例讲解如何自定义 Starters。

【例 4-8】　自定义一个 Starter（spring_boot_mystarters）。要求：当类路径中存在 MyService 类时，自动配置该类的 Bean，并可以在 application.properties 中配置相应 Bean 的属性。

具体实现步骤如下。

1. 新建 Spring Boot 项目 spring_boot_mystarters

首先，选择菜单 File|New|Spring Starter Project，打开如图 4.14 所示的 New Spring

Starter Project 对话框。其次,在对话框中输入项目名称 spring_boot_mystarters。最后,单击 Next 与 Finish 按钮。

图 4.14 新建 Spring Boot 项目 spring_boot_mystarters

2. 修改 pom 文件

修改 Spring Boot 项目 spring_boot_mystarters 的 pom 文件,增加 Spring Boot 自身的自动配置作为依赖,pom 文件代码如下:

```
<?xml version="1.0" encoding="UTF-8"?>
<project xmlns="http://maven.apache.org/POM/4.0.0"
xmlns:xsi="http://www.w3.org/2001/XMLSchema-instance"
    xsi:schemaLocation="http://maven.apache.org/POM/4.0.0
http://maven.apache.org/xsd/maven-4.0.0.xsd">
    <modelVersion>4.0.0</modelVersion>
    <parent>
        <groupId>org.springframework.boot</groupId>
        <artifactId>spring-boot-starter-parent</artifactId>
        <version>2.1.4.RELEASE</version>
        <relativePath/> <!-- lookup parent from repository -->
    </parent>
    <groupId>com.ch</groupId>
    <artifactId>spring_boot_mystarters</artifactId>
    <version>0.0.1-SNAPSHOT</version>
    <name>spring_boot_mystarters</name>
    <description>Demo project for Spring Boot</description>
    <properties>
        <java.version>11</java.version>
    </properties>
```

```xml
<dependencies>
    <dependency>
        <groupId>org.springframework.boot</groupId>
        <artifactId>spring-boot-autoconfigure</artifactId>
    </dependency>
    <dependency>
        <groupId>org.springframework.boot</groupId>
        <artifactId>spring-boot-configuration-processor</artifactId>
        <optional>true</optional>
    </dependency>
    <dependency>
        <groupId>org.springframework.boot</groupId>
        <artifactId>spring-boot-starter-test</artifactId>
        <scope>test</scope>
    </dependency>
</dependencies>
<build>
    <plugins>
        <plugin>
            <groupId>org.springframework.boot</groupId>
            <artifactId>spring-boot-maven-plugin</artifactId>
        </plugin>
    </plugins>
</build>
</project>
```

3. 创建属性配置类 MyProperties

在项目 spring_boot_mystarters 的 com.ch.spring_boot_mystarters 包中，创建属性配置类 MyProperties。在使用 spring_boot_mystarters 的 Spring Boot 项目的 application.properties 中，可以使用 my.msg=设置属性；若不设置，默认为 my.msg=默认值。属性配置类 MyProperties 的代码如下：

```java
package com.ch.spring_boot_mystarters;
import org.springframework.boot.context.properties.ConfigurationProperties;
//在 application.properties 中通过 my.msg=设置属性
@ConfigurationProperties(prefix = "my")
public class MyProperties {
    private String msg = "默认值";
    public String getMsg() {
        return msg;
    }
    public void setMsg(String msg) {
        this.msg = msg;
    }
}
```

4. 创建判断依据类 MyService

在项目 spring_boot_mystarters 的 com.ch.spring_boot_mystarters 包中，创建判断依

据类 MyService。本例自定义的 Starters 将根据该类的存在与否来创建该类的 Bean，该类可以是第三方类库的类。判断依据类 MyService 的代码如下：

```java
package com.ch.spring_boot_mystarters;
public class MyService {
    private String msg;
    public String sayMsg() {
        return "my " + msg;
    }
    public String getMsg() {
        return msg;
    }
    public void setMsg(String msg) {
        this.msg = msg;
    }
}
```

5. 创建自动配置类 MyAutoConfiguration

在项目 spring_boot_mystarters 的 com.ch.spring_boot_mystarters 包中，创建自动配置类 MyAutoConfiguration。在该类中使用@EnableConfigurationProperties 注解开启属性配置类 MyProperties 提供参数；使用@ConditionalOnClass 注解判断类加载器（类路径）中是否存在 MyService 类；使用@ConditionalOnMissingBean 注解判断当容器中不存在 MyService 的 Bean 时，自动配置这个 Bean。自动配置类 MyAutoConfiguration 的代码如下：

```java
package com.ch.spring_boot_mystarters;
import org.springframework.beans.factory.annotation.Autowired;
import org.springframework.boot.autoconfigure.condition.ConditionalOnClass;
import org.springframework.boot.autoconfigure.condition.ConditionalOnMissingBean;
import org.springframework.boot.autoconfigure.condition.ConditionalOnProperty;
import org.springframework.boot.context.properties.EnableConfigurationProperties;
import org.springframework.context.annotation.Bean;
import org.springframework.context.annotation.Configuration;
@Configuration
//开启属性配置类 MyProperties 提供参数
@EnableConfigurationProperties(MyProperties.class)
//类加载器（类路径）中是否存在对应的类
@ConditionalOnClass(MyService.class)
//应用环境中属性是否存在指定的值
@ConditionalOnProperty(prefix = "my", value = "enabled", matchIfMissing = true)
public class MyAutoConfiguration {
    @Autowired
    private MyProperties myProperties;
    @Bean
    //当容器中不存在 MyService 的 Bean 时，自动配置这个 Bean
    @ConditionalOnMissingBean(MyService.class)
    public MyService myService() {
        MyService myService = new MyService();
```

```
        myService.setMsg(myProperties.getMsg());
        return myService;
    }
}
```

6. 注册配置

在项目 spring_boot_mystarters 的 src/main/resources 目录下新建文件夹 META-INF，并在该文件夹下创建名为 spring.factories 的文件。在 spring.factories 文件中添加如下内容注册自动配置类 MyAutoConfiguration：

```
org.springframework.boot.autoconfigure.EnableAutoConfiguration = \
com.ch.spring_boot_mystarters.MyAutoConfiguration
```

上述文件内容中，若有多个自动配置，则使用","分开，此处"\"是为了换行后仍然能读到属性值。

至此，经过上述 6 个步骤后，自定义 Starters(spring_boot_mystarters)已经完成。可以将 spring_boot_mystarters 安装到 Maven 的本地库，或者将 jar 包发布到 Maven 的私服上。

【例 4-9】 在该例中创建 Spring Boot 的 Web 应用 ch4_2，并在 ch4_2 中使用 spring_boot_mystarters。

具体实现步骤如下。

1. 创建 Spring Boot 的 Web 应用 ch4_2

使用 STS 快速创建 Spring Boot 的 Web 应用 ch4_2。

2. 添加 spring_boot_mystarters 的依赖

在 Web 应用 ch4_2 的 pom.xml 文件中添加 spring_boot_mystarters 的依赖，具体代码如下：

```xml
<dependency>
    <groupId>com.ch</groupId>
    <artifactId>spring_boot_mystarters</artifactId>
    <version>0.0.1-SNAPSHOT</version>
</dependency>
```

添加依赖后，可以在 Maven 的依赖中查看 spring_boot_mystarters 依赖，如图 4.15 所示。

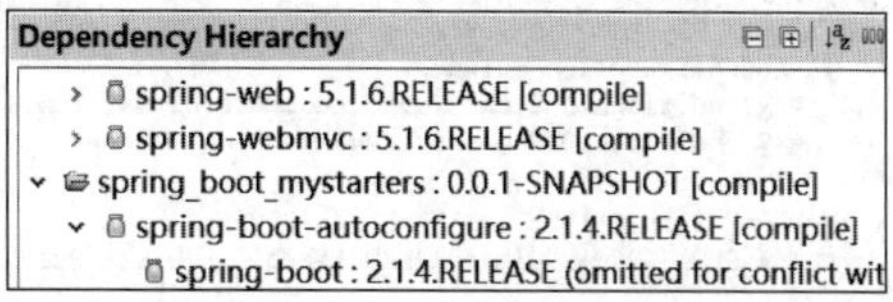

图 4.15 查看 spring_boot_mystarters 依赖

3. 修改程序入口类 Ch42Application，测试 spring_boot_mystarters

类 Ch42Application 修改后的代码如下：

```java
package com.ch.ch4_2;
import org.springframework.beans.factory.annotation.Autowired;
import org.springframework.boot.SpringApplication;
import org.springframework.boot.autoconfigure.SpringBootApplication;
```

```
import org.springframework.web.bind.annotation.RequestMapping;
import org.springframework.web.bind.annotation.RestController;
import com.ch.spring_boot_mystarters.MyService;
@RestController
@SpringBootApplication
public class Ch42Application {
    @Autowired MyService myService;
    public static void main(String[] args) {
        SpringApplication.run(Ch42Application.class, args);
    }
    @RequestMapping("/testStarters")
    public String index() {
        return myService.sayMsg();
    }
}
```

运行 Ch42Application 应用程序，启动 Web 应用。通过访问 http://localhost:8080/testStarters 测试 spring_boot_mystarters，运行效果如图 4.16 所示。

这时，在 Web 应用 ch4_2 的 application.properties 文件中配置 msg 的内容：

my.msg = starter pom

然后，运行 Ch42Application 应用程序，重新启动 Web 应用。再次访问 http://localhost:8080/testStarters，运行效果如图 4.17 所示。

图 4.16　访问 http://localhost:8080/testStarters　　　图 4.17　配置 msg 后的效果

另外，可以在 Web 应用 ch4_2 的 application.properties 文件中配置 debug 属性（debug=true），查看自动配置报告。重新启动 Web 应用，可以在控制台中查看到如图 4.18 所示的自定义的自动配置。

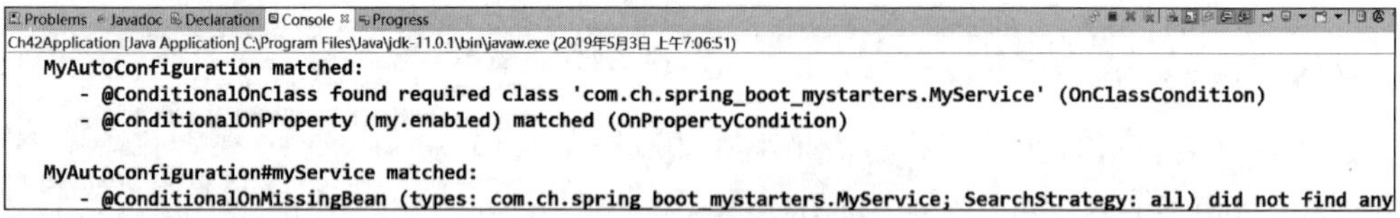

图 4.18　查看自动配置报告

4.6　本章小结

本章重点讲解了 Spring Boot 的基本配置、自动配置原理以及条件注解。通过本章的学习，开发者可以利用 Spring Boot 的自动配置与条件注解贡献自己的 Starters。

习题 4

1. 如何读取 Spring Boot 项目的应用配置？请举例说明。

2. 参考例 4-7，编写 Spring Boot 应用程序 practice4_2。要求如下：以不同的操作系统作为条件，若在 Windows 操作系统下运行程序，则输出列表命令为 dir；若在 Linux 操作系统下运行程序，则输出列表命令为 ls。

3. 参考例 4-8 与例 4-9，自定义一个 Starter(spring_boot_addstarters)和 Spring Boot 的 Web 应用 practice4_3。在 practice4_3 中，使用 spring_boot_addstarters 计算两个整数的和，通过访问 http://localhost:8080/testAddStarters 返回两个整数的和。在 spring_boot_addstarters 中，首先创建属性配置类 AddProperties（有 Integer 类型的 number1 与 number2 两个属性），在该属性配置类中使用@ConfigurationProperties(prefix="add")注解设置属性前缀为 add；其次，创建判断依据类 AddService（有 Integer 类型的 number1 与 number2 两个属性），在 AddService 类中提供 add 方法(计算 number1 与 number2 的和)；再次，创建自动配置类 AddAutoConfiguration，当类路径中存在 AddService 类时，自动配置该类的 Bean，并可以将相应 Bean 的属性在 application.properties 中配置；最后，注册自动配置类 AddAutoConfiguration。

Spring Boot 的 Web 开发

学习目的与要求

本章首先介绍 Spring Boot 的 Web 开发支持，然后介绍 Thymeleaf 视图模板引擎技术，最后介绍 Spring Boot 的 Web 开发技术（JSON 数据交互、文件上传与下载、异常统一处理以及对 JSP 的支持）。通过本章的学习，掌握 Spring Boot 的 Web 开发技术。

本章主要内容

- Thymeleaf 模板引擎。
- Spring Boot 处理 JSON 数据。
- Spring Boot 的文件上传与下载。
- Spring Boot 的异常处理。

Web 开发是一种基于 B/S 架构（即浏览器/服务器）的应用软件开发技术，分为前端（用户接口）和后端（业务逻辑和数据），前端的可视化及用户交互由浏览器实现，即以浏览器作为客户端，实现客户与服务器远程的数据交互。Spring Boot 的 Web 开发内容主要包括内嵌 Servlet 容器和 Spring MVC。

5.1 Spring Boot 的 Web 开发支持

Spring Boot 提供了 spring-boot-starter-web 依赖模块，该依赖模块包含 Spring Boot 预定义的 Web 开发常用依赖包，为 Web 开发者提供了内嵌的 Servlet 容器（Tomcat）以及 Spring MVC 的依赖。如果开发者希望开发 Spring Boot 的 Web 应用程序，可以在 Spring

Boot 项目的 pom.xml 文件中添加如下依赖配置：

```
<dependency>
    <groupId>org.springframework.boot</groupId>
    <artifactId>spring-boot-starter-web</artifactId>
</dependency>
```

视频讲解

Spring Boot 将自动关联 Web 开发的相关依赖，如 tomcat、spring-webmvc 等，进而对 Web 开发提供支持，并对相关技术的配置实现自动配置。

另外，开发者也可以使用 Spring Tool Suite 集成开发工具快速创建 Spring Starter Project，在 New Spring Starter Project Dependencies 对话框中添加 Spring Boot 的 Web 依赖，如图 5.1 所示。

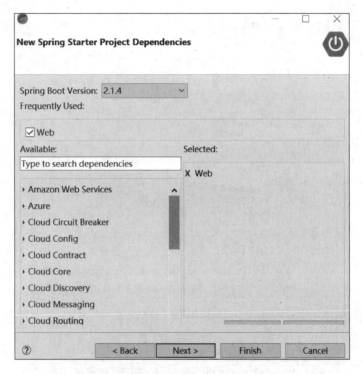

图 5.1 添加 Spring Boot 的 Web 依赖

5.2 Thymeleaf 模板引擎

在 Spring Boot 的 Web 应用中，建议开发者使用 HTML 完成动态页面。Spring Boot 提供了许多模板引擎，主要包括 FreeMarker、Groovy、Thymeleaf、Velocity 和 Mustache。因为 Thymeleaf 提供了完美的 Spring MVC 支持，所以在 Spring Boot 的 Web 应用中推荐使用 Thymeleaf 作为模板引擎。

Thymeleaf 是一个 Java 类库，是一个 XML/XHTML/HTML5 的模板引擎，能够处理 HTML、XML、JavaScript 以及 CSS，可以作为 MVC Web 应用的 View 层显示数据。

5.2.1　Spring Boot 的 Thymeleaf 支持

在 Spring Boot 1.X 版本中，spring-boot-starter-thymeleaf 依赖包含了 spring-boot-starter-web 模块。但是，在 Spring 5 中，WebFlux 的出现对于 Web 应用的解决方案将不再唯一。所以，spring-boot-starter-thymeleaf 依赖不再包含 spring-boot-starter-web 模块，需要开发人员自己选择 spring-boot-starter-web 模块依赖。下面通过一个实例，讲解如何创建基于 Thymeleaf 模板引擎的 Spring Boot Web 应用 ch5_1。

【例 5-1】　创建基于 Thymeleaf 模板引擎的 Spring Boot Web 应用 ch5_1。

具体实现步骤如下。

1. 创建 Spring Starter Project

选择菜单 File|New|Spring Starter Project，打开 New Spring Starter Project 对话框，在该对话框中选择和输入相关信息，如图 5.2 所示。

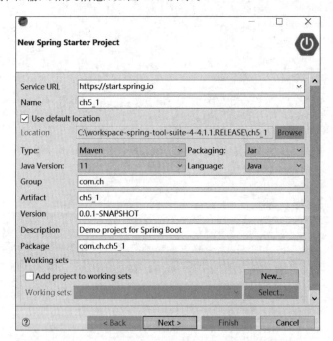

图 5.2　创建基于 Thymeleaf 模板引擎的 Spring Boot Web 应用 ch5_1

2. 选择依赖

单击图 5.2 中的 Next 按钮，打开 New Spring Starter Project Dependencies 对话框，选择 Spring Web Starter 和 Thymeleaf 依赖，如图 5.3 所示。

3. 打开项目目录

单击图 5.3 中的 Finish 按钮，创建如图 5.4 所示的基于 Thymeleaf 模板引擎的 Spring Boot Web 应用 ch5_1。

第5章 Spring Boot 的 Web 开发

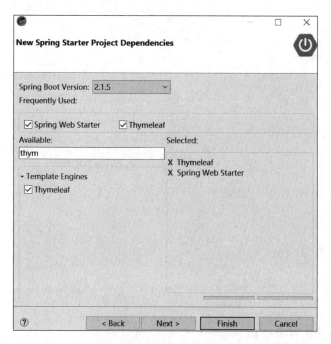

图 5.3　选择 Spring Web Starter 和 Thymeleaf 依赖

图 5.4　基于 Thymeleaf 模板引擎的 Spring Boot Web 应用 ch5_1

Tymeleaf 模板默认将 JS 脚本、CSS 样式、图片等静态文件放置在 src/main/resources/static 目录下，将视图页面放在 src/main/resources/templates 目录下。

4. 创建控制器类

创建一个名为 com.ch.ch5_1.controller 的包。并在该包中创建控制器类 TestThymeleafController，代码如下：

```
package com.ch.ch5_1.controller;
import org.springframework.stereotype.Controller;
@Controller
public class TestThymeleafController {
    @RequestMapping("/")
    public String test(){
        //根据 Tymeleaf 模板，默认将返回 src/main/resources/templates/index.html
        return "index";
    }
}
```

5. 新建 index.html 页面

在 src/main/resources/templates 目录下新建 index.html 页面，代码如下：

```
<!DOCTYPE html>
<html>
<head>
```

145

```
<meta charset="UTF-8">
<title>Insert title here</title>
</head>
<body>
测试Spring Boot的Thymeleaf支持
</body>
</html>
```

6. 运行测试

首先，运行Ch51Application主类。然后，访问http://localhost:8080/ch5_1/（因为配置文件中配置了Web应用的上下文路径为ch5_1）。运行效果如图5.5所示。

图5.5 例5-1运行结果

5.2.2 Thymeleaf基础语法

1. 引入Thymeleaf

首先，将View层页面文件的html标签修改如下：

`<html xmlns:th="http://www.thymeleaf.org">`

然后，在View层页面文件的其他标签里，使用th:*动态处理页面，示例代码如下：

`<img th:src="'images/' + ${aBook.picture}"/>`

其中，${aBook.picture}获得数据对象aBook的picture属性。

2. 输出内容

使用th:text和th:utext(对HTML标签解析)将文本内容输出到所在标签的body中。假如在国际化资源文件messages_en_US.properties中有消息文本"test.myText = Test International Message"，那么在页面中可以使用如下两种方式获得消息文本：

```
<p th:text="#{test.myText}"></p>
<!-- 不对HTML标签解析，即输出<strong>Test International Message</strong> -->
<p th:utext="#{test.myText}"></p>
<!-- 对HTML标签解析，即输出加粗的"Test International Message" -->
```

3. 基本表达式

1) 变量表达式：${…}

变量表达式用于访问容器上下文环境中的变量，示例代码如下：

`<span th:text="${information}">`

2) 选择变量表达式：*{…}

选择变量表达式计算的是选定的对象(th:object属性绑定的对象)，示例代码如下：

```
<div th:object = "${session.user}">
    name: <span th: text = "*{firstName}"></span><br>
    <!-- firstName 为 user 对象的属性 -->
    surname: <span th: text = "*{lastName}"></span><br>
    nationality: <span th: text = "*{nationality}"></span><br>
</div>
```

3) 信息表达式：#{…}

信息表达式一般用于显示页面静态文本,将可能需要根据需求而整体变动的静态文本放在 properties 文件中以便维护(如国际化),通常与 th:text 属性一起使用,示例代码如下：

```
<p th:text = "#{test.myText}"></p>
```

4. 引入 URL

Thymeleaf 模板通过@{…}表达式引入 URL,示例代码如下：

```
<!-- 默认访问 src/main/resources/static 下的 css 文件夹 -->
<link rel = "stylesheet" th:href = "@{css/bootstrap.min.css}" />
<!-- 访问相对路径 -->
<a th:href = "@{/}">去看看</a>
<!-- 访问绝对路径 -->
<a th:href = "@{http://www.tup.tsinghua.edu.cn/index.html(param1 = '传参')}">去清华大学出版社</a>
<!-- 默认访问 src/main/resources/static 下的 images 文件夹 -->
<img th:src = "'images/' + ${aBook.picture}"/>
```

5. 访问 WebContext 对象中的属性

Thymeleaf 模板通过一些专门的表达式从模板的 WebContext 获取请求参数、请求、会话和应用程序中的属性,具体如下：

${xxx}将返回存储在 Thymeleaf 模板上下文中的变量 xxx 或请求 request 作用域中的属性 xxx。

${param.xxx}将返回一个名为 xxx 的请求参数(可能是多个值)。

${session.xxx}将返回一个名为 xxx 的 HttpSession 作用域中的属性。

${application.xxx}将返回一个名为 xxx 的全局 ServletContext 上下文作用中的属性。

与 EL 表达式一样,使用${xxx}获得变量值,使用${对象变量名.属性名}获取 JavaBean 属性值。但需要注意的是,${}表达式只能在 th 标签内部有效。

6. 运算符

在 Thymeleaf 模板的表达式中可以使用+、-、*、/、%等各种算术运算符,也可以使用>、<、<=、>=、==、!=等各种逻辑运算符。示例代码如下：

```
<tr th:class = "(${row} == 'even')? 'even' : 'odd'">…</tr>
```

7. 条件判断

1) if 和 unless

只有在 th:if 条件成立时才显示标签内容；th:unless 与 th:if 相反,只有在条件不成立时才显示标签内容。示例代码如下：

```html
<a href="success.html" th:if="${user != nul}">成功</a>
<a href="success.html" th:unless="${user = nul}">成功</a>
```

2）switch 语句

Thymeleaf 模板也支持多路选择 switch 语句结构，默认属性 default 可用"*"表示。示例代码如下：

```html
<div th:switch="${user.role}">
    <p th:case="'admin'">User is an administrator</p>
    <p th:case="'teacher'">User is a teacher</p>
    <p th:case="*">User is a student</p>
</div>
```

8. 循环

1）基本循环

Thymeleaf 模板使用 th:each="obj,iterStat:${objList}" 标签进行迭代循环，迭代对象可以是 java.util.List、java.util.Map 或数组等。示例代码如下：

```html
<!-- 循环取出集合数据 -->
<div class="col-md-4 col-sm-6" th:each="book:${books}">
    <a href="">
        <img th:src="'images/'+${book.picture}" alt="图书封面" style="height:180px;width:40%;"/>
    </a>
    <div class="caption">
        <h4 th:text="${book.bname}"></h4>
        <p th:text="${book.author}"></p>
        <p th:text="${book.isbn}"></p>
        <p th:text="${book.price}"></p>
        <p th:text="${book.publishing}"></p>
    </div>
</div>
```

2）循环状态的使用

在 th:each 标签中可以使用循环状态变量，该变量有如下属性。

index：当前迭代对象的 index（从 0 开始计数）。

count：当前迭代对象的 index（从 1 开始计数）。

size：迭代对象的大小。

current：当前迭代变量。

even/odd：布尔值，当前循环是否是偶数/奇数（从 0 开始计数）。

first：布尔值，当前循环是否是第一个。

last：布尔值，当前循环是否是最后一个。

使用循环状态变量的示例代码如下：

```html
<!-- 循环取出集合数据 -->
<div class="col-md-4 col-sm-6" th:each="book,bookStat:${books}">
    <a href="">
        <img th:src="'images/' + ${book.picture}" alt="图书封面" style="height:180px;width:40%;"/>
    </a>
    <div class="caption">
        <!-- 循环状态 bookStat -->
        <h3 th:text="${bookStat.count}"></h3>
        <h4 th:text="${book.bname}"></h4>
        <p th:text="${book.author}"></p>
        <p th:text="${book.isbn}"></p>
        <p th:text="${book.price}"></p>
        <p th:text="${book.publishing}"></p>
    </div>
</div>
```

9. 内置对象

在实际 Web 项目开发中，经常传递列表、日期等数据。所以，Thymeleaf 模板提供了很多内置对象，可以通过#直接访问。这些内置对象一般都以 s 结尾，如 dates、lists、numbers、strings 等。Thymeleaf 模板通过 ${#…} 表达式访问内置对象。常见的内置对象如下。

#dates：日期格式化的内置对象，操作的方法是 java.util.Date 类的方法。

#calendars：类似于#dates，但操作的方法是 java.util.Calendar 类的方法。

#numbers：数字格式化的内置对象。

#strings：字符串格式化的内置对象，操作的方法参照 java.lang.String。

#objects：参照 java.lang.Object。

#bools：判断 boolean 类型的内置对象。

#arrays：数组操作的内置对象。

#lists：列表操作的内置对象，参照 java.util.List。

#sets：Set 操作的内置对象，参照 java.util.Set。

#maps：Map 操作的内置对象，参照 java.util.Map。

#aggregates：创建数组或集合的聚合的内置对象。

#messages：在变量表达式内部获取外部消息的内置对象。

假如有如下控制器方法：

```java
@RequestMapping("/testObject")
public String testObject(Model model) {
    //系统时间 new Date()
    model.addAttribute("nowTime", new Date());
    //系统日历对象
    model.addAttribute("nowCalendar", Calendar.getInstance());
    //创建 BigDecimal 对象
    BigDecimal money = new BigDecimal(2019.613);
    model.addAttribute("myMoney", money);
```

```java
//字符串
String tsts = "Test strings";
model.addAttribute("str", tsts);
//boolean 类型
boolean b = false;
model.addAttribute("bool", b);
//数组(这里不能使用 int 定义数组)
Integer aint[] = {1,2,3,4,5};
model.addAttribute("mya", aint);
//List 列表 1
List<String> nameList1 = new ArrayList<String>();
nameList1.add("陈恒 1");
nameList1.add("陈恒 3");
nameList1.add("陈恒 2");
model.addAttribute("myList1", nameList1);
//Set 集合
Set<String> st = new HashSet<String>();
st.add("set1");
st.add("set2");
model.addAttribute("mySet", st);
//Map 集合
Map<String, Object> map = new HashMap<String, Object>();
map.put("key1", "value1");
map.put("key2", "value2");
model.addAttribute("myMap", map);
//List 列表 2
List<String> nameList2 = new ArrayList<String>();
nameList2.add("陈恒 6");
nameList2.add("陈恒 5");
nameList2.add("陈恒 4");
model.addAttribute("myList2", nameList2);
return "showObject";
}
```

那么,可以在 src/main/resources/templates/showObject.html 视图页面文件中使用内置对象操作数据。showObject.html 的代码如下:

```html
<!DOCTYPE html>
<html xmlns:th="http://www.thymeleaf.org">
<head>
<meta charset="UTF-8">
<title>Insert title here</title>
</head>
<body>
    格式化控制器传递过来的系统时间 nowTime
    <span th:text="${#dates.format(nowTime, 'yyyy/MM/dd')}"></span>
    <br>
    创建一个日期对象
    <span th:text="${#dates.create(2019,6,13)}"></span>
    <br>
    格式化控制器传递过来的系统日历 nowCalendar:
```

```html
<span th:text="${#calendars.format(nowCalendar,'yyyy-MM-dd')}"></span>
<br>
格式化控制器传递过来的 BigDecimal 对象 myMoney：
<span th:text="${#numbers.formatInteger(myMoney,3)}"></span>
<br>
计算控制器传递过来的字符串 str 的长度：
<span th:text="${#strings.length(str)}"></span>
<br>
返回对象，当控制器传递过来的 BigDecimal 对象 myMoney 为空时，返回默认值 9999：
<span th:text="${#objects.nullSafe(myMoney,9999)}"></span>
<br>
判断 boolean 数据是否是 false：
<span th:text="${#bools.isFalse(bool)}"></span>
<br>
判断数组 mya 中是否包含元素 5：
<span th:text="${#arrays.contains(mya,5)}"></span>
<br>
排序列表 myList1 的数据：
<span th:text="${#lists.sort(myList1)}"></span>
<br>
判断集合 mySet 中是否包含元素 set2：
<span th:text="${#sets.contains(mySet,'set2')}"></span>
<br>
判断 myMap 中是否包含 key1 关键字：
<span th:text="${#maps.containsKey(myMap,'key1')}"></span>
<br>
将数组 mya 中的元素求和：
<span th:text="${#aggregates.sum(mya)}"></span>
<br>
将数组 mya 中的元素求平均：
<span th:text="${#aggregates.avg(mya)}"></span>
<br>
如果未找到消息，则返回默认消息（如"??msgKey_zh_CN??"）：
<span th:text="${#messages.msg('msgKey')}"></span>
</body>
</html>
```

5.2.3 Thymeleaf 的常用属性

通过 5.2.2 节的学习，我们发现 Thymeleaf 语法的使用都是通过在 html 页面的标签中添加 th:xxx 关键字来实现模板套用，且其属性与 html 页面标签基本类似。常用属性有以下几种。

1. th:action

th:action 定义后台控制器路径，类似于 <form> 标签的 action 属性。示例代码如下：

```html
<form th:action="@{/login}">…</form>
```

2. th:each

th:each 用于集合对象遍历,功能类似于 JSTL 标签<c:forEach>。示例代码如下:

```html
<div class="col-md-4 col-sm-6" th:each="gtype:${gtypes}">
    <div class="caption">
        <p th:text="${gtype.id}"></p>
        <p th:text="${gtype.typename}"></p>
    </div>
</div>
```

3. th:field

th:field 常用于表单参数绑定,通常与 th:object 一起使用。示例代码如下:

```html
<form th:action="@{/login}" th:object="${user}">
    <input type="text" value="" th:field="*{username}"></input>
    <input type="text" value="" th:field="*{role}"></input>
</form>
```

4. th:href

th:href 用于定义超链接,类似于<a>标签的 href 属性。value 形式为@{/logout},示例代码如下:

```html
<a th:href="@{/gogo}"></a>
```

5. th:id

th:id 用于 id 的声明,类似于 html 标签中的 id 属性。示例代码如下:

```html
<div th:id="stu+(${rowStat.index}+1)"></div>
```

6. th:if

th:if 用于条件判断,如果为否则标签不显示。示例代码如下:

```html
<div th:if="${rowStat.index} == 0">… do something …</div>
```

7. th:fragment

th:fragment 声明定义该属性的 div 为模板片段,常用于头文件、页尾文件的引入,常与 th:include、th:replace 一起使用。

假如在 ch5_1 的 src/main/resources/templates 目录下声明模板片段文件 footer.html,具体代码如下:

```html
<!DOCTYPE html>
<html xmlns:th="http://www.thymeleaf.org">
<head>
<meta charset="UTF-8">
<title>Insert title here</title>
```

```html
</head>
<body>
    <!-- 声明片段 content -->
    <div th:fragment="content">
        主体内容
    </div>
    <!-- 声明片段 copy -->
    <div th:fragment="copy">
        ©清华大学出版社
    </div>
</body>
</html>
```

那么,可以在 ch5_1 的 src/main/resources/templates/index.html 文件中引入模板片段,具体代码如下:

```html
<!DOCTYPE html>
<html xmlns:th="http://www.thymeleaf.org">
<head>
<meta charset="UTF-8">
<title>Insert title here</title>
</head>
<body>
    测试 Spring Boot 的 Thymeleaf 支持<br>
    引入主体内容模板片段:
    <div th:include="footer::content"></div>
    引入版权所有模板片段:
    <div th:replace="footer::copy"></div>
</body>
</html>
```

8. th:object

th:object 用于表单数据对象绑定,将表单绑定到后台 controller 的一个 JavaBean 参数,常与 th:field 一起使用,进行表单数据绑定。下面通过实例讲解表单提交及数据绑定的实现过程。

【例 5-2】 表单提交及数据绑定的实现过程。

具体实现步骤如下。

1) 创建实体类

在 Web 应用 ch5_1 的 src/main/java 目录下,创建 com.ch.ch5_1.model 包,并在该包中创建实体类 LoginBean,代码如下:

```java
package com.ch.ch5_1.model;
public class LoginBean {
    String uname;
    String urole;
    //省略 set 和 get 方法
}
```

2) 创建控制器类

在 Web 应用 ch5_1 的 com.ch.ch5_1.controller 包中,创建控制器类 LoginController,具体代码如下:

```java
package com.ch.ch5_1.controller;
import org.springframework.stereotype.Controller;
import org.springframework.ui.Model;
import org.springframework.web.bind.annotation.ModelAttribute;
import org.springframework.web.bind.annotation.RequestMapping;
import com.ch.ch5_1.model.LoginBean;
@Controller
public class LoginController {
    @RequestMapping("/toLogin")
    public String toLogin(Model model) {
        /* loginBean 与 login.html 页面中的 th:object = "${loginBean}"相同,类似于 Spring
        MVC 的表单绑定。*/
        model.addAttribute("loginBean", new LoginBean());
        return "login";
    }
    @RequestMapping("/login")
    public String greetingSubmit(@ModelAttribute LoginBean loginBean) {
        /* @ModelAttribute LoginBean loginBean 接收 login.html 页面中的表单数据,并将
        loginBean 对象保存到 model 中返回给 result.html 页面显示。*/
        System.out.println("测试提交的数据:" + loginBean.getUname());
        return "result";
    }
}
```

3) 创建页面表示层

在 Web 应用 ch5_1 的 src/main/resources/templates 目录下,创建页面 login.html 和 result.html。

页面 login.html 的代码如下:

```html
<!DOCTYPE html>
<html xmlns:th="http://www.thymeleaf.org">
<head>
<meta charset="UTF-8">
<title>Insert title here</title>
</head>
<body>
    <h1>Form</h1>
    <form action="#" th:action="@{/login}" th:object="${loginBean}" method="post">
        <!-- th:field = "*{uname}"的 uname 与实体类的属性相同,即绑定 loginBean 对象  -->
        <p>Uname: <input type="text" th:field="*{uname}" th:placeholder="请输入用户名" /></p>
        <p>Urole: <input type="text" th:field="*{urole}" th:placeholder="请输入角色" /></p>
        <p><input type="submit" value="Submit" /><input type="reset" value="Reset" /></p>
    </form>
```

```
</body>
</html>
```

页面 result.html 的代码如下：

```
<!DOCTYPE html>
<html xmlns:th = "http://www.thymeleaf.org">
<head>
<meta charset = "UTF-8">
<title>Insert title here</title>
</head>
<body>
    <h1>Result</h1>
    <p th:text = "'Uname: ' + ${loginBean.uname}" />
    <p th:text = "'Urole: ' + ${loginBean.urole}" />
    <a href = "toLogin">继续提交</a>
</body>
</html>
```

4）运行

首先，运行 Ch51Application 主类。然后，访问 http://localhost:8080/ch5_1/toLogin。运行结果如图 5.6 所示。

在图 5.6 的文本框中输入信息后，单击 Sumbit 按钮，打开如图 5.7 所示的页面。

图 5.6　页面 login.html 的运行结果

图 5.7　页面 result.html 的运行结果

9. th:src

th:src 用于外部资源引入，类似于 <script> 标签的 src 属性。示例代码如下：

```
<img th:src = "'images/' + ${aBook.picture}" />
```

10. th:text

th:text 用于文本显示，将文本内容显示到所在标签的 body 中。示例代码如下：

```
<td th:text = "${username}"></td>
```

11. th:value

th:value 用于标签赋值，类似于标签的 value 属性。示例代码如下：

```
<option th:value="Adult">Adult</option>
<input type="hidden" th:value="${msg}"/>
```

12. th:style

th:style 用于修改标签 style。示例代码如下：

```
<span th:style="'display:' + @{(${sitrue} ? 'none' : 'inline-block')}"></span>
```

13. th:onclick

th:onclick 用于修改单击事件，示例代码如下：

```
<button th:onclick="'getCollect()'"></button>
```

5.2.4 Spring Boot 与 Thymeleaf 实现页面信息国际化

在 Spring Boot 的 Web 应用中实现页面信息国际化非常简单，下面通过实例讲解国际化的实现过程。

【例 5-3】 国际化的实现过程。

具体实现步骤如下。

1. 编写国际化资源属性文件

1) 编写管理员模块的国际化信息

在 ch5_1 的 src/main/resources 目录下创建 i18n/admin 文件夹，并在该文件夹下创建 adminMessages.properties、adminMessages_en_US.properties 和 adminMessages_zh_CN.properties 资源属性文件。adminMessages.properties 表示默认加载的信息；adminMessages_en_US.properties 表示英文信息（en 代表语言代码，US 代表国家地区）；adminMessages_zh_CN.properties 表示中文信息。

adminMessages.properties 的内容如下：

```
test.admin = \u6D4B\u8BD5\u540E\u53F0
admin = \u540E\u53F0\u9875\u9762
```

adminMessages_en_US.properties 的内容如下：

```
test.admin = test admin
admin = admin
```

adminMessages_zh_CN.properties 的内容如下：

```
test.admin = \u6D4B\u8BD5\u540E\u53F0
admin = \u540E\u53F0\u9875\u9762
```

2) 编写用户模块的国际化信息

在 ch5_1 的 src/main/resources 目录下创建 i18n/before 文件夹，并在该文件夹下创建 beforeMessages.properties、beforeMessages_en_US.properties 和 beforeMessages_zh_CN

.properties 资源属性文件。

beforeMessages.properties 的内容如下：

```
test.before = \u6D4B\u8BD5\u524D\u53F0
before = \u524D\u53F0\u9875\u9762
```

beforeMessages_en_US.properties 的内容如下：

```
test.before = test before
before = before
```

beforeMessages_zh_CN.properties 的内容如下：

```
test.before = \u6D4B\u8BD5\u524D\u53F0
before = \u524D\u53F0\u9875\u9762
```

3）编写公共模块的国际化信息

在 ch5_1 的 src/main/resources 目录下创建 i18n/common 文件夹，并在该文件夹下创建 commonMessages.properties、commonMessages_en_US.properties 和 commonMessages_zh_CN.properties 资源属性文件。

commonMessages.properties 的内容如下：

```
chinese.key = \u4E2D\u6587\u7248
english.key = \u82F1\u6587\u7248
return = \u8FD4\u56DE\u9996\u9875
```

commonMessages_en_US.properties 的内容如下：

```
chinese.key = chinese
english.key = english
return = return
```

commonMessages_zh_CN.properties 的内容如下：

```
chinese.key = \u4E2D\u6587\u7248
english.key = \u82F1\u6587\u7248
return = \u8FD4\u56DE\u9996\u9875
```

2. 添加配置文件内容，引入资源属性文件

在 ch5_1 应用的配置文件中，添加如下内容，引入资源属性文件：

```
spring.messages.basename = i18n/admin/adminMessages,i18n/before/beforeMessages,
i18n/common/commonMessages
```

3. 重写 localeResolver 方法配置语言区域选择

在 ch5_1 应用的 com.ch.ch5_1 包中，创建配置类 LocaleConfig，该配置类实现 WebMvcConfigurer 接口，并配置语言区域选择。LocaleConfig 的代码如下：

```
package com.ch.ch5_1;
import java.util.Locale;
import org.springframework.boot.autoconfigure.EnableAutoConfiguration;
```

```java
import org.springframework.context.annotation.Bean;
import org.springframework.context.annotation.Configuration;
import org.springframework.web.servlet.LocaleResolver;
import org.springframework.web.servlet.config.annotation.InterceptorRegistry;
import org.springframework.web.servlet.config.annotation.WebMvcConfigurer;
import org.springframework.web.servlet.i18n.LocaleChangeInterceptor;
import org.springframework.web.servlet.i18n.SessionLocaleResolver;
@Configuration
@EnableAutoConfiguration
public class LocaleConfig implements WebMvcConfigurer {
    /**
     * 根据用户本次会话过程中的语义设定语言区域
     * (如用户进入首页时选择的语言种类)
     * @return
     */
    @Bean
    public LocaleResolver localeResolver() {
        SessionLocaleResolver slr = new SessionLocaleResolver();
        //默认语言
        slr.setDefaultLocale(Locale.CHINA);
        return slr;
    }
    /**
     * 使用SessionLocaleResolver存储语言区域时,
     * 必须配置localeChangeInterceptor拦截器
     * @return
     */
    @Bean
    public LocaleChangeInterceptor localeChangeInterceptor() {
        LocaleChangeInterceptor lci = new LocaleChangeInterceptor();
        //选择语言的参数名
        lci.setParamName("locale");
        return lci;
    }
    /**
     * 注册拦截器
     */
    @Override
    public void addInterceptors(InterceptorRegistry registry) {
        registry.addInterceptor(localeChangeInterceptor());
    }
}
```

4. 创建控制器类I18nTestController

在ch5_1应用的com.ch.ch5_1.controller包中,创建控制器类I18nTestController,具体代码如下:

```java
package com.ch.ch5_1.controller;
import org.springframework.stereotype.Controller;
import org.springframework.web.bind.annotation.RequestMapping;
@Controller
@RequestMapping("/i18n")
public class I18nTestController {
```

```
    @RequestMapping("/first")
    public String testI18n(){
        return "/i18n/first";
    }
    @RequestMapping("/admin")
    public String admin(){
        return "/i18n/admin";
    }
    @RequestMapping("/before")
    public String before(){
        return "/i18n/before";
    }
}
```

5．创建视图页面，并获得国际化信息

在 ch5_1 应用的 src/main/resources/templates 目录下，创建文件夹 i18n；并在该文件夹中创建 admin.html、before.html 和 first.html 视图页面，并在这些视图页面中使用 th:text="#{xxx}"获得国际化信息。

admin.html 的代码如下：

```
<!DOCTYPE html>
<html xmlns:th="http://www.thymeleaf.org">
<head>
<meta charset="UTF-8">
<title>Insert title here</title>
</head>
<body>
    <span th:text="#{admin}"></span><br>
    <a th:href="@{/i18n/first}" th:text="#{return}"></a>
</body>
</html>
```

before.html 的代码如下：

```
<!DOCTYPE html>
<html xmlns:th="http://www.thymeleaf.org">
<head>
<meta charset="UTF-8">
<title>Insert title here</title>
</head>
<body>
    <span th:text="#{before}"></span><br>
    <a th:href="@{/i18n/first}" th:text="#{return}"></a>
</body>
</html>
```

first.html 的代码如下：

```
<!DOCTYPE html>
<html xmlns:th="http://www.thymeleaf.org">
<head>
<meta charset="UTF-8">
```

```html
        <title>Insert title here</title>
    </head>
    <body>
        <a th:href = "@{/i18n/first(locale = 'zh_CN')}" th:text = "#{chinese.key}"></a>
        <a th:href = "@{/i18n/first(locale = 'en_US')}" th:text = "#{english.key}"></a>
        <br>
        <a th:href = "@{/i18n/admin}" th:text = "#{test.admin}"></a><br>
        <a th:href = "@{/i18n/before}"  th:text = "#{test.before}"></a><br>
    </body>
</html>
```

6. 运行

首先，运行 Ch51Application 主类。然后，访问 http://localhost:8080/ch5_1/i18n/first。运行效果如图 5.8 所示。

单击图 5.8 中的"英文版"，打开如图 5.9 所示的页面效果。

图 5.8　程序入口页面

图 5.9　英文版效果

5.2.5　Spring Boot 与 Thymeleaf 的表单验证

本节使用 Hibernate Validator 对表单进行验证，注意它和 Hibernate 无关，只是使用它进行数据验证。在 Spring MVC 的 Web 应用中，需要加载 Hibernate Validator 所依赖的 jar 包。而在 Spring Boot 的 Web 应用中，不再需要加载 Hibernate Validator 所依赖的 jar 包。这是因为 spring-boot-starter-web 中已经依赖了 hibernate-validator 的 jar 包。

使用 Hibernate Validator 验证表单时，需要利用它的标注类型在实体模型的属性上嵌入约束。

1. 空检查

@Null：验证对象是否为 null。
@NotNull：验证对象是否不为 null，无法检查长度为 0 的字符串。
@NotBlank：检查约束字符串是不是 null，还有被 trim 后的长度是否大于 0，只针对字符串，且会去掉前后空格。
@NotEmpty：检查约束元素是否为 null 或者是 empty。
示例代码如下：

```
@NotBlank(message = "{goods.gname.required}") //goods.gname.required 为属性文件的错误代码
private String gname;
```

2. booelan 检查

@AssertTrue：验证 boolean 属性是否为 true。
@AssertFalse：验证 boolean 属性是否为 false。
示例代码如下：

```
@AssertTrue
private boolean isLogin;
```

3. 长度检查

@Size(min＝,max＝)：验证对象(Array、Collection、Map、String)长度是否在给定的范围之内。
@Length(min＝,max＝)：验证字符串长度是否在给定的范围之内。
示例代码如下：

```
@Length(min=1,max=100)
private String gdescription;
```

4. 日期检查

@Past：验证 Date 和 Calendar 对象是否在当前时间之前。
@Future：验证 Date 和 Calendar 对象是否在当前时间之后。
@Pattern：验证 String 对象是否符合正则表达式的规则。
示例代码如下：

```
@Past(message="{gdate.invalid}")
private Date gdate;
```

5. 数值检查

@Min：验证 Number 和 String 对象是否大于等于指定的值。
@Max：验证 Number 和 String 对象是否小于等于指定的值。
@DecimalMax：被标注的值必须不大于约束中指定的最大值，这个约束的参数是一个通过 BigDecimal 定义的最大值的字符串表示，小数存在精度。
@DecimalMin：被标注的值必须不小于约束中指定的最小值，这个约束的参数是一个通过 BigDecimal 定义的最小值的字符串表示，小数存在精度。
@Digits：验证 Number 和 String 的构成是否合法。
@Digits(integer＝,fraction＝)：验证字符串是否符合指定格式的数字，integer 指定整数精度，fraction 指定小数精度。
@Range(min＝,max＝)：检查数字是否介于 min 和 max 之间。
@Valid：对关联对象进行校验，如果关联对象是一个集合或者数组，那么对其中的元素进行校验；如果关联对象是一个 map，则对其中的值部分进行校验。
@CreditCardNumber：信用卡验证。
@Email：验证是否是邮件地址，如果为 null，不进行验证，通过验证。

示例代码如下：

```
@Range(min = 0, max = 100, message = "{gprice.invalid}")
private double gprice;
```

下面通过实例讲解使用 Hibernate Validator 验证表单的过程。

【例 5-4】 使用 Hibernate Validator 验证表单的过程。

具体实现步骤如下。

1. 创建表单实体模型

在 ch5_1 应用的 com.ch.ch5_1.model 包中，创建表单实体模型类 Goods，在该类使用 Hibernate Validator 的标注类型进行表单验证。代码如下：

```java
package com.ch.ch5_1.model;
import javax.validation.constraints.NotBlank;
import org.hibernate.validator.constraints.Length;
import org.hibernate.validator.constraints.Range;
public class Goods {
    @NotBlank(message = "商品名必须输入")
    @Length(min = 1, max = 5, message = "商品名长度在 1 到 5 之间")
    private String gname;
    @Range(min = 0, max = 100, message = "商品价格在 0 到 100 之间")
    private double gprice;
    //省略 set 和 get 方法
}
```

2. 创建控制器

在 ch5_1 应用的 com.ch.ch5_1.controller 包中，创建控制器类 TestValidatorController。在该类中有两个处理方法，一个是界面初始化处理方法 testValidator，一个是添加请求处理方法 add。在 add 方法中，使用@Validated 注解使验证生效。代码如下：

```java
package com.ch.ch5_1.controller;
import org.springframework.stereotype.Controller;
import org.springframework.validation.BindingResult;
import org.springframework.validation.annotation.Validated;
import org.springframework.web.bind.annotation.ModelAttribute;
import org.springframework.web.bind.annotation.RequestMapping;
import com.ch.ch5_1.model.Goods;
@Controller
public class TestValidatorController {
    @RequestMapping("/testValidator")
    public String testValidator(@ModelAttribute("goodsInfo") Goods goods){
        goods.setGname("商品名初始化");
        goods.setGprice(0.0);
        return "testValidator";
    }
```

```
    @RequestMapping(value = "/add")
    public String add(@ModelAttribute("goodsInfo") @Validated Goods goods,BindingResult rs){
        //@ModelAttribute("goodsInfo")与th:object = "${goodsInfo}"相对应
        if(rs.hasErrors()){              //验证失败
            return "testValidator";
        }
        //验证成功,可以到任意地方,在这里直接到testValidator界面
        return "testValidator";
    }
}
```

3. 创建视图页面

在ch5_1应用的src/main/resources/templates目录下,创建视图页面testValidator.html。在视图页面中,直接读取到ModelAttribute里面注入的数据,然后通过th:errors="*{xxx}"获得验证错误信息。代码如下：

```
<!DOCTYPE html>
<html xmlns:th="http://www.thymeleaf.org">
<head>
<meta charset="UTF-8">
<title>Insert title here</title>
</head>
<body>
    <h2>通过th:object访问对象的方式</h2>
    <div th:object="${goodsInfo}">
        <p th:text="*{gname}"></p>
        <p th:text="*{gprice}"></p>
    </div>
    <h1>表单提交</h1>
    <!-- 表单提交用户信息,注意表单参数的设置,直接是*{} -->
    <form th:action="@{/add}" th:object="${goodsInfo}" method="post">
        <div><span>商品名</span><input type="text" th:field="*{gname}"/><span th:errors="*{gname}"></span></div>
        <div><span>商品价格</span><input type="text" th:field="*{gprice}"/><span th:errors="*{gprice}"></span></div>
        <input type="submit"/>
    </form>
</body>
</html>
```

4. 运行

首先,运行Ch51Application主类。然后,访问http://localhost:8080/ch5_1/testValidator。测试效果如图5.10所示。

图 5.10 表单验证

5.2.6 基于 Thymeleaf 与 BootStrap 的 Web 开发实例

在本书的后续 Web 应用开发中，尽量使用前端开发工具包 BootStrap、JavaScript 框架 jQuery 和 Spring MVC 框架。BootStrap 和 jQuery 的相关知识，请读者自行学习。下面通过一个实例，讲解如何创建基于 Thymeleaf 模板引擎的 Spring Boot Web 应用 ch5_2。

【例 5-5】 创建基于 Thymeleaf 模板引擎的 Spring Boot Web 应用 ch5_2。

具体实现步骤如下。

1. 创建基于 Thymeleaf 模板引擎的 Spring Boot Web 应用 ch5_2

选择菜单 File|New|Spring Starter Project，打开 New Spring Starter Project 对话框，在该对话框中选择和输入相关信息，如图 5.11 所示。

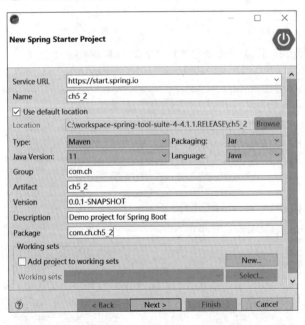

图 5.11 创建基于 Thymeleaf 模板引擎的 Spring Boot Web 应用 ch5_2

单击图 5.11 中的 Next 按钮,打开 New Spring Starter Project Dependencies 对话框,选择 Spring Web Starter 和 Thymeleaf 依赖,如图 5.12 所示。

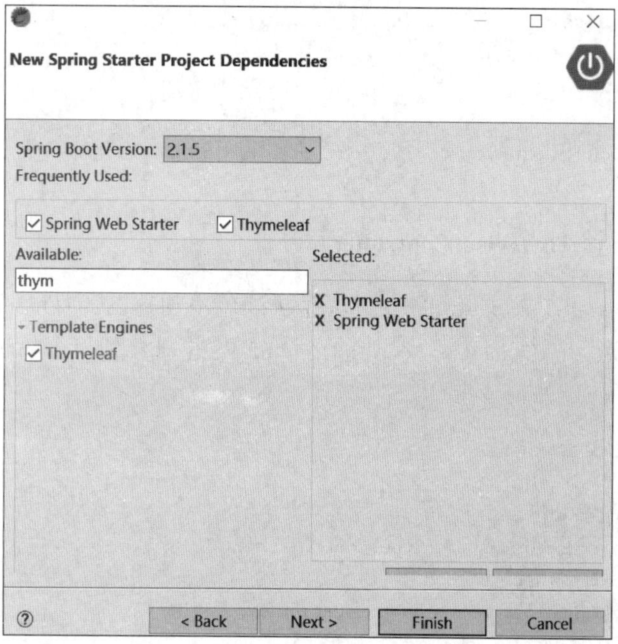

图 5.12　选择 Spring Web Starter 和 Thymeleaf 依赖

单击图 5.12 中的 Finish 按钮,完成创建基于 Thymeleaf 模板引擎的 Spring Boot Web 应用 ch5_2。

2. 设置 Web 应用 ch5_2 的上下文路径

在 ch5_2 的 application.properties 文件中配置如下内容:

```
server.servlet.context-path = /ch5_2
```

3. 创建实体类 Book

创建名为 com.ch.ch5_2.model 的包,并在该包中创建名为 Book 的实体类,此实体类用在模板页面展示数据。代码如下:

```java
package com.ch.ch5_2.model;
public class Book {
    String isbn;
    Double price;
    String bname;
    String publishing;
    String author;
    String picture;
    public Book(String isbn, Double price, String bname, String publishing, String author, String picture) {
        super();
```

```
        this.isbn = isbn;
        this.price = price;
        this.bname = bname;
        this.publishing = publishing;
        this.author = author;
        this.picture = picture;
    }
    //省略 set 和 get 方法
}
```

4. 创建控制器类 ThymeleafController

创建名为 com.ch.ch5_2.controller 的包,并在该包中创建名为 ThymeleafController 的控制器类。在该控制器类中,实例化 Book 类的多个对象,并保存到集合 ArrayList<Book>中。代码如下:

```
package com.ch.ch5_2.controller;
import java.util.ArrayList;
import java.util.List;
import org.springframework.stereotype.Controller;
import org.springframework.ui.Model;
import org.springframework.web.bind.annotation.RequestMapping;
import com.ch.ch5_1.model.Book;
@Controller
public class ThymeleafController {
    @RequestMapping("/")
    public String index(Model model) {
        Book teacherGeng = new Book(
                "9787302464259",
                59.5,
                "Java 2 实用教程(第 5 版)",
                "清华大学出版社",
                "耿祥义",
                "073423-02.jpg"
                );
        List<Book> chenHeng = new ArrayList<Book>();
        Book b1 = new Book(
                "9787302529118",
                69.8,
                "Java Web 开发从入门到实战(微课版)",
                "清华大学出版社",
                "陈恒",
                "082526-01.jpg"
                );
        chenHeng.add(b1);
        Book b2 = new Book(
                "9787302502968",
                69.8,
                "Java EE 框架整合开发入门到实战——Spring+Spring MVC+MyBatis(微课版)",
                "清华大学出版社",
```

```
                "陈恒",
                "079720-01.jpg");
        chenHeng.add(b2);
        model.addAttribute("aBook", teacherGeng);
        model.addAttribute("books", chenHeng);
        //根据Thymeleaf模板,默认将返回src/main/resources/templates/index.html
        return "index";
    }
}
```

5. 整理脚本样式静态文件

JS 脚本、CSS 样式、图片等静态文件默认放置在 src/main/resources/static 目录下,ch5_2 应用引入了 BootStrap 和 jQuery,结构如图 5.13 所示。

6. View 视图页面

Thymeleaf 模板默认将视图页面放在 src/main/resources/templates 目录下。因此,在 src/main/resources/templates 目录下新建 html 页面文件 index.html。在该页面中,使用 Thymeleaf 模板显示控制器类 TestThymeleafController 中的 model 对象数据。代码如下:

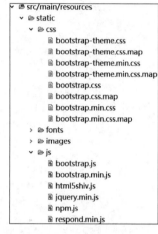

图 5.13 静态文件位置

```html
<!DOCTYPE html>
<html xmlns:th="http://www.thymeleaf.org">
<head>
<meta charset="UTF-8">
<title>Insert title here</title>
<link rel="stylesheet" th:href="@{css/bootstrap.min.css}"/>
<link rel="stylesheet" th:href="@{css/bootstrap-theme.min.css}"/>
</head>
<body>
    <!-- 面板 -->
    <div class="panel panel-primary">
        <!-- 面板头信息 -->
        <div class="panel-heading">
            <!-- 面板标题 -->
            <h3 class="panel-title">第一个基于Thymeleaf模板引擎的Spring Boot Web 应用
            </h3>
        </div>
    </div>
    <!-- 容器 -->
    <div class="container">
        <div>
            <h4>图书列表</h4>
        </div>
        <div class="row">
            <!-- col-md针对桌面显示器,col-sm针对平板 -->
```

```html
                <div class = "col-md-4 col-sm-6">
                    <a href = "">
                        <img th:src = "'images/' + ${aBook.picture}" alt = "图书封面"
                        style = "height: 180px; width: 40%;"/>
                    </a>
                    <!-- caption容器中放置其他基本信息,例如标题、文本描述等 -->
                    <div class = "caption">
                        <h4 th:text = "${aBook.bname}"></h4>
                        <p th:text = "${aBook.author}"></p>
                        <p th:text = "${aBook.isbn}"></p>
                        <p th:text = "${aBook.price}"></p>
                        <p th:text = "${aBook.publishing}"></p>
                    </div>
                </div>
                <!-- 循环取出集合数据 -->
                <div class = "col-md-4 col-sm-6" th:each = "book: ${books}">
                    <a href = "">
                        <img th:src = "'images/' + ${book.picture}" alt = "图书封面"
                        style = "height: 180px; width: 40%;"/>
                    </a>
                    <div class = "caption">
                        <h4 th:text = "${book.bname}"></h4>
                        <p th:text = "${book.author}"></p>
                        <p th:text = "${book.isbn}"></p>
                        <p th:text = "${book.price}"></p>
                        <p th:text = "${book.publishing}"></p>
                    </div>
                </div>
            </div>
        </div>
    </body>
</html>
```

7. 运行

首先,运行 Ch52Application 主类。然后,访问 http://localhost:8080/ch5_2/。运行效果如图 5.14 所示。

图 5.14 例 5-5 运行结果

5.3 Spring Boot 处理 JSON 数据

视频讲解

在 Spring Boot 的 Web 应用中，内置了 JSON 数据的解析功能，默认使用 Jackson 自动完成解析（不需要加载 Jackson 依赖包）。当控制器返回一个 Java 对象或集合数据时，Spring Boot 自动将其转换成 JSON 数据，使用起来很方便简洁。

Spring Boot 处理 JSON 数据时，需要用到两个重要的 JSON 格式转换注解，分别是 @RequestBody 和 @ResponseBody。

- @RequestBody：用于将请求体中的数据绑定到方法的形参中，该注解应用在方法的形参上。
- @ResponseBody：用于直接返回 JSON 对象，该注解应用在方法上。

下面通过一个实例讲解 Spring Boot 处理 JSON 数据的过程，该实例针对返回实体对象、ArrayList 集合、Map<String, Object>集合以及 List<Map<String, Object>>集合分别处理。

【例 5-6】 Spring Boot 处理 JSON 数据的过程。

具体实现步骤如下。

1. 创建实体类

在 ch5_2 应用的 com.ch.ch5_2.model 包中，创建实体类 Person。具体代码如下：

```
package com.ch.ch5_2.model;
public class Person {
    private String pname;
    private String password;
    private Integer page;
    //省略 set 和 get 方法
}
```

2. 创建视图页面

在 ch5_2 应用的 src/main/resources/templates 目录下，创建视图页面 input.html。在 input.html 页面中，引入 jQuery 框架，并使用它的 ajax 方法进行异步请求。具体代码如下：

```
<!DOCTYPE html>
<html xmlns:th="http://www.thymeleaf.org">
<head>
<meta charset="UTF-8">
<title>Insert title here</title>
<link rel="stylesheet" th:href="@{css/bootstrap.min.css}" />
<!-- 默认访问 src/main/resources/static 下的 css 文件夹 -->
<link rel="stylesheet" th:href="@{css/bootstrap-theme.min.css}" />
<!-- 引入 jQuery -->
<script type="text/javascript" th:src="@{js/jquery.min.js}"></script>
<script type="text/javascript">
```

```javascript
            function testJson() {
                //获取输入的值 pname 为 id
                var pname = $("#pname").val();
                var password = $("#password").val();
                var page = $("#page").val();
                alert(password);
                $.ajax({
                    //发送请求的 URL 字符串
                    url : "testJson",
                    //定义回调响应的数据格式为 JSON 字符串,该属性可以省略
                    dataType : "json",
                    //请求类型
                    type : "post",
                    //定义发送请求的数据格式为 JSON 字符串
                    contentType : "application/json",
                    //data 表示发送的数据
                    data : JSON.stringify({pname:pname,password:password,page:page}),
                    //成功响应的结果
                    success : function(data){
                        if(data != null){
                            //返回一个 Person 对象
                            //alert("输入的用户名:" + data.pname + ",密码:" + data.password +
                            ",年龄:" +  data.page);
                            //ArrayList<Person>对象
                            /** for(var i = 0; i < data.length; i++){
                                alert(data[i].pname);
                            } **/
                            //返回一个 Map<String, Object>对象
                            //alert(data.pname);       //pname 为 key
                            //返回一个 List<Map<String, Object>>对象
                            for(var i = 0; i < data.length; i++){
                                alert(data[i].pname);
                            }
                        }
                    },
                    //请求出错
                    error:function(){
                        alert("数据发送失败");
                    }
                });
            }
        </script>
    </head>
    <body>
        <div class="panel panel-primary">
            <div class="panel-heading">
                <h3 class="panel-title">处理 JSON 数据</h3>
            </div>
        </div>
        <div class="container">
            <div>
```

```html
            <h4>添加用户</h4>
        </div>
        <div class="row">
            <div class="col-md-6 col-sm-6">
                <form class="form-horizontal" action="">
                    <div class="form-group">
                        <div class="input-group col-md-6">
                            <span class="input-group-addon">
                                <i class="glyphicon glyphicon-pencil"></i>
                            </span>
                            <input class="form-control" type="text"
                                id="pname" th:placeholder="请输入用户名"/>
                        </div>
                    </div>
                    <div class="form-group">
                        <div class="input-group col-md-6">
                            <span class="input-group-addon">
                                <i class="glyphicon glyphicon-pencil"></i>
                            </span>
                            <input class="form-control" type="text"
                                id="password" th:placeholder="请输入密码"/>
                        </div>
                    </div>
                    <div class="form-group">
                        <div class="input-group col-md-6">
                            <span class="input-group-addon">
                                <i class="glyphicon glyphicon-pencil"></i>
                            </span>
                            <input class="form-control" type="text"
                                id="page" th:placeholder="请输入年龄"/>
                        </div>
                    </div>
                    <div class="form-group">
                        <div class="col-md-6">
                            <div class="btn-group btn-group-justified">
                                <div class="btn-group">
                                    <button type="button" onclick="testJson()"
                                        class="btn btn-success">
                                        <span class="glyphicon glyphicon-share"></span>
                                         测试
                                    </button>
                                </div>
                            </div>
                        </div>
                    </div>
                </form>
            </div>
        </div>
    </div>
</body>
</html>
```

3. 创建控制器

在 ch5_2 应用的 com.ch.ch5_2.controller 包中，创建控制器类 TestJsonController。在该类中有两个处理方法，一个是界面导航方法 input，一个是接收页面请求的方法。具体代码如下：

```java
package com.ch.ch5_2.controller;
import java.util.ArrayList;
import java.util.HashMap;
import java.util.List;
import java.util.Map;
import org.springframework.stereotype.Controller;
import org.springframework.web.bind.annotation.RequestBody;
import org.springframework.web.bind.annotation.RequestMapping;
import org.springframework.web.bind.annotation.ResponseBody;
import com.ch.ch5_2.model.Person;
@Controller
public class TestJsonController {
    /**
     * 进入视图页面
     */
    @RequestMapping("/input")
    public String input() {
        return "input";
    }
    /**
     * 接收页面请求的 JSON 数据
     */
    @RequestMapping("/testJson")
    @ResponseBody
    /* @RestController 注解相当于@ResponseBody ＋ @Controller 合在一起的作用。
      ①如果只是使用@RestController 注解 Controller,则 Controller 中的方法无法返回 jsp 页面
    或者 html,返回的内容就是 return 的内容。
      ②如果需要返回到指定页面,则需要用@Controller 注解。如果需要返回 JSON,XML 或自定义
    mediaType 内容到页面,则需要在对应的方法上加上@ResponseBody 注解。
    */
    public List<Map<String, Object>> testJson(@RequestBody Person user) {
        //打印接收的 JSON 格式数据
        System.out.println("pname = " + user.getPname() +
                ", password = " + user.getPassword() + ",page = " + user.getPage());
        //返回 Person 对象
        //return user;
        /** ArrayList<Person> allp = new ArrayList<Person>();
        Person p1 = new Person();
        p1.setPname("陈恒 1");
        p1.setPassword("123456");
        p1.setPage(80);
        allp.add(p1);
```

```
            Person p2 = new Person();
            p2.setPname("陈恒 2");
            p2.setPassword("78910");
            p2.setPage(90);
            allp.add(p2);
            //返回 ArrayList<Person>对象
            return allp;
            **/
            Map<String, Object> map = new HashMap<String, Object>();
            map.put("pname", "陈恒 2");
            map.put("password", "123456");
            map.put("page", 25);
            //返回一个 Map<String, Object>对象
            //return map;
            //返回一个 List<Map<String, Object>>对象
            List<Map<String, Object>> allp = new ArrayList<Map<String, Object>>();
            allp.add(map);
            Map<String, Object> map1 = new HashMap<String, Object>();
            map1.put("pname", "陈恒 3");
            map1.put("password", "54321");
            map1.put("page", 55);
            allp.add(map1);
            return allp;
      }
}
```

4. 运行

首先,运行 Ch52Application 主类。然后,访问 http://localhost:8080/ch5_2/input。运行效果如图 5.15 所示。

图 5.15　input.html 运行效果

5.4 Spring Boot 文件上传与下载

视频讲解

文件上传与下载是 Web 应用开发中常用的功能之一。本节将讲解如何在 Spring Boot 的 Web 应用开发中实现文件的上传与下载。

在实际的 Web 应用开发中,为了成功上传文件,必须将表单的 method 设置为 post,并将 enctype 设置为 multipart/form-data。只有这样设置,浏览器才能将所选文件的二进制数据发送给服务器。

从 Servlet 3.0 开始,就提供了处理文件上传的方法,但这种文件上传需要在 Java Servlet 中完成,而 Spring MVC 提供了更简单的封装。Spring MVC 是通过 Apache Commons FileUpload 技术实现一个 MultipartResolver 的实现类 CommonsMultipartResolver 完成文件上传的。因此,Spring MVC 的文件上传需要依赖 Apache Commons FileUpload 组件。

Spring MVC 将上传文件自动绑定到 MultipartFile 对象中,MultipartFile 提供了获取上传文件内容、文件名等方法,并通过 transferTo 方法将文件上传到服务器的磁盘中。MultipartFile 的常用方法如下。

- byte[] getBytes():获取文件数据。
- String getContentType():获取文件 MIME 类型,如 image/jpeg 等。
- InputStream getInputStream():获取文件流。
- String getName():获取表单中文件组件的名字。
- String getOriginalFilename():获取上传文件的原名。
- long getSize():获取文件的字节大小,单位为 b(byte)。
- boolean isEmpty():是否有(选择)上传文件。
- void transferTo(File dest):将上传文件保存到一个目标文件中。

Spring Boot 的 spring-boot-starter-web 已经集成了 Spring MVC,所以使用 Spring Boot 实现文件上传更加便捷,只需引入 Apache Commons FileUpload 组件依赖即可。

下面通过一个实例讲解 Spring Boot 文件上传与下载的实现过程。

【例 5-7】 Spring Boot 文件上传与下载。

具体实现步骤如下。

1. 引入 Apache Commons FileUpload 组件依赖

在 Web 应用 ch5_2 的 pom.xml 文件中,添加 Apache Commons FileUpload 组件依赖。代码如下:

```
<dependency>
    <groupId>commons-fileupload</groupId>
    <artifactId>commons-fileupload</artifactId>
    <!-- 由于 commons-fileupload 组件不属于 Spring Boot,所以需要加上版本 -->
    <version>1.3.3</version>
</dependency>
```

2. 设置上传文件大小限制

在 Web 应用 ch5_2 的配置文件 application.properties 中,添加如下配置限制上传文件大小。

```
#上传文件时,默认单个上传文件大小是1MB,max-file-size设置单个上传文件大小
spring.servlet.multipart.max-file-size=50MB
#默认总文件大小是10MB,max-request-size设置总上传文件大小
spring.servlet.multipart.max-request-size=500MB
```

3. 创建选择文件视图页面

在 ch5_2 应用的 src/main/resources/templates 目录下,创建选择文件视图页面 uploadFile.html。该页面中有一个 enctype 属性值为 multipart/form-data 的 form 表单。具体代码如下:

```html
<!DOCTYPE html>
<html xmlns:th="http://www.thymeleaf.org">
<head>
<meta charset="UTF-8">
<title>Insert title here</title>
<link rel="stylesheet" th:href="@{css/bootstrap.min.css}" />
<!-- 默认访问 src/main/resources/static 下的 css 文件夹 -->
<link rel="stylesheet" th:href="@{css/bootstrap-theme.min.css}" />
</head>
<body>
<div class="panel panel-primary">
        <div class="panel-heading">
            <h3 class="panel-title">文件上传示例</h3>
        </div>
    </div>
    <div class="container">
        <div class="row">
            <div class="col-md-6 col-sm-6">
                <form class="form-horizontal" action="upload"
                method="post" enctype="multipart/form-data">
                    <div class="form-group">
                        <div class="input-group col-md-6">
                            <span class="input-group-addon">
                                <i class="glyphicon glyphicon-pencil"></i>
                            </span>
                            <input class="form-control" type="text"
                             name="description" th:placeholder="文件描述"/>
                        </div>
                    </div>
                    <div class="form-group">
                        <div class="input-group col-md-6">
                            <span class="input-group-addon">
                                <i class="glyphicon glyphicon-search"></i>
                            </span>
```

```html
                    <input class="form-control" type="file"
                        name="myfile" th:placeholder="请选择文件"/>
                </div>
            </div>
            <div class="form-group">
                <div class="col-md-6">
                    <div class="btn-group btn-group-justified">
                        <div class="btn-group">
                            <button type="submit" class="btn btn-success">
                                <span class="glyphicon glyphicon-share"></span>
                                 上传文件
                            </button>
                        </div>
                    </div>
                </div>
            </div>
        </form>
    </div>
</div>
</body>
</html>
```

4. 创建控制器

在 ch5_2 应用的 com.ch.ch5_2.controller 包中，创建控制器类 TestFileUpload。在该类中有 4 个处理方法，一个是界面导航方法 uploadFile，一个是实现文件上传的 upload 方法，一个是显示将要被下载文件的 showDownLoad 方法，一个是实现下载功能的 download 方法。具体代码如下：

```java
package com.ch.ch5_2.controller;
import java.io.File;
import java.io.IOException;
import java.net.URLEncoder;
import javax.servlet.http.HttpServletRequest;
import org.apache.commons.io.FileUtils;
import org.springframework.http.MediaType;
import org.springframework.http.ResponseEntity;
import org.springframework.http.ResponseEntity.BodyBuilder;
import org.springframework.stereotype.Controller;
import org.springframework.ui.Model;
import org.springframework.web.bind.annotation.RequestHeader;
import org.springframework.web.bind.annotation.RequestMapping;
import org.springframework.web.bind.annotation.RequestParam;
import org.springframework.web.multipart.MultipartFile;
@Controller
public class TestFileUpload {
    /**
     * 进入文件选择页面
     */
```

```java
@RequestMapping("/uploadFile")
public String uploadFile() {
    return "uploadFile";
}
/**
 * 上传文件自动绑定到 MultipartFile 对象中,
 * 在这里使用处理方法的形参接收请求参数。
 */
@RequestMapping("/upload")
public String upload(
        HttpServletRequest request,
        @RequestParam("description") String description,
        @RequestParam("myfile") MultipartFile myfile)
        throws IllegalStateException, IOException {
    System.out.println("文件描述:" + description);
    //如果选择了上传文件,将文件上传到指定的目录 uploadFiles
    if(!myfile.isEmpty()) {
        //上传文件路径
        String path = request.getServletContext().getRealPath("/uploadFiles/");
        //获得上传文件原名
        String fileName = myfile.getOriginalFilename();
        File filePath = new File(path + File.separator + fileName);
        //如果文件目录不存在,创建目录
        if(!filePath.getParentFile().exists()) {
            filePath.getParentFile().mkdirs();
        }
        //将上传文件保存到一个目标文件中
        myfile.transferTo(filePath);
    }
    //转发到一个请求处理方法,查询将要下载的文件
    return "forward:/showDownLoad";
}
/**
 * 显示要下载的文件
 */
@RequestMapping("/showDownLoad")
public String showDownLoad(HttpServletRequest request, Model model) {
    String path = request.getServletContext().getRealPath("/uploadFiles/");
    File fileDir = new File(path);
    //从指定目录获得文件列表
    File filesList[] = fileDir.listFiles();
    model.addAttribute("filesList", filesList);
    return "showFile";
}
/**
 * 实现下载功能
 */
@RequestMapping("/download")
public ResponseEntity<byte[]> download(
        HttpServletRequest request,
        @RequestParam("filename") String filename,
```

```
        @RequestHeader("User-Agent") String userAgent) throws IOException {
    //下载文件路径
    String path = request.getServletContext().getRealPath("/uploadFiles/");
    //构建将要下载的文件对象
    File downFile = new File(path + File.separator + filename);
    //ok 表示 HTTP 中的状态是 200
    BodyBuilder builder =  ResponseEntity.ok();
    //内容长度
    builder.contentLength(downFile.length());
    //application/octet-stream:二进制流数据(最常见的文件下载)
    builder.contentType(MediaType.APPLICATION_OCTET_STREAM);
    //使用 URLEncoder.encode 对文件名进行编码
    filename = URLEncoder.encode(filename,"UTF-8");
    /**
     * 设置实际的响应文件名,告诉浏览器文件要用于"下载"和"保存".
     * 不同的浏览器,处理方式不同,根据浏览器的实际情况区别对待.
     */
    if(userAgent.indexOf("MSIE") > 0) {
        //IE 浏览器,只需要用 UTF-8 字符集进行 URL 编码
        builder.header("Content-Disposition", "attachment; filename=" + filename);
    }else {
        /** 非 IE 浏览器,如 FireFox、Chrome 等浏览器,则需要说明编码的字符集
         * filename 后面有一个 * 号,在 UTF-8 后面有两个单引号
         */
        builder.header("Content-Disposition", "attachment; filename*=UTF-8''" + filename);
    }
    return builder.body(FileUtils.readFileToByteArray(downFile));
    }
}
```

5. 创建文件下载视图页面

在 ch5_2 应用的 src/main/resources/templates 目录下,创建文件下载视图页面 showFile.html。具体代码如下:

```
<!DOCTYPE html>
<html xmlns:th="http://www.thymeleaf.org">
<head>
<meta charset="UTF-8">
<title>Insert title here</title>
<link rel="stylesheet" th:href="@{css/bootstrap.min.css}" />
<!-- 默认访问 src/main/resources/static 下的 css 文件夹 -->
<link rel="stylesheet" th:href="@{css/bootstrap-theme.min.css}" />
<body>
    <div class="panel panel-primary">
        <div class="panel-heading">
            <h3 class="panel-title">文件下载示例</h3>
        </div>
    </div>
    <div class="container">
```

```html
                <div class = "panel panel - primary">
                    <div class = "panel - heading">
                        <h3 class = "panel - title">文件列表</h3>
                    </div>
                    <div class = "panel - body">
                        <div class = "table table - responsive">
                            <table class = "table table - bordered table - hover">
                                <tbody class = "text - center">
                                    <tr th:each = "file,fileStat: ${filesList}">
                                        <td>
                                            <span th:text = "${fileStat.count}"></span>
                                        </td>
                                        <td>
                                        <!-- file.name 相当于调用 getName()方法获得文件名称    -->
                                            <a th:href = "@{download(filename = ${file.name})}">
                                                <span th:text = "${file.name}"></span>
                                            </a>
                                        </td>
                                    </tr>
                                </tbody>
                            </table>
                        </div>
                    </div>
                </div>
            </div>
        </body>
</html>
```

6. 运行

首先，运行 Ch52Application 主类。然后，访问 http://localhost:8080/ch5_2/uploadFile。运行效果如图 5.16 所示。

图 5.16　文件选择界面

在图 5.16 中输入文件描述，并选择上传文件后，单击"上传文件"按钮，实现文件上传。文件上传成功后，打开如图 5.17 所示的下载文件列表页面。

单击图 5.17 中的文件名即可下载文件，至此，文件上传与下载示例演示完毕。

图 5.17 下载文件列表页面

5.5 Spring Boot 的异常统一处理

视频讲解

在 Spring Boot 应用的开发中,不管是对底层数据库操作,对业务层操作,还是对控制层操作,都会不可避免地遇到各种可预知的、不可预知的异常需要处理。如果每个过程都单独处理异常,那么系统的代码耦合度高、工作量大且不好统一,以后维护的工作量也很大。

如果能将所有类型的异常处理从各层中解耦出来,则既保证了相关处理过程的功能较单一,也实现了异常信息的统一处理和维护。幸运的是,Spring 框架支持这样的实现。本节将从自定义 error 页面、@ExceptionHandler 注解以及 @ControllerAdvice 3 种方式讲解 Spring Boot 应用的异常统一处理。

5.5.1 自定义 error 页面

在 Spring Boot Web 应用的 src/main/resources/templates 目录下添加 error.html 页面,访问发生错误或异常时,Spring Boot 将自动找到该页面作为错误页面。Spring Boot 为错误页面提供了以下属性。

- timestamp:错误发生时间;
- status:HTTP 状态码;
- error:错误原因;
- exception:异常的类名;
- message:异常消息(如果这个错误是由异常引起的);
- errors:BindingResult 异常里的各种错误(如果这个错误是由异常引起的);
- trace:异常跟踪信息(如果这个错误是由异常引起的);
- path:错误发生时请求的 URL 路径。

下面通过一个实例讲解在 Spring Boot 应用的开发中,如何使用自定义 error 页面。

【例 5-8】 自定义 error 页面。

具体实现步骤如下。

1. 创建基于 Thymeleaf 模板引擎的 Spring Boot Web 应用 ch5_3

参照 5.2.6 节的例 5-5，创建基于 Thymeleaf 模板引擎的 Spring Boot Web 应用 ch5_3。

2. 设置 Web 应用 ch5_3 的上下文路径

在 ch5_3 的 application.properties 文件中配置如下内容：

server.servlet.context-path=/ch5_3

3. 创建自定义异常类 MyException

创建名为 com.ch.ch5_3.exception 的包，并在该包中创建名为 MyException 的异常类。具体代码如下：

```
package com.ch.ch5_3.exception;
public class MyException extends Exception {
    private static final long serialVersionUID = 1L;
    public MyException() {
        super();
    }
    public MyException(String message) {
        super(message);
    }
}
```

4. 创建控制器类 TestHandleExceptionController

创建名为 com.ch.ch5_3.controller 的包，并在该包中创建名为 TestHandleExceptionController 的控制器类。在该控制器类中，有 4 个请求处理方法，一个是导航到 index.html，另外 3 个分别抛出不同的异常（并没有处理异常）。具体代码如下：

```
package com.ch.ch5_3.controller;
import java.sql.SQLException;
import org.springframework.stereotype.Controller;
import org.springframework.web.bind.annotation.RequestMapping;
import com.ch.ch5_3.exception.MyException;
@Controller
public class TestHandleExceptionController {
    @RequestMapping("/")
    public String index() {
        return "index";
    }
    @RequestMapping("/db")
    public void db() throws SQLException {
        throw new SQLException("数据库异常");
    }
    @RequestMapping("/my")
    public void my() throws MyException {
        throw new MyException("自定义异常");
```

```
    }
    @RequestMapping("/no")
    public void no() throws Exception {
        throw new Exception("未知异常");
    }
}
```

5. 整理脚本样式静态文件

JS 脚本、CSS 样式、图片等静态文件默认放置在 src/main/resources/static 目录下，ch5_3 应用引入了与 ch5_2 一样的 BootStrap 和 jQuery。

6. View 视图页面

Thymeleaf 模板默认将视图页面放在 src/main/resources/templates 目录下。因此，我们在 src/main/resources/templates 目录下新建 html 页面文件 index.html 和 error.html。

在 index.html 页面中，有 4 个超链接请求，3 个请求在控制器中有对应处理，另一个请求是 404 错误。具体代码如下：

```html
<!DOCTYPE html>
<html xmlns:th="http://www.thymeleaf.org">
<head>
<meta charset="UTF-8">
<title>index</title>
<link rel="stylesheet" th:href="@{css/bootstrap.min.css}" />
<!-- 默认访问 src/main/resources/static 下的 css 文件夹 -->
<link rel="stylesheet" th:href="@{css/bootstrap-theme.min.css}" />
</head>
<body>
    <div class="panel panel-primary">
        <div class="panel-heading">
            <h3 class="panel-title">异常处理示例</h3>
        </div>
    </div>
    <div class="container">
        <div class="row">
            <div class="col-md-4 col-sm-6">
                <a th:href="@{db}">处理数据库异常</a><br>
                <a th:href="@{my}">处理自定义异常</a><br>
                <a th:href="@{no}">处理未知错误</a>
                <hr>
                <a th:href="@{nofound}">404 错误</a>
            </div>
        </div>
    </div>
</body>
</html>
```

在 error.html 页面中，使用 Spring Boot 为错误页面提供的属性显示错误消息。具体代码如下：

```html
<!DOCTYPE html>
<html xmlns:th="http://www.thymeleaf.org">
<head>
<meta charset="UTF-8">
<title>error</title>
<link rel="stylesheet" th:href="@{css/bootstrap.min.css}" />
<!-- 默认访问src/main/resources/static下的css文件夹 -->
<link rel="stylesheet" th:href="@{css/bootstrap-theme.min.css}" />
</head>
<body>
    <div class="panel-l container clearfix">
        <div class="error">
            <p class="title"><span class="code" th:text="${status}"></span>非常抱歉,没有找到您要查看的页面</p>
            <div class="common-hint-word">
                <div th:text="${#dates.format(timestamp,'yyyy-MM-dd HH:mm:ss')}"></div>
                <div th:text="${message}"></div>
                <div th:text="${error}"></div>
            </div>
        </div>
    </div>
</body>
</html>
```

7. 运行

首先,运行 Ch53Application 主类。然后,访问 http://localhost:8080/ch5_3/打开 index.html 页面,运行效果如图 5.18 所示。

单击图 5.18 中的超链接时,Spring Boot 应用将根据链接请求,到控制器中找对应的处理。例如,单击图 5.18 中的"处理数据库异常"链接时,将执行控制器中的 public void db() throws SQLException 方法,而该方法仅仅抛出了 SQLException 异常,并没有处理异常。当 Spring Boot 发现有异常抛出且没有处理时,将自动在 src/main/resources/templates 目录下找到 error.html 页面显示异常信息,效果如图 5.19 所示。

图 5.18 index.html 页面

图 5.19 error.html 页面

从上述例 5-8 的运行结果可以看出,使用自定义 error 页面并没有真正处理异常,只是将异常或错误信息显示给客户端,因为在服务器控制台上同样抛出了异常,如图 5.20 所示。

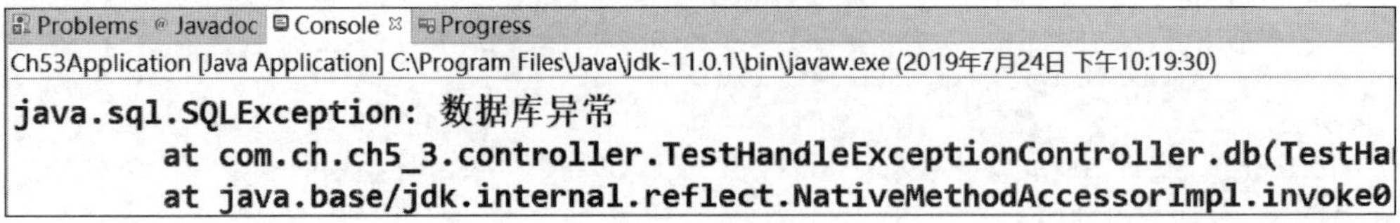

```
Problems  Javadoc  Console  Progress
Ch53Application [Java Application] C:\Program Files\Java\jdk-11.0.1\bin\javaw.exe (2019年7月24日 下午10:19:30)
java.sql.SQLException: 数据库异常
        at com.ch.ch5_3.controller.TestHandleExceptionController.db(TestHa
        at java.base/jdk.internal.reflect.NativeMethodAccessorImpl.invoke0
```

图 5.20　异常信息

5.5.2　@ExceptionHandler 注解

在 5.5.1 节中使用自定义 error 页面并没有真正处理异常，在本节可以使用 @ExceptionHandler 注解处理异常。如果在 Controller 中有一个使用 @ExceptionHandler 注解修饰的方法，那么当 Controller 的任何方法抛出异常时，都由该方法处理异常。

下面通过实例讲解如何使用 @ExceptionHandler 注解处理异常。

【例 5-9】　使用 @ExceptionHandler 注解处理异常。

具体实现步骤如下。

1. 在控制器类中添加使用 @ExceptionHandler 注解修饰的方法

在例 5-8 的控制器类 TestHandleExceptionController 中，添加一个使用 @ExceptionHandler 注解修饰的方法，具体代码如下：

```java
@ExceptionHandler(value = Exception.class)
public String handlerException(Exception e) {
    //数据库异常
    if (e instanceof SQLException) {
        return "sqlError";
    } else if (e instanceof MyException) {      //自定义异常
        return "myError";
    } else {                                     //未知异常
        return "noError";
    }
}
```

2. 创建 sqlError、myError 和 noError 页面

在 ch5_3 的 src/main/resources/templates 目录下，创建 sqlError、myError 和 noError 页面。当发生 SQLException 异常时，Spring Boot 处理后，显示 sqlError 页面；当发生 MyException 异常时，Spring Boot 处理后，显示 myError 页面；当发生未知异常时，Spring Boot 处理后，显示 noError 页面。具体代码略。

3. 运行

再次运行 Ch53Application 主类后，访问 http://localhost:8080/ch5_3/打开 index .html 页面，单击"处理数据库异常"链接时，执行控制器中的 public void db() throws SQLException 方法，该方法抛出了 SQLException，这时 Spring Boot 会自动执行使用 @ExceptionHandler 注解修饰的方法 public String handlerException(Exception e) 进行异常处理并打开 sqlError.html 页面，同时观察控制台有没有抛出异常信息。注意单击"404

错误"链接时,还是由自定义 error 页面显示错误信息,这是因为没有执行控制器中抛出异常的方法,进而不会执行使用@ExceptionHandler 注解修饰的方法。

从例 5-9 可以看出,在控制器中添加使用@ExceptionHandler 注解修饰的方法才能处理异常。而一个 Spring Boot 应用中往往存在多个控制器,不太适合在每个控制器中添加使用@ExceptionHandler 注解修饰的方法进行异常处理。可以将使用@ExceptionHandler 注解修饰的方法放到一个父类中,然后所有需要处理异常的控制器继承该类即可。例如,可以将例 5-9 中使用@ExceptionHandler 注解修饰的方法移到一个父类 BaseController 中,然后让控制器类 TestHandleExceptionController 继承该父类即可处理异常。

5.5.3　@ControllerAdvice 注解

使用 5.5.2 节中父类 Controller 进行异常处理,也有其自身的缺点,那就是代码耦合性太高。可以使用@ControllerAdvice 注解降低这种父子耦合关系。

@ControllerAdvice 注解,顾名思义,是一个增强的 Controller。使用该 Controller,可以实现 3 个方面的功能:全局异常处理、全局数据绑定以及全局数据预处理。本节将学习如何使用@ControllerAdvice 注解进行全局异常处理。

使用@ControllerAdvice 注解的类是当前 Spring Boot 应用中所有类的统一异常处理类,该类中使用@ExceptionHandler 注解的方法统一处理异常,不需要在每个 Controller 中逐一定义异常处理方法,这是因为对所有注解了@RequestMapping 的控制器方法有效。

下面通过实例讲解如何使用@ControllerAdvice 注解进行全局异常处理。

【例 5-10】　使用@ControllerAdvice 注解进行全局异常处理。

具体实现步骤如下。

1. 创建使用@ControllerAdvice 注解的类

在 ch5_3 的 com.ch.ch5_3.controller 包中,创建名为 GlobalExceptionHandlerController 的类。使用@ControllerAdvice 注解修饰该类,并将例 5-9 中使用@ExceptionHandler 注解修饰的方法移到该类中,具体代码如下:

```java
package com.ch.ch5_3.controller;
import java.sql.SQLException;
import org.springframework.web.bind.annotation.ControllerAdvice;
import org.springframework.web.bind.annotation.ExceptionHandler;
import com.ch.ch5_3.exception.MyException;
@ControllerAdvice
public class GlobalExceptionHandlerController {
    @ExceptionHandler(value = Exception.class)
    public String handlerException(Exception e) {
        //数据库异常
        if (e instanceof SQLException) {
            return "sqlError";
        } else if (e instanceof MyException) {     //自定义异常
            return "myError";
        } else {                                    //未知异常
            return "noError";
        }
```

        }
    }

2. 运行

再次运行 Ch53Application 主类后,访问 http://localhost:8080/ch5_3/打开 index.html页面测试即可。

5.6 Spring Boot 对 JSP 的支持

视频讲解

尽管 Spring Boot 建议使用 HTML 完成动态页面,但也有部分 Java Web 应用使用 JSP 完成动态页面。遗憾的是 Spring Boot 官方不推荐使用 JSP 技术,但考虑到是常用的技术,本节将介绍 Spring Boot 如何集成 JSP 技术。

下面通过实例讲解 Spring Boot 如何集成 JSP 技术。

【例 5-11】 Spring Boot 集成 JSP 技术。

具体实现步骤如下。

1. 创建 Spring Boot Web 应用 ch5_4

选择菜单 File|New|Spring Starter Project,打开 New Spring Starter Project 对话框,在该对话框中选择和输入相关信息,如图 5.21 所示。

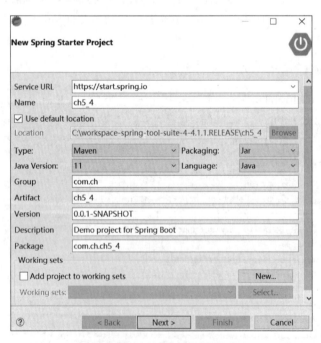

图 5.21 创建 Spring Boot Web 应用 ch5_4

单击图 5.21 中的 Next 按钮,打开 New Spring Starter Project Dependencies 对话框,在该对话框中,选择 Spring Web Starter 依赖,如图 5.22 所示。

第5章 Spring Boot 的 Web 开发

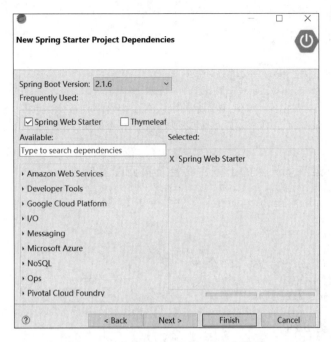

图 5.22 选择 Spring Web Starter 依赖

单击图 5.22 中的 Finish 按钮，完成创建 Spring Boot Web 应用 ch5_4。

2. 修改 pom.xml 文件，添加 Servlet、Tomcat 和 JSTL 依赖

因为在 JSP 页面中使用 EL 和 JSTL 标签显示数据，所以在 pom.xml 文件中，除了添加 Servlet 和 Tomcat 依赖外，还需要添加 JSTL 依赖。具体代码如下：

```
<!-- 添加 Servlet 依赖 -->
<dependency>
    <groupId>javax.servlet</groupId>
    <artifactId>javax.servlet-api</artifactId>
    <scope>provided</scope>
    <!-- provided 被依赖包理论上可以参与编译、测试、运行等阶段，相当于 compile，但是在打包阶
    段做了 exclude 的动作。适用场景：例如，如果我们在开发一个 Web 应用，在编译时需要依赖
    servlet-api.jar，但是在运行时不需要该 jar 包，因为这个 jar 包已由应用服务器提供，此时需
    要使用 provided 进行范围修饰。-->
</dependency>
<!-- 添加 Tomcat 依赖 -->
<dependency>
    <groupId>org.springframework.boot</groupId>
    <artifactId>spring-boot-starter-tomcat</artifactId>
    <scope>provided</scope>
</dependency>
<!-- Jasper 是 Tomcat 使用的引擎，使用 tomcat-embed-jasper 可以将 Web 应用在内嵌的 Tomcat
下运行 -->
<dependency>
    <groupId>org.apache.tomcat.embed</groupId>
    <artifactId>tomcat-embed-jasper</artifactId>
```

```xml
        <scope>provided</scope>
    </dependency>
    <!-- 添加 JSTL 依赖 -->
    <dependency>
        <groupId>javax.servlet</groupId>
        <artifactId>jstl</artifactId>
        <!-- 如果没有指定 scope 值,该元素的默认值为 compile。被依赖包需要参与到当前项目的编
        译、测试、打包、运行等阶段。打包的时候通常会包含被依赖包。-->
    </dependency>
```

3. 设置 Web 应用 ch5_4 的上下文路径及页面配置信息

在 ch5_4 的 application.properties 文件中配置如下内容:

```
server.servlet.context-path=/ch5_4
#设置页面前缀目录
spring.mvc.view.prefix=/WEB-INF/jsp/
#设置页面后缀
spring.mvc.view.suffix=.jsp
```

4. 创建实体类 Book

创建名为 com.ch.ch5_4.model 的包,并在该包中创建名为 Book 的实体类。此实体类用在模板页面展示数据,代码与例 5-5 中的 Book 一样,不再赘述。

5. 创建控制器类 ThymeleafController

创建名为 com.ch.ch5_4.controller 的包,并在该包中创建名为 ThymeleafController 的控制器类。在该控制器类中,实例化 Book 类的多个对象,并保存到集合 ArrayList<Book>中。代码与例 5-5 中的 ThymeleafController 一样,不再赘述。

6. 整理脚本样式静态文件

JS 脚本、CSS 样式、图片等静态文件默认放置在 src/main/resources/static 目录下,ch5_4 应用引入的 BootStrap 和 jQuery 与例 5-5 中的一样,不再赘述。

7. View 视图页面

从 application.properties 配置文件中可知,将 JSP 文件路径指定到/WEB-INF/jsp/目录。因此,我们需要在 src/main 目录下创建目录 webapp/WEB-INF/jsp/,并在该目录下创建 JSP 文件 index.jsp。代码如下:

```jsp
<%@ page language="java" contentType="text/html; charset=UTF-8" pageEncoding="UTF-8"%>
<!-- 引入 JSTL 标签 -->
<%@ taglib prefix="c" uri="http://java.sun.com/jsp/jstl/core" %>
<%
    String path = request.getContextPath();
    String basePath = request.getScheme() + "://" + request.getServerName() + ":" + request.getServerPort() + path + "/";
%>
```

```html
<!DOCTYPE html>
<html>
<head>
<base href="<%=basePath%>">
<meta charset="UTF-8">
<title>JSP 测试</title>
<link href="css/bootstrap.min.css" rel="stylesheet">
<link href="css/bootstrap-theme.min.css" rel="stylesheet">
</head>
<body>
    <div class="panel panel-primary">
        <div class="panel-heading">
            <h3 class="panel-title">第一个基于 JSP 技术的 Spring Boot Web 应用</h3>
        </div>
    </div>
    <div class="container">
        <div>
        <h4>图书列表</h4>
    </div>
        <div class="row">
            <div class="col-md-4 col-sm-6">
                <!-- 使用 EL 表达式 -->
                <a href="">
    <img src="images/${aBook.picture}" alt="图书封面" style="height:180px; width:40%;"/>
                </a>
                <div class="caption">
                    <h4>${aBook.bname}</h4>
                    <p>${aBook.author}</p>
                    <p>${aBook.isbn}</p>
                    <p>${aBook.price}</p>
                    <p>${aBook.publishing}</p>
                </div>
            </div>
            <!-- 使用 JSTL 标签 forEach 循环取出集合数据 -->
            <c:forEach var="book" items="${books}">
                <div class="col-md-4 col-sm-6">
                <a href="">
    <img src="images/${book.picture}" alt="图书封面" style="height:180px; width:40%;"/>
                </a>
                <div class="caption">
                    <h4>${book.bname}</h4>
                    <p>${book.author}</p>
                    <p>${book.isbn}</p>
                    <p>${book.price}</p>
                    <p>${book.publishing}</p>
                </div>
            </div>
            </c:forEach>
        </div>
    </div>
</body>
</html>
```

8. 运行

首先,运行 Ch54Application 主类。然后,访问 http://localhost:8080/ch5_4/。运行效果如图 5.23 所示。

图 5.23 例 5-11 运行结果

5.7 本章小结

本章首先介绍了 Spring Boot 的 Web 开发支持,然后详细讲述了 Spring Boot 推荐使用的 Thymeleaf 模板引擎,包括 Thymeleaf 的基础语法、常用属性以及国际化。同时,本章还介绍了 Spring Boot 对 JSON 数据的处理、文件上传下载、异常统一处理和对 JSP 的支持等 Web 应用开发的常用功能。

习题 5

使用 Hibernate Validator 验证如图 5.24 所示的表单信息,具体要求如下:
(1) 用户名必须输入,并且长度范围为 5～20。
(2) 年龄范围为 18～60。
(3) 工作日期在系统时间之前。

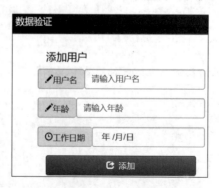

图 5.24 输入页面

第 6 章

Spring Boot 的数据访问

学习目的与要求

本章详细介绍 Spring Boot 访问数据库的解决方案。通过本章的学习，掌握 Spring Boot 访问关系型数据库及非关系型数据库的解决方案。

本章主要内容

- Spring Data JPA。
- Spring Boot 整合 MyBatis。
- Spring Boot 的事务管理。
- Spring Boot 整合 REST。
- Spring Boot 整合 MongoDB。
- Spring Boot 整合 Redis。
- 数据缓存 Cache。

Spring Data 是 Spring 访问数据库的一揽子解决方案，是一个伞形项目，包含大量关系型数据库及非关系型数据库的数据访问解决方案。本章将详细介绍 Spring Data JPA、Spring Data REST、Spring Data MongoDB、Spring Data Redis 等 Spring Data 的子项目。

6.1 Spring Data JPA

Spring Data JPA 是 Spring Data 的子项目，在讲解 Spring Data JPA 之前，先了解一下 Hibernate。这是因为 Spring Data JPA 是由 Hibernate 默认实现的。

Hibernate是一个开源的对象关系映射框架,它对JDBC进行了非常轻量级的对象封装,它将POJO(Plain Ordinary Java Object)简单的Java对象与数据库表建立映射关系,是一个全自动的ORM(Object Relational Mapping)框架。Hibernate可以自动生成SQL语句、自动执行,使得Java开发人员可以随心所欲地使用对象编程思维来操纵数据库。

JPA(Java Persistence API)是官方提出的Java持久化规范。JPA通过注解或XML描述对象—关系(表)的映射关系,并将内存中的实体对象持久化到数据库。

Spring Data JPA通过提供基于JPA的Repository极大地简化了JPA的写法,在几乎不写实现的情况下,实现数据库的访问和操作。使用Spring Data JPA建立数据访问层十分方便,只需要定义一个继承JpaRepository接口的接口即可。

继承了JpaRepository接口的自定义数据访问接口,具有JpaRepository接口的所有数据访问操作方法。JpaRepository接口的源代码如下:

```
package org.springframework.data.jpa.repository;
import java.util.List;
import javax.persistence.EntityManager;
import org.springframework.data.domain.Example;
import org.springframework.data.domain.Sort;
import org.springframework.data.repository.NoRepositoryBean;
import org.springframework.data.repository.PagingAndSortingRepository;
import org.springframework.data.repository.query.QueryByExampleExecutor;
@NoRepositoryBean
public interface JpaRepository<T, ID> extends PagingAndSortingRepository<T, ID>,
QueryByExampleExecutor<T> {
    List<T> findAll();
    List<T> findAll(Sort sort);
    List<T> findAllById(Iterable<ID> ids);
    <S extends T> List<S> saveAll(Iterable<S> entities);
    void flush();
    <S extends T> S saveAndFlush(S entity);
    void deleteInBatch(Iterable<T> entities);
    void deleteAllInBatch();
    T getOne(ID id);
    <S extends T> List<S> findAll(Example<S> example);
    <S extends T> List<S> findAll(Example<S> example, Sort sort);
}
```

JpaRepository接口提供的常用方法如下。

List<T> findAll():查询所有实体对象数据,返回一个List集合。

List<T> findAll(Sort sort):按照指定的排序规则查询所有实体对象数据,返回一个List集合。

List<T> findAllById(Iterable<ID> ids):根据所提供的实体对象id(多个),将对应的实体全部查询出来,并返回一个List集合。

<S extends T> List<S> saveAll(Iterable<S> entities):将提供的集合中的实体对象数据保存到数据库。

void flush():将缓存的对象数据操作更新到数据库。

＜S extends T＞S saveAndFlush(S entity)：保存对象的同时立即更新到数据库。
void deleteInBatch(Iterable＜T＞entities)：批量删除提供的实体对象。
void deleteAllInBatch()：批量删除所有的实体对象。
T getOne(ID id)：根据 id 获得对应的实体对象。
＜S extends T＞List＜S＞findAll(Example＜S＞example)：根据提供的 example 实例查询实体对象数据。
＜S extends T＞List＜S＞findAll(Example＜S＞example，Sort sort)：根据提供的 example 实例，并按照指定规则，查询实体对象数据。

6.1.1 Spring Boot 的支持

在 Spring Boot 应用中，如果需要使用 Spring Data JPA 访问数据库，那么可以通过 STS 创建 Spring Boot 应用时选择 Spring Data JPA 模块依赖，如图 6.1 所示。

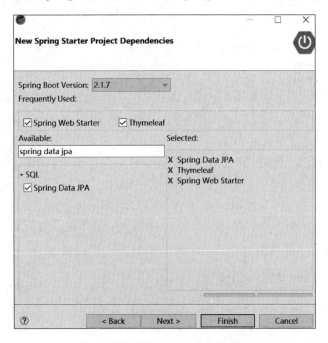

图 6.1 选择 Spring Data JPA 模块

如果在 Maven 项目中需要使用 Spring Data JPA 访问数据库，那么可以在 pom.xml 文件中通过如下配置添加 Spring Data JPA 模块依赖：

```
<dependency>
    <groupId>org.springframework.boot</groupId>
    <artifactId>spring-boot-starter-data-jpa</artifactId>
</dependency>
```

1. JDBC 的自动配置

Spring Data JPA 模块的依赖关系如图 6.2 所示，从图中可知，spring-boot-starter-data-jpa 依赖于 spring-boot-starter-jdbc，而 Spring Boot 对 spring-boot-starter-jdbc 做了自动配置。JDBC 自动配置源码位于 org.springframework.boot.autoconfigure.jdbc 包下，如图 6.3 所示。

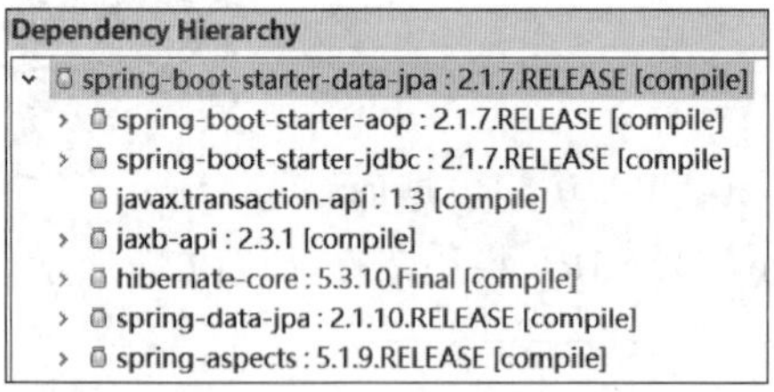

图 6.2 spring-boot-starter-data-jpa 依赖关系

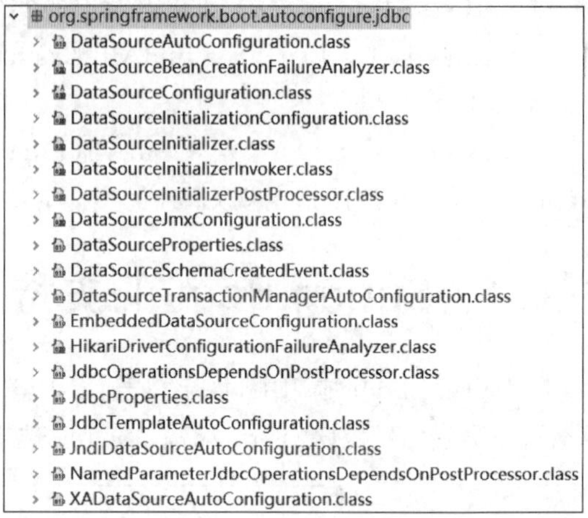

图 6.3 spring-boot-starter-jdbc 自动配置源码位置

从 DataSourceProperties 类可以看出，可以使用 spring.datasource 为前缀的属性在 application.properties 配置文件中配置 datasource。

2. JPA 的自动配置

Spring Boot 对 JPA 的自动配置位于 org.springframework.boot.autoconfigure.orm.jpa 包下，如图 6.4 所示。

从 HibernateJpaAutoConfiguration 类可以看出，Spring Boot 对 JPA 的默认实现是 Hibernate。从 JpaProperties 类可以看出，可以使用 spring.jpa 为前缀的属性在 application.properties 配置文件中配置 JPA。

3. Spring Data JPA 的自动配置

Spring Boot 对 Spring Data JPA 的自动配置位于 org.springframework.boot.autoconfigure.data.jpa 包下，如图 6.5 所示。

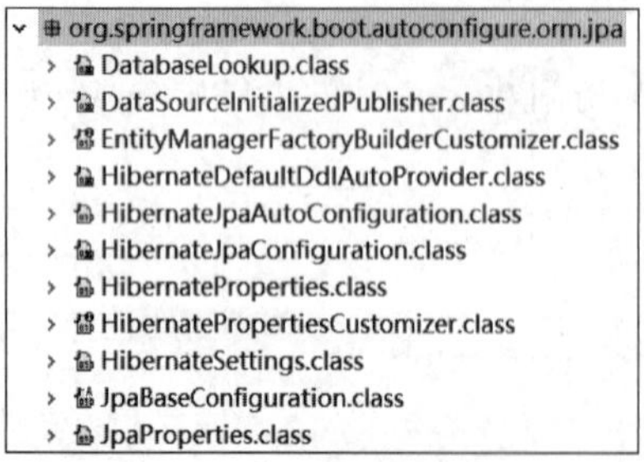

图 6.4 JPA 的自动配置源码位置

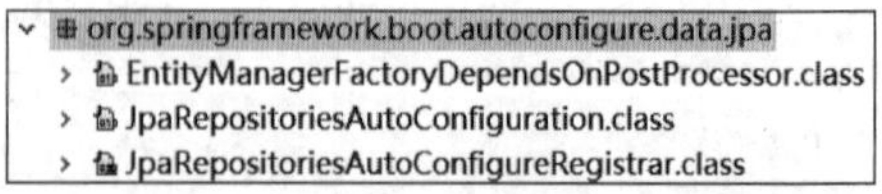

图 6.5 Spring Data JPA 的自动配置源码位置

从 JpaRepositoriesAutoConfiguration 类可以看出，JpaRepositoriesAutoConfiguration 依赖于 HibernateJpaAutoConfiguration 配置；从 JpaRepositoriesAutoConfigureRegistrar 类可以看出，Spring Boot 自动开启了对 Spring Data JPA 的支持，即开发人员无须在配置类中显式声明@EnableJpaRepositories。

4. Spring Boot 应用的 Spring Data JPA

从上述分析可知，在 Spring Boot 应用中使用 Spring Data JPA 访问数据库时，除了添加 spring-boot-starter-data-jpa 依赖外，只需定义 DataSource、持久化实体类和数据访问层，并在需要使用数据访问的地方（如 Service 层）依赖注入数据访问层即可。

6.1.2 简单条件查询

从前面的学习可知，只需定义一个继承 JpaRepository 接口的接口即可使用 Spring Data JPA 建立数据访问层。因此，自定义的数据访问接口完全继承了 JpaRepository 的接口方法。但更重要的是，在自定义的数据访问接口中可以根据查询关键字定义查询方法，这些查询方法需要符合它的命名规则，一般是根据持久化实体类的属性名来确定的。

1. 查询关键字

目前，Spring Data JPA 支持的查询关键字如表 6.1 所示。

表 6.1 查询关键字

关 键 字	示 例	JPQL 代码段
And	findByLastnameAndFirstname	… where x.lastname = ?1 and x.firstname = ?2
Or	findByLastnameOrFirstname	… where x.lastname = ?1 or x.firstname = ?2
Is,Equals	findByFirstname,findByFirstnameIs, findByFirstnameEquals	… where x.firstname = ?1
Between	findByStartDateBetween	… where x.startDate between ?1 and ?2
LessThan	findByAgeLessThan	… where x.age < ?1
LessThanEqual	findByAgeLessThanEqual	… where x.age <= ?1
GreaterThan	findByAgeGreaterThan	… where x.age > ?1
GreaterThanEqual	findByAgeGreaterThanEqual	… where x.age >= ?1
After	findByStartDateAfter	… where x.startDate > ?1
Before	findByStartDateBefore	… where x.startDate < ?1
IsNull	findByAgeIsNull	… where x.age is null
IsNotNull,NotNull	findByAge(Is)NotNull	… where x.age not null
Like	findByFirstnameLike	… where x.firstname like ?1
NotLike	findByFirstnameNotLike	… where x.firstname not like ?1
StartingWith	findByFirstnameStartingWith	… where x.firstname like ?1 参数后加%，即以参数开头的模糊查询
EndingWith	findByFirstnameEndingWith	… where x.firstname like ?1 参数前加%，即以参数结尾的模糊查询

续表

关 键 字	示 例	JPQL 代码段
Containing	findByFirstnameContaining	…where x.firstname like ?1 参数两边加%,即包含参数的模糊查询
OrderBy	findByAgeOrderByLastnameDesc	…where x.age = ?1 order by x.lastname desc
Not	findByLastnameNot	…where x.lastname <> ?1
In	findByAgeIn(Collection<Age> ages)	…where x.age in ?1
NotIn	findByAgeNotIn(Collection<Age> ages)	…where x.age not in ?1
True	findByActiveTrue()	…where x.active = true
False	findByActiveFalse()	…where x.active = false
IgnoreCase	findByFirstnameIgnoreCase	…where UPPER(x.firstname) = UPPER(?1)

2. 限制查询结果数量

在 Spring Data JPA 中,使用 Top 和 First 关键字限制查询结果数量。示例代码如下:

```
public interface UserRepository extends JpaRepository<MyUser, Integer>{
    /**
     * 获得符合查询条件的前 10 条
     */
    public List<MyUser> findTop10ByUnameLike(String uname);
    /**
     * 获得符合查询条件的前 15 条
     */
    public List<MyUser> findFirst15ByUnameLike(String uname);
}
```

3. 简单条件查询示例

下面通过实例讲解在 Spring Boot Web 应用中如何使用 Spring Data JPA 进行简单条件查询。

【例 6-1】 使用 Spring Data JPA 进行简单条件查询。

具体实现步骤如下。

1) 创建数据库

本书采用的关系型数据库是 MySQL 5.x,为了演示本例,首先通过命令"CREATE DATABASE springbootjpa;"创建名为 springbootjpa 的数据库。

2) 创建基于 Thymeleaf 和 Spring Data JPA 的 Spring Boot Web 应用 ch6_1

参考图 6.1 创建基于 Thymeleaf 和 Spring Data JPA 的 Spring Boot Web 应用 ch6_1。

3) 修改 pom.xml 文件,添加 MySQL 依赖

在 pom.xml 文件中添加如下依赖:

```
<dependency>
    <groupId>mysql</groupId>
    <artifactId>mysql-connector-java</artifactId>
    <version>5.1.45</version>
    <!-- 使用 MySQL8 时,连接器请使用 8.x -->
</dependency>
```

4）设置 Web 应用 ch6_1 的上下文路径及数据源配置信息

在 ch6_1 的 application.properties 文件中配置如下内容：

```
server.servlet.context-path=/ch6_1
###
## 数据源信息配置
###
# 数据库地址
spring.datasource.url=jdbc:mysql://localhost:3306/springbootjpa?characterEncoding=utf8
# 数据库 MySQL 为 8.x 时，url 为
# jdbc:mysql://localhost:3306/springbootjpa?useSSL=false&serverTimezone=Asia/Beijing&ch
# aracterEncoding=utf-8
# 数据库用户名
spring.datasource.username=root
# 数据库密码
spring.datasource.password=root
# 数据库驱动
spring.datasource.driver-class-name=com.mysql.jdbc.Driver
# 数据库 MySQL 为 8.x 时，驱动类为 com.mysql.cj.jdbc.Driver
####
# JPA 持久化配置
####
# 指定数据库类型
spring.jpa.database=MYSQL
# 指定是否在日志中显示 SQL 语句
spring.jpa.show-sql=true
# 指定自动创建、更新数据库表等配置，update 表示如果数据库中存在持久化类对应的表就不创建，
# 不存在就创建
spring.jpa.hibernate.ddl-auto=update
# 让控制器输出的 JSON 字符串格式更美观
spring.jackson.serialization.indent-output=true
```

5）创建持久化实体类 MyUser

创建名为 com.ch.ch6_1.entity 的包，并在该包中创建名为 MyUser 的持久化实体类。具体代码如下：

```java
package com.ch.ch6_1.entity;
import java.io.Serializable;
import javax.persistence.Entity;
import javax.persistence.GeneratedValue;
import javax.persistence.GenerationType;
import javax.persistence.Id;
import javax.persistence.Table;
@Entity
@Table(name = "user_table")
public class MyUser implements Serializable{
    private static final long serialVersionUID = 1L;
    @Id
    @GeneratedValue(strategy = GenerationType.IDENTITY)
    private int id;                              //主键
```

```
    /** 使用@Column 注解,可以配置列相关属性(列名,长度等),
     * 可以省略,默认为属性名小写,如果属性名是词组,将在中间加上"_"。
     */
    private String uname;
    private String usex;
    private int age;
    //省略 get 和 set 方法
}
```

在持久化类中,@Entity 注解表明该实体类是一个与数据库表映射的实体类。@Table 表示实体类与哪个数据库表映射,如果没有通过 name 属性指定表名,默认为小写的类名。如果类名为词组,将在中间加上"_"(如 MyUser 类对应的表名为 my_user)。@Id 注解的属性表示该属性映射为数据库表的主键。@GeneratedValue 注解默认使用主键生成方式为自增。如果是 MySQL、SQL Server 等关系型数据库,可映射成一个递增的主键;如果是 Oracle 等关系型数据库 hibernate,将自动生成一个名为 HIBERNATE_SEQUENCE 的序列。

6) 创建数据访问层

创建名为 com.ch.ch6_1.repository 的包,并在该包中创建名为 UserRepository 的接口,该接口继承 JpaRepository 接口。具体代码如下:

```
package com.ch.ch6_1.repository;
import java.util.List;
import org.springframework.data.jpa.repository.JpaRepository;
import com.ch.ch6_1.entity.MyUser;
/**
 * 这里不需要使用@Repository 注解数据访问层,
 * 因为 Spring Boot 自动配置了 JpaRepository
 */
public interface UserRepository extends JpaRepository<MyUser, Integer>{
    public MyUser findByUname(String uname);
    public List<MyUser> findByUnameLike(String uname);
}
```

由于 UserRepository 接口继承了 JpaRepository 接口,因此 UserRepository 接口中除了上述自定义的两个接口方法外(方法名命名规范参照表 6.1),还拥有 JpaRepository 的接口方法。

7) 创建业务层

创建名为 com.ch.ch6_1.service 的包,并在该包中创建 UserService 接口和接口的实现类 UserServiceImpl。

UserService 接口的具体代码如下:

```
package com.ch.ch6_1.service;
import java.util.List;
import com.ch.ch6_1.entity.MyUser;
public interface UserService {
    public void saveAll();
    public List<MyUser> findAll();
```

```java
    public MyUser findByUname(String uname);
    public List<MyUser> findByUnameLike(String uname);
    public MyUser getOne(int id);
}
```

UserServiceImpl 实现类的具体代码如下:

```java
package com.ch.ch6_1.service;
import java.util.ArrayList;
import java.util.List;
import org.springframework.beans.factory.annotation.Autowired;
import org.springframework.stereotype.Service;
import com.ch.ch6_1.entity.MyUser;
import com.ch.ch6_1.repository.UserRepository;
@Service
public class UserServiceImpl implements UserService{
    @Autowired                              //依赖注入数据访问层
    private UserRepository userRepository;
    @Override
    public void saveAll() {
        MyUser mu1 = new MyUser();
        mu1.setUname("陈恒1");
        mu1.setUsex("男");
        mu1.setAge(88);
        MyUser mu2 = new MyUser();
        mu2.setUname("陈恒2");
        mu2.setUsex("女");
        mu2.setAge(18);
        MyUser mu3 = new MyUser();
        mu3.setUname("陈恒3");
        mu3.setUsex("男");
        mu3.setAge(99);
        List<MyUser> users = new ArrayList<MyUser>();
        users.add(mu1);
        users.add(mu2);
        users.add(mu3);
        //调用父接口中的方法 saveAll
        userRepository.saveAll(users);
    }
    @Override
    public List<MyUser> findAll() {
        //调用父接口中的方法 findAll
        return userRepository.findAll();
    }
    @Override
    public MyUser findByUname(String uname) {
        return userRepository.findByUname(uname);
    }
    @Override
    public List<MyUser> findByUnameLike(String uname) {
        return userRepository.findByUnameLike("%" + uname + "%");
```

```java
    }
    @Override
    public MyUser getOne(int id) {
        //调用父接口中的方法 getOne
        return userRepository.getOne(id);
    }
}
```

8) 创建控制器类 UserTestController

创建名为 com.ch.ch6_1.controller 的包,并在该包中创建名为 UserTestController 的控制器类。具体代码如下:

```java
package com.ch.ch6_1.controller;
import org.springframework.beans.factory.annotation.Autowired;
import org.springframework.stereotype.Controller;
import org.springframework.ui.Model;
import org.springframework.web.bind.annotation.RequestMapping;
import org.springframework.web.bind.annotation.ResponseBody;
import com.ch.ch6_1.service.UserService;
@Controller
public class UserTestController {
    @Autowired
    private UserService userService;
    @RequestMapping("/save")
    @ResponseBody
    public String save() {
        userService.saveAll();
        return "保存用户成功!";
    }
    @RequestMapping("/findByUname")
    public String findByUname(String uname, Model model) {
        model.addAttribute("title", "根据用户名查询一个用户");
        model.addAttribute("auser", userService.findByUname(uname));
        return "showAuser";
    }
    @RequestMapping("/getOne")
    public String getOne(int id, Model model) {
        model.addAttribute("title", "根据用户 id 查询一个用户");
        model.addAttribute("auser", userService.getOne(id));
        return "showAuser";
    }
    @RequestMapping("/findAll")
    public String findAll(Model model){
        model.addAttribute("title", "查询所有用户");
        model.addAttribute("allUsers", userService.findAll());
        return "showAll";
    }
    @RequestMapping("/findByUnameLike")
    public String findByUnameLike(String uname, Model model){
```

```
            model.addAttribute("title","根据用户名模糊查询所有用户");
            model.addAttribute("allUsers",userService.findByUnameLike(uname));
            return "showAll";
        }
    }
```

9) 整理脚本样式静态文件

JS 脚本、CSS 样式、图片等静态文件默认放置在 src/main/resources/static 目录下，ch6_1 应用引入的 BootStrap 和 jQuery 与例 5-5 中的一样，不再赘述。

10) 创建 View 视图页面

在 src/main/resources/templates 目录下，创建视图页面 showAll.html 和 showAuser.html。

showAll.html 的具体代码如下：

```html
<!DOCTYPE html>
<html xmlns:th="http://www.thymeleaf.org">
<head>
<meta charset="UTF-8">
<title>显示查询结果</title>
<link rel="stylesheet" th:href="@{css/bootstrap.min.css}" />
<link rel="stylesheet" th:href="@{css/bootstrap-theme.min.css}" />
</head>
<body>
    <div class="panel panel-primary">
        <div class="panel-heading">
            <h3 class="panel-title">Spring Data JPA 简单查询</h3>
        </div>
    </div>
    <div class="container">
        <div class="panel panel-primary">
            <div class="panel-heading">
                <h3 class="panel-title"><span th:text="${title}"></span></h3>
            </div>
            <div class="panel-body">
                <div class="table table-responsive">
                    <table class="table table-bordered table-hover">
                        <tbody class="text-center">
                            <tr th:each="user:${allUsers}">
                                <td>
                                    <span th:text="${user.id}"></span>
                                </td>
                                <td>
                                    <span th:text="${user.uname}"></span>
                                </td>
                                <td>
                                    <span th:text="${user.usex}"></span>
                                </td>
                                <td>
                                    <span th:text="${user.age}"></span>
```

```
                    </td>
                </tr>
            </tbody>
        </table>
    </div>
  </div>
 </div>
</div>
</body>
</html>
```

showAuser.html 的具体代码如下：

```
<!DOCTYPE html>
<html xmlns:th="http://www.thymeleaf.org">
<head>
<meta charset="UTF-8">
<title>显示查询结果</title>
<link rel="stylesheet" th:href="@{css/bootstrap.min.css}" />
<link rel="stylesheet" th:href="@{css/bootstrap-theme.min.css}" />
</head>
<body>
    <div class="panel panel-primary">
        <div class="panel-heading">
            <h3 class="panel-title">Spring Data JPA 简单查询</h3>
        </div>
    </div>
    <div class="container">
        <div class="panel panel-primary">
            <div class="panel-heading">
                <h3 class="panel-title"><span th:text="${title}"></span></h3>
            </div>
            <div class="panel-body">
                <div class="table table-responsive">
                    <table class="table table-bordered table-hover">
                        <tbody class="text-center">
                            <tr>
                                <td>
                                    <span th:text="${auser.id}"></span>
                                </td>
                                <td>
                                    <span th:text="${auser.uname}"></span>
                                </td>
                                <td>
                                    <span th:text="${auser.usex}"></span>
                                </td>
                                <td>
                                    <span th:text="${auser.age}"></span>
                                </td>
                            </tr>
                        </tbody>
```

```
                </table>
            </div>
        </div>
    </div>
</body>
</html>
```

11）运行

首先，运行 Ch61Application 主类。然后，访问"http://localhost:8080/ch6_1/save/"。运行效果如图 6.6 所示。

"http://localhost:8080/ch6_1/save/"成功运行后，在 MySQL 的 springbootjpa 数据库中创建一张名为 user_table 的数据库表，并插入三条记录。

通过访问"http://localhost:8080/ch6_1/findAll"查询所有用户，运行效果如图 6.7 所示。

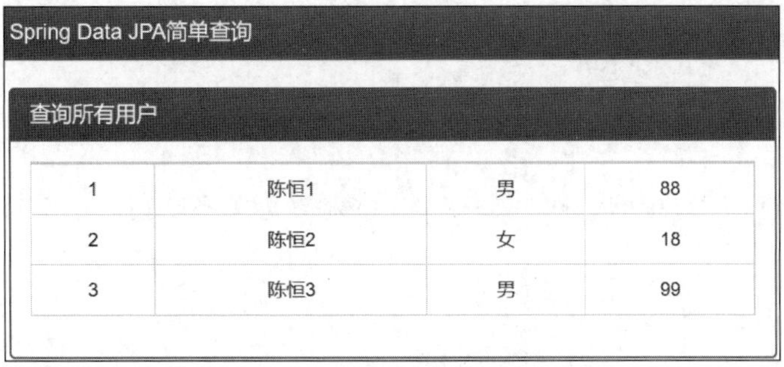

图 6.6　保存用户

图 6.7　查询所有用户

通过访问"http://localhost:8080/ch6_1/findByUnameLike?uname=陈"模糊查询所有陈姓用户，运行效果如图 6.8 所示。

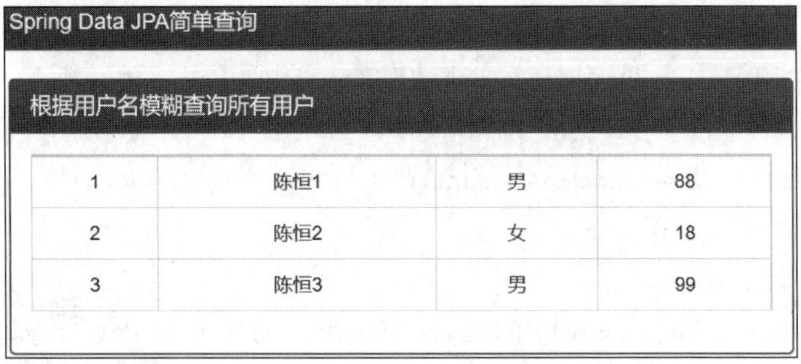

图 6.8　模糊查询所有陈姓用户

通过访问"http://localhost:8080/ch6_1/findByUname?uname=陈恒 2"查询一个名为陈恒 2 的用户信息，运行效果如图 6.9 所示。

通过访问"http://localhost:8080/ch6_1/getOne?id=1"查询一个 id 为 1 的用户信息，运行效果如图 6.10 所示。

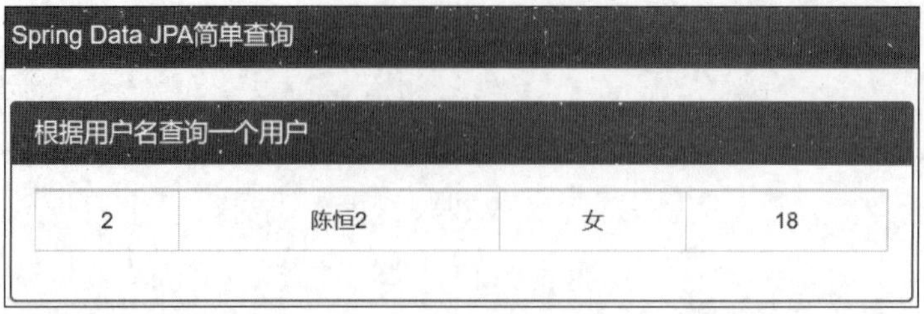

图 6.9　查询一个名为陈恒 2 的用户信息

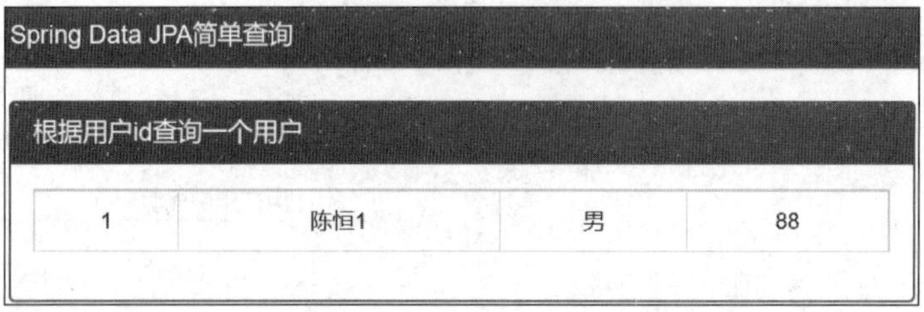

图 6.10　查询一个 id 为 1 的用户信息

6.1.3　关联查询

在 Spring Data JPA 中有一对一、一对多、多对多等关系映射。本节针对这些关系映射进行讲解。

1. @OneToOne

一对一关系在现实生活中是十分常见的。例如一个大学生只有一张一卡通，一张一卡通只属于一个大学生。再如人与身份证的关系也是一对一的关系。

在 Spring Data JPA 中，可用两种方式描述一对一关系映射：一种是通过外键的方式（一个实体通过外键关联到另一个实体的主键）；一种是通过一张关联表来保存两个实体一对一的关系。下面通过外键的方式讲解一对一关系映射。

【例 6-2】　使用 Spring Data JPA 实现人与身份证的一对一关系映射。

首先，为例 6-2 创建基于 Spring Data JPA 的 Spring Boot Web 应用 ch6_2。ch6_2 应用的数据库、pom.xml 以及 application.properties 与 ch6_1 应用基本一样，不再赘述。

其他内容具体实现步骤如下：

1）创建持久化实体类

创建名为 com.ch.ch6_2.entity 的包，并在该包中创建名为 Person 和 IdCard 的持久化实体类。

Person 的具体代码如下：

```
package com.ch.ch6_2.entity;
import java.io.Serializable;
import javax.persistence.CascadeType;
import javax.persistence.Entity;
```

视频讲解

```java
import javax.persistence.FetchType;
import javax.persistence.GeneratedValue;
import javax.persistence.GenerationType;
import javax.persistence.Id;
import javax.persistence.JoinColumn;
import javax.persistence.OneToOne;
import javax.persistence.Table;
import com.fasterxml.jackson.annotation.JsonIgnore;
import com.fasterxml.jackson.annotation.JsonIgnoreProperties;
@Entity
@Table(name = "person_table")
/** 解决 No serializer found for class org.hibernate.proxy.pojo.bytebuddy.ByteBuddyInterceptor
异常 */
@JsonIgnoreProperties(value = {"hibernateLazyInitializer"})
public class Person implements Serializable{
    private static final long serialVersionUID = 1L;
    @Id
    @GeneratedValue(strategy = GenerationType.IDENTITY)
    private int id;                          //自动递增的主键
    private String pname;
    private String psex;
    private int page;
    @OneToOne(
            optional = true,
            fetch = FetchType.LAZY,
            targetEntity = IdCard.class,
            cascade = CascadeType.ALL
            )
    /**
     * 指明 Person 对应表的 id_Card_id 列作为外键与 IdCard 对应表的 id 列进行关联
     * unique = true 指明 id_Card_id 列的值不可重复
     */
    @JoinColumn(
            name = "id_Card_id",
            referencedColumnName = "id",
            unique = true
            )
    @JsonIgnore
    //如果 A 对象持有 B 的引用,B 对象持有 A 的引用,这样就形成了循环引用
    //如果直接使用 json 转换会报错,使用@JsonIgnore 解决该错误
    private IdCard idCard;
    //省略 get 和 set 方法
}
```

上述实体类 Person 中,@OneToOne 注解有 5 个属性: targetEntity、cascade、fetch、optional 和 mappedBy。

targetEntity 属性: class 类型属性。定义关系类的类型,默认是该成员属性对应的类类型,所以通常不需要提供定义。

cascade 属性: CascadeType[]类型。该属性定义类和类之间的级联关系。定义的级联

关系将被容器视为对当前类对象及其关联类对象采取相同的操作,而且这种关系是递归调用的。cascade 的值只能从 CascadeType.PERSIST(级联新建)、CascadeType.REMOVE(级联删除)、CascadeType.REFRESH(级联刷新)、CascadeType.MERGE(级联更新)中选择一个或多个。还有一个选择是使用 CascadeType.ALL,表示选择全部四项。

fetch 属性:分为两种。FetchType.LAZY:懒加载,加载一个实体时,定义懒加载的属性不会立即从数据库中加载。FetchType.EAGER:急加载,加载一个实体时,定义急加载的属性会立即从数据库中加载。

optional 属性:optional = true,表示 idCard 属性可以为 null,也就是允许没有身份证,如未成年人没有身份证。

mappedBy 属性:mappedBy 标签一定是定义在关系的被维护端,它指向关系的维护端;只有 @OneToOne、@OneToMany、@ManyToMany 才有 mappedBy 属性,ManyToOne 不存在该属性。拥有 mappedBy 注解的实体类为关系的被维护端。

IdCard 的代码如下:

```java
package com.ch.ch6_2.entity;
import java.io.Serializable;
import java.util.Calendar;
import javax.persistence.CascadeType;
import javax.persistence.Entity;
import javax.persistence.FetchType;
import javax.persistence.GeneratedValue;
import javax.persistence.GenerationType;
import javax.persistence.Id;
import javax.persistence.OneToOne;
import javax.persistence.Table;
import javax.persistence.Temporal;
import javax.persistence.TemporalType;
import com.fasterxml.jackson.annotation.JsonIgnore;
import com.fasterxml.jackson.annotation.JsonIgnoreProperties;
@Entity
@Table(name = "idcard_table")
@JsonIgnoreProperties(value = { "hibernateLazyInitializer"})
public class IdCard implements Serializable{
    private static final long serialVersionUID = 1L;
    @Id
    @GeneratedValue(strategy = GenerationType.IDENTITY)
    private int id;                              //自动递增的主键
    private String code;
    /**
     * @Temporal 主要是用来指明 java.util.Date 或 java.util.Calendar 类型的属性
     * 具体与数据库(date、time、timestamp)3 个类型中的哪一个进行映射
     */
    @Temporal(value = TemporalType.DATE)
    private Calendar birthday;
    private String address;
    /**
     * optional = false 设置 person 属性值不能为 null,也就是身份证必须有对应的主人
```

```
 * mappedBy = "idCard"与 Person 类中的 idCard 属性一致
 * 拥有 mappedBy 注解的实体类为关系的被维护端
 */
@OneToOne(
        optional = false,
        fetch = FetchType.LAZY,
        targetEntity = Person.class,
        mappedBy = "idCard",
        cascade = CascadeType.ALL
        )
private Person person;                          //对应的人
//省略 get 和 set 方法
}
```

2) 创建数据访问层

创建名为 com.ch.ch6_2.repository 的包，并在该包中创建名为 IdCardRepository 和 PersonRepository 的接口。

IdCardRepository 的代码如下：

```
package com.ch.ch6_2.repository;
import java.util.List;
import org.springframework.data.jpa.repository.JpaRepository;
import com.ch.ch6_2.entity.IdCard;
public interface IdCardRepository extends JpaRepository<IdCard, Integer>{
    /**
     * 根据人员 ID 查询身份信息(关联查询,根据 person 属性的 id)
     * 相当于 JPQL 语句:select ic from IdCard ic where ic.person.id = ?1
     */
    public IdCard findByPerson_id(Integer id);
    /**
     * 根据地址和身份证号查询身份信息
     * 相当于 JPQL 语句:select ic from IdCard ic where ic.address = ?1 and ic.code =?2
     */
    public List<IdCard> findByAddressAndCode(String address, String code);
}
```

按照 Spring Data JPA 的规则，查询两个有关联关系的对象，可以通过方法名中的下画线"_"来标识。如根据人员 ID 查询身份信息 findByPerson_id。JPQL(Java Persistence Query Language)是一种和 SQL 非常类似的中间性和对象化查询语言，它最终被编译成针对不同底层数据库的 SQL 查询，从而屏蔽不同数据库的差异。JPQL 语句可以是 select 语句、update 语句或 delete 语句，它们都通过 Query 接口封装执行。JPQL 的具体学习内容不是本书的重点，需要学习的读者请参考相关内容学习。

PersonRepository 的代码如下：

```
package com.ch.ch6_2.repository;
import java.util.List;
import org.springframework.data.jpa.repository.JpaRepository;
import com.ch.ch6_2.entity.Person;
public interface PersonRepository extends JpaRepository<Person, Integer>{
```

```java
/**
 * 根据身份 ID 查询人员信息(关联查询,根据 idCard 属性的 id)
 * 相当于 JPQL 语句:select p from Person p where p.idCard.id = ?1
 */
public Person findByIdCard_id(Integer id);
/**
 * 根据人名和性别查询人员信息
 * 相当于 JPQL 语句:select p from Person p where p.pname = ?1 and p.psex = ?2
 */
public List<Person> findByPnameAndPsex(String pname, String psex);
}
```

3)创建业务层

创建名为 com.ch.ch6_2.service 的包,并在该包中创建名为 PersonAndIdCardService 的接口和接口实现类 PersonAndIdCardServiceImpl。

PersonAndIdCardService 的代码如下:

```java
package com.ch.ch6_2.service;
import java.util.List;
import com.ch.ch6_2.entity.IdCard;
import com.ch.ch6_2.entity.Person;
public interface PersonAndIdCardService {
    public void saveAll();
    public List<Person> findAllPerson();
    public List<IdCard> findAllIdCard();
    public IdCard findByPerson_id(Integer id);
    public List<IdCard> findByAddressAndCode(String address, String code);
    public Person findByIdCard_id(Integer id);
    public List<Person> findByPnameAndPsex(String pname, String psex);
    public IdCard getOneIdCard(Integer id);
    public Person getOnePerson(Integer id);
}
```

PersonAndIdCardServiceImpl 的代码如下:

```java
package com.ch.ch6_2.service;
import java.util.ArrayList;
import java.util.Calendar;
import java.util.List;
import org.springframework.beans.factory.annotation.Autowired;
import org.springframework.stereotype.Service;
import com.ch.ch6_2.entity.IdCard;
import com.ch.ch6_2.entity.Person;
import com.ch.ch6_2.repository.IdCardRepository;
import com.ch.ch6_2.repository.PersonRepository;
@Service
public class PersonAndIdCardServiceImpl implements PersonAndIdCardService{
    @Autowired
    private IdCardRepository idCardRepository;
    @Autowired
    private PersonRepository personRepository;
```

```java
@Override
public void saveAll() {
    //保存身份证
    IdCard ic1 = new IdCard();
    ic1.setCode("123456789");
    ic1.setAddress("北京");
    Calendar c1 = Calendar.getInstance();
    c1.set(2019, 8, 13);
    ic1.setBirthday(c1);
    IdCard ic2 = new IdCard();
    ic2.setCode("000123456789");
    ic2.setAddress("上海");
    Calendar c2 = Calendar.getInstance();
    c2.set(2019, 8, 14);
    ic2.setBirthday(c2);
    IdCard ic3 = new IdCard();
    ic3.setCode("1111123456789");
    ic3.setAddress("广州");
    Calendar c3 = Calendar.getInstance();
    c3.set(2019, 8, 15);
    ic3.setBirthday(c3);
    List<IdCard> idCards = new ArrayList<IdCard>();
    idCards.add(ic1);
    idCards.add(ic2);
    idCards.add(ic3);
    idCardRepository.saveAll(idCards);
    //保存人员
    Person p1 = new Person();
    p1.setPname("陈恒1");
    p1.setPsex("男");
    p1.setPage(88);
    p1.setIdCard(ic1);
    Person p2 = new Person();
    p2.setPname("陈恒2");
    p2.setPsex("女");
    p2.setPage(99);
    p2.setIdCard(ic2);
    Person p3 = new Person();
    p3.setPname("陈恒3");
    p3.setPsex("女");
    p3.setPage(18);
    p3.setIdCard(ic3);
    List<Person> persons = new ArrayList<Person>();
    persons.add(p1);
    persons.add(p2);
    persons.add(p3);
    personRepository.saveAll(persons);
}
@Override
public List<Person> findAllPerson() {
    return personRepository.findAll();
```

```java
    }
    @Override
    public List<IdCard> findAllIdCard() {
        return idCardRepository.findAll();
    }
    /**
     * 根据人员 ID 查询身份信息(关联查询)
     */
    @Override
    public IdCard findByPerson_id(Integer id) {
        return idCardRepository.findByPerson_id(id);
    }
    @Override
    public List<IdCard> findByAddressAndCode(String address, String code) {
        return idCardRepository.findByAddressAndCode(address, code);
    }
    /**
     * 根据身份 ID 查询人员信息(关联查询)
     */
    @Override
    public Person findByIdCard_id(Integer id) {
        return personRepository.findByIdCard_id(id);
    }
    @Override
    public List<Person> findByPnameAndPsex(String pname, String psex) {
        return personRepository.findByPnameAndPsex(pname, psex);
    }
    @Override
    public IdCard getOneIdCard(Integer id) {
        return idCardRepository.getOne(id);
    }
    @Override
    public Person getOnePerson(Integer id) {
        return personRepository.getOne(id);
    }
}
```

4)创建控制器类

创建名为 com.ch.ch6_2.controller 的包,并在该包中创建名为 TestOneToOneController 的控制器类。

TestOneToOneController 的代码如下:

```java
package com.ch.ch6_2.controller;
import java.util.List;
import org.springframework.beans.factory.annotation.Autowired;
import org.springframework.web.bind.annotation.RequestMapping;
import org.springframework.web.bind.annotation.RestController;
import com.ch.ch6_2.entity.IdCard;
import com.ch.ch6_2.entity.Person;
import com.ch.ch6_2.service.PersonAndIdCardService;
```

```java
@RestController
public class TestOneToOneController {
    @Autowired
    private PersonAndIdCardService personAndIdCardService;
    @RequestMapping("/save")
    public String save() {
        personAndIdCardService.saveAll();
        return "人员和身份保存成功!";
    }
    @RequestMapping("/findAllPerson")
    public List<Person> findAllPerson() {
        return  personAndIdCardService.findAllPerson();
    }
    @RequestMapping("/findAllIdCard")
    public List<IdCard>  findAllIdCard() {
        return personAndIdCardService.findAllIdCard();
    }
    /**
     * 根据人员ID查询身份信息(关联查询)
     */
    @RequestMapping("/findByPerson_id")
    public IdCard findByPerson_id(Integer id) {
        return personAndIdCardService.findByPerson_id(id);
    }
    @RequestMapping("/findByAddressAndCode")
    public List<IdCard> findByAddressAndCode(String address, String code){
        return personAndIdCardService.findByAddressAndCode(address, code);
    }
    /**
     * 根据身份ID查询人员信息(关联查询)
     */
    @RequestMapping("/findByIdCard_id")
    public Person findByIdCard_id(Integer id) {
        return personAndIdCardService.findByIdCard_id(id);
    }
    @RequestMapping("/findByPnameAndPsex")
    public List<Person> findByPnameAndPsex(String pname, String psex) {
        return personAndIdCardService.findByPnameAndPsex(pname, psex);
    }
    @RequestMapping("/getOneIdCard")
    public IdCard getOneIdCard(Integer id) {
        return personAndIdCardService.getOneIdCard(id);
    }
    @RequestMapping("/getOnePerson")
    public Person getOnePerson(Integer id) {
        return personAndIdCardService.getOnePerson(id);
    }
}
```

5) 运行

首先,运行Ch62Application主类。然后,访问"http://localhost:8080/ch6_2/save/"。

运行效果如图 6.11 所示。

图 6.11 保存数据

"http://localhost:8080/ch6_2/save/"成功运行后，在 MySQL 的 springbootjpa 数据库中创建名为 idcard_table 和 person_table 的数据库表（实体类成功加载后就已创建好数据表），并分别插入三条记录。

通过"http://localhost:8080/ch6_2/findByIdCard_id?id=1"查询身份证 id 为 1 的人员信息（关联查询），运行效果如图 6.12 所示。

通过"http://localhost:8080/ch6_2/findByPerson_id?id=1"查询人员 id 为 1 的身份证信息（关联查询），运行效果如图 6.13 所示。

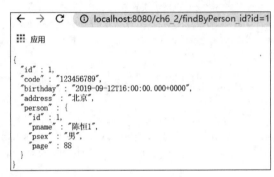

图 6.12 查询身份证 id 为 1 的人员信息　　图 6.13 查询人员 id 为 1 的身份证信息

2. @OneToMany 和 @ManyToOne

在实际生活中，作者和文章是一对多的双向关系。那么在 Spring Data JPA 中，如何描述一对多的双向关系呢？

在 Spring Data JPA 中，使用@OneToMany 和@ManyToOne 来表示一对多的双向关联。例如，一端（Author）使用@OneToMany，多端（Article）使用@ManyToOne。

在 JPA 规范中，一对多的双向关系由多端（如 Article）来维护。就是说多端为关系的维护端，负责关系的增删改查；一端则为关系的被维护端，不能维护关系。

一端（Author）使用 @OneToMany 注解的 mappedBy = "author" 属性表明一端（Author）是关系的被维护端。多端（Article）使用@ManyToOne 和@JoinColumn 来注解属性 author，@ManyToOne 表明 Article 是多端，@JoinColumn 设置在 article 表的关联字段（外键）上。

【例 6-3】 使用 Spring Data JPA 实现 Author 与 Article 的一对多关系映射。

在 ch6_2 应用中实现例 6-3，具体实现步骤如下。

1) 创建持久化实体类

在 com.ch.ch6_2.entity 包中，创建名为 Author 和 Article 的持久化实体类。

Author 的代码如下：

```
package com.ch.ch6_2.entity;
import java.io.Serializable;
```

```java
import java.util.List;
import javax.persistence.CascadeType;
import javax.persistence.Entity;
import javax.persistence.FetchType;
import javax.persistence.GeneratedValue;
import javax.persistence.GenerationType;
import javax.persistence.Id;
import javax.persistence.OneToMany;
import javax.persistence.Table;
import com.fasterxml.jackson.annotation.JsonIgnoreProperties;
@Entity
@Table(name = "author_table")
@JsonIgnoreProperties(value = { "hibernateLazyInitializer"})
public class Author implements Serializable{
    private static final long serialVersionUID = 1L;
    @Id
    @GeneratedValue(strategy = GenerationType.IDENTITY)
    private int id;
    //作者名
    private String aname;
    //文章列表,作者与文章是一对多的关系
    @OneToMany(
            mappedBy = "author",
            cascade = CascadeType.ALL,
            targetEntity = Article.class,
            fetch = FetchType.LAZY
            )
    private List<Article> articleList;
    //省略set和get方法
}
```

Article 的代码如下:

```java
package com.ch.ch6_2.entity;
import java.io.Serializable;
import javax.persistence.Basic;
import javax.persistence.CascadeType;
import javax.persistence.Column;
import javax.persistence.Entity;
import javax.persistence.FetchType;
import javax.persistence.GeneratedValue;
import javax.persistence.GenerationType;
import javax.persistence.Id;
import javax.persistence.JoinColumn;
import javax.persistence.Lob;
import javax.persistence.ManyToOne;
import javax.persistence.Table;
import javax.validation.constraints.NotEmpty;
import javax.validation.constraints.Size;
import com.fasterxml.jackson.annotation.JsonIgnore;
import com.fasterxml.jackson.annotation.JsonIgnoreProperties;
```

```java
@Entity
@Table(name = "article_table")
@JsonIgnoreProperties(value = { "hibernateLazyInitializer"})
public class Article    implements Serializable{
    private static final long serialVersionUID = 1L;
    @Id
    @GeneratedValue(strategy = GenerationType.IDENTITY)
    private int id;
    //标题
    @NotEmpty(message = "标题不能为空")
    @Size(min = 2, max = 50)
    @Column(nullable = false, length = 50)
    private String title;
    //文章内容
    @Lob    //大对象,映射为MySQL的Long文本类型
    @Basic(fetch = FetchType.LAZY)
    @NotEmpty(message = "内容不能为空")
    @Size(min = 2)
    @Column(nullable = false)
    private String content;
    //所属作者,文章与作者是多对一的关系
    @ManyToOne(cascade = {CascadeType.MERGE,CascadeType.REFRESH},optional = false)
    //可选属性optional = false,表示author不能为空。删除文章,不影响用户
    @JoinColumn(name = "id_author_id")          //设置在article表中的关联字段(外键)
    @JsonIgnore
    private Author author;
    //省略set和get方法
}
```

2)创建数据访问层

在com.ch.ch6_2.repository包中,创建名为AuthorRepository和ArticleRepository的接口。

AuthorRepository的代码如下:

```
package com.ch.ch6_2.repository;
import org.springframework.data.jpa.repository.JpaRepository;
import com.ch.ch6_2.entity.Author;
public interface AuthorRepository extends JpaRepository<Author, Integer>{
    /**
     * 根据文章标题包含的内容,查询作者(关联查询)
     * 相当于JPQL语句:select a from Author a   inner join   a.articleList t where t.title like %?1%
     */
    public Author findByArticleList_titleContaining(String title);
}
```

ArticleRepository的代码如下:

```
package com.ch.ch6_2.repository;
import java.util.List;
import org.springframework.data.jpa.repository.JpaRepository;
```

```java
import com.ch.ch6_2.entity.Article;
public interface ArticleRepository extends JpaRepository<Article, Integer>{
    /**
     * 根据作者id查询文章信息(关联查询,根据author属性的id)
     * 相当于JPQL语句:select a from Article a where a.author.id = ?1
     */
    public List<Article> findByAuthor_id(Integer id);
    /**
     * 根据作者名查询文章信息(关联查询,根据author属性的aname)
     * 相当于JPQL语句:select a from Article a where a.author.aname = ?1
     */
    public List<Article> findByAuthor_aname(String aname);
}
```

3）创建业务层

在 com.ch.ch6_2.service 包中，创建名为 AuthorAndArticleService 的接口和接口实现类 AuthorAndArticleServiceImpl。

AuthorAndArticleService 的代码如下：

```java
package com.ch.ch6_2.service;
import java.util.List;
import com.ch.ch6_2.entity.Article;
public interface AuthorAndArticleService {
    public void saveAll();
    public List<Article> findByAuthor_id(Integer id);
    public List<Article> findByAuthor_aname(String aname);
    public Author findByArticleList_titleContaining(String title);
}
```

AuthorAndArticleServiceImpl 的代码如下：

```java
package com.ch.ch6_2.service;
import java.util.ArrayList;
import java.util.List;
import org.springframework.beans.factory.annotation.Autowired;
import org.springframework.stereotype.Service;
import com.ch.ch6_2.entity.Article;
import com.ch.ch6_2.entity.Author;
import com.ch.ch6_2.repository.ArticleRepository;
import com.ch.ch6_2.repository.AuthorRepository;
@Service
public class AuthorAndArticleServiceImpl implements AuthorAndArticleService{
    @Autowired
    private AuthorRepository authorRepository;
    @Autowired
    private ArticleRepository articleRepository;
    @Override
    public void saveAll() {
        //保存作者(先保存一的一端)
        Author a1 = new Author();
        a1.setAname("陈恒1");
```

```java
        Author a2 = new Author();
        a2.setAname("陈恒 2");
        ArrayList<Author> allAuthor = new ArrayList<Author>();
        allAuthor.add(a1);
        allAuthor.add(a2);
        authorRepository.saveAll(allAuthor);
        //保存文章
        Article at1 = new Article();
        at1.setTitle("JPA 的一对多 111");
        at1.setContent("其实一对多映射关系很常见 111。");
        //设置关系
        at1.setAuthor(a1);
        Article at2 = new Article();
        at2.setTitle("JPA 的一对多 222");
        at2.setContent("其实一对多映射关系很常见 222。");
        //设置关系
        at2.setAuthor(a1);                        //文章 2 与文章 1 作者相同
        Article at3 = new Article();
        at3.setTitle("JPA 的一对多 333");
        at3.setContent("其实一对多映射关系很常见 333。");
        //设置关系
        at3.setAuthor(a2);
        Article at4 = new Article();
        at4.setTitle("JPA 的一对多 444");
        at4.setContent("其实一对多映射关系很常见 444。");
        //设置关系
        at4.setAuthor(a2);                        //文章 3 与文章 4 作者相同
        ArrayList<Article> allAt = new ArrayList<Article>();
        allAt.add(at1);
        allAt.add(at2);
        allAt.add(at3);
        allAt.add(at4);
        articleRepository.saveAll(allAt);
    }
    @Override
    public List<Article> findByAuthor_id(Integer id) {
        return articleRepository.findByAuthor_id(id);
    }
    @Override
    public List<Article> findByAuthor_aname(String aname) {
        return articleRepository.findByAuthor_aname(aname);
    }
    @Override
    public Author findByArticleList_titleContaining(String title) {
        return authorRepository.findByArticleList_titleContaining(title);
    }
}
```

4) 创建控制器类

在 com.ch.ch6_2.controller 包中,创建名为 TestOneToManyController 的控制器类。

TestOneToManyController 的代码如下：

```java
package com.ch.ch6_2.controller;
import java.util.List;
import org.springframework.beans.factory.annotation.Autowired;
import org.springframework.web.bind.annotation.RequestMapping;
import org.springframework.web.bind.annotation.RestController;
import com.ch.ch6_2.entity.Article;
import com.ch.ch6_2.service.AuthorAndArticleService;
@RestController
public class TestOneToManyController {
    @Autowired
    private AuthorAndArticleService authorAndArticleService;
    @RequestMapping("/saveOneToMany")
    public String save() {
        authorAndArticleService.saveAll();
        return "作者和文章保存成功!";
    }
    @RequestMapping("/findArticleByAuthor_id")
    public List<Article> findByAuthor_id(Integer id) {
        return authorAndArticleService.findByAuthor_id(id);
    }
    @RequestMapping("/findArticleByAuthor_aname")
    public List<Article> findByAuthor_aname(String aname){
        return authorAndArticleService.findByAuthor_aname(aname);
    }
    @RequestMapping("/findByArticleList_titleContaining")
    public Author findByArticleList_titleContaining(String title) {
        return authorAndArticleService.findByArticleList_titleContaining(title);
    }
}
```

5）运行

首先，运行 Ch62Application 主类。然后，访问"http://localhost：8080/ch6_2/saveOneToMany/"，运行效果如图 6.14 所示。

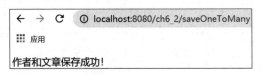

图 6.14　保存数据

"http://localhost：8080/ch6_2/saveOneToMany/"成功运行后，在 MySQL 的 springbootjpa 数据库中创建名为 author_table 和 article_table 的数据库表，并在 author_table 表中插入两条记录，同时在 article_table 表中插入四条记录。

通过"http://localhost:8080/ch6_2/findArticleByAuthor_id?id=2"查询作者 id 为 2 的文章列表（关联查询），运行效果如图 6.15 所示。

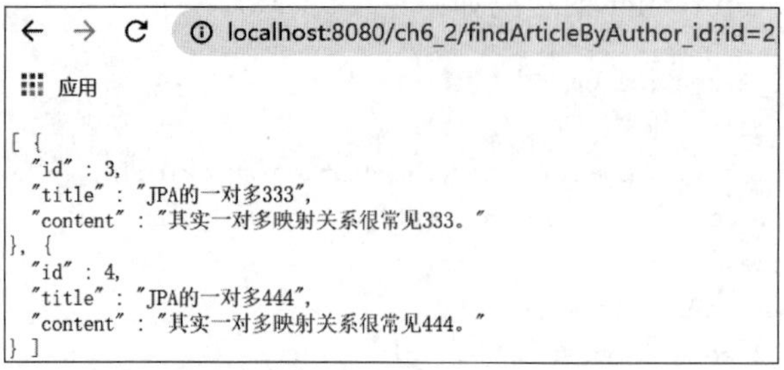

图 6.15　查询作者 id 为 2 的文章列表

通过"http://localhost:8080/ch6_2/findArticleByAuthor_aname?aname=陈恒1"查询作者名为陈恒1的文章列表（关联查询），运行效果如图 6.16 所示。

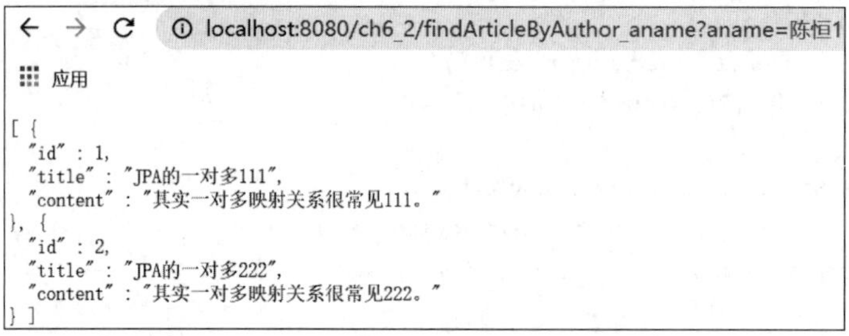

图 6.16　查询作者名为陈恒1的文章列表

通过"http://localhost:8080/ch6_2/findByArticleList_titleContaining?title=对多1"查询文章标题包含"对多1"的作者（关联查询），运行效果如图 6.17 所示。

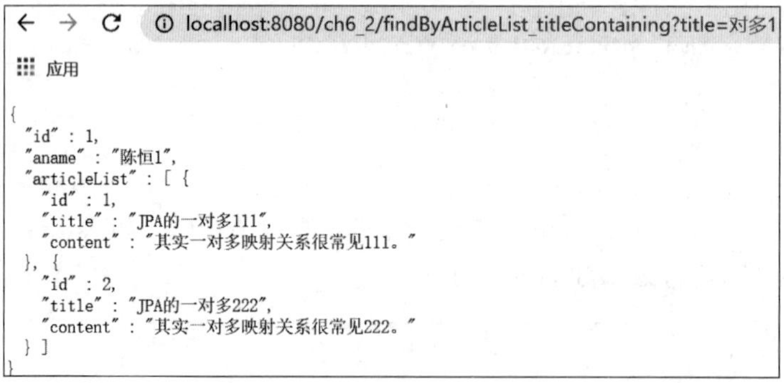

图 6.17　查询文章标题包含"对多1"的作者

3．@ManyToMany

在实际生活中，用户和权限是多对多的关系。一个用户可以有多个权限，一个权限也可以被很多用户拥有。

在 Spring Data JPA 中使用@ManyToMany 来注解多对多的映射关系，由一个关联表来维护。关联表的表名默认是：主表名＋下画线＋从表名（主表是指关系维护端对应的表，从表是指关系被维护端对应的表）。关联表只有两个外键字段，分别指向主表 ID 和从表

ID。字段的名称默认为：主表名+下画线+主表中的主键列名,从表名+下画线+从表中的主键列名。需要注意的是,多对多关系中一般不设置级联保存、级联删除、级联更新等操作。

【例 6-4】 使用 Spring Data JPA 实现用户(User)与权限(Authority)的多对多关系映射。

在 ch6_2 应用中实现例 6-4,具体实现步骤如下。

1) 创建持久化实体类

在 com.ch.ch6_2.entity 包中,创建名为 User 和 Authority 的持久化实体类。
User 的代码如下：

```java
package com.ch.ch6_2.entity;
import java.io.Serializable;
import java.util.List;
import javax.persistence.Entity;
import javax.persistence.GeneratedValue;
import javax.persistence.GenerationType;
import javax.persistence.Id;
import javax.persistence.JoinColumn;
import javax.persistence.JoinTable;
import javax.persistence.ManyToMany;
import javax.persistence.Table;
import com.fasterxml.jackson.annotation.JsonIgnoreProperties;
@Entity
@Table(name = "user")
@JsonIgnoreProperties(value = {"hibernateLazyInitializer"})
public class User implements Serializable{
    private static final long serialVersionUID = 1L;
    @Id
    @GeneratedValue(strategy = GenerationType.IDENTITY)
    private int id;
    private String username;
    private String password;
    @ManyToMany
    @JoinTable(name = "user_authority",joinColumns = @JoinColumn(name = "user_id"),
    inverseJoinColumns = @JoinColumn(name = "authority_id"))
    /** 1. 关系维护端,负责多对多关系的绑定和解除
    2. @JoinTable 注解的 name 属性指定关联表的名字,joinColumns 指定外键的名字,关联到关系维护端(User)
    3. inverseJoinColumns 指定外键的名字,需要关联的关系称为被维护端(Authority)
    4. 其实可以不使用@JoinTable 注解,默认生成的关联表名称为主表表名+下画线+从表表名,即表名为 user_authority
    关联到主表的外键名:主表名+下画线+主表中的主键列名,即 user_id
    关联到从表的外键名:主表中用于关联的属性名+下画线+从表的主键列名,即 authority_id。
    主表就是关系维护端对应的表,从表就是关系被维护端对应的表
    */
    private List<Authority> authorityList;
    //省略 get 和 set 方法
}
```

Authority 的代码如下：

```java
package com.ch.ch6_2.entity;
import java.io.Serializable;
import java.util.List;
import javax.persistence.Column;
import javax.persistence.Entity;
import javax.persistence.GeneratedValue;
import javax.persistence.GenerationType;
import javax.persistence.Id;
import javax.persistence.ManyToMany;
import javax.persistence.Table;
import com.fasterxml.jackson.annotation.JsonIgnore;
import com.fasterxml.jackson.annotation.JsonIgnoreProperties;
@Entity
@Table(name = "authority")
@JsonIgnoreProperties(value = { "hibernateLazyInitializer"})
public class Authority implements Serializable{
    private static final long serialVersionUID = 1L;
    @Id
    @GeneratedValue(strategy = GenerationType.IDENTITY)
    private int id;
    @Column(nullable = false)
    private String name;
    @ManyToMany(mappedBy = "authorityList")
    @JsonIgnore
    private List<User> userList;
    //省略 get 和 set 方法
}
```

2）创建数据访问层

在 com.ch.ch6_2.repository 包中，创建名为 UserRepository 和 AuthorityRepository 的接口。

UserRepository 的代码如下：

```java
package com.ch.ch6_2.repository;
import java.util.List;
import org.springframework.data.jpa.repository.JpaRepository;
import com.ch.ch6_2.entity.User;
public interface UserRepository extends JpaRepository<User, Integer>{
    /**
     * 根据权限 id 查询拥有该权限的用户(关联查询)
     * 相当于 JPQL 语句:select u from User u inner join u.authorityList a where a.id = ?1
     */
    public List<User> findByAuthorityList_id(int id);
    /**
     * 根据权限名查询拥有该权限的用户(关联查询)
     * 相当于 JPQL 语句:select u from User u inner join u.authorityList a where a.name = ?1
     */
    public List<User> findByAuthorityList_name(String name);
}
```

AuthorityRepository 的代码如下：

```java
package com.ch.ch6_2.repository;
import java.util.List;
import org.springframework.data.jpa.repository.JpaRepository;
import com.ch.ch6_2.entity.Authority;
public interface AuthorityRepository extends JpaRepository<Authority, Integer>{
    /**
     * 根据用户id查询用户所拥有的权限(关联查询)
     * 相当于JPQL语句:select a from Authority a inner join a.userList u where u.id = ?1
     */
    public List<Authority> findByUserList_id(int id);
    /**
     * 根据用户名查询用户所拥有的权限(关联查询)
     * 相当于JPQL语句:select a from Authority a inner join a.userList u where u.username = ?1
     */
    public List<Authority> findByUserList_Username(String username);
}
```

3）创建业务层

在 com.ch.ch6_2.service 包中，创建名为 UserAndAuthorityService 的接口和接口实现类 UserAndAuthorityServiceImpl。

UserAndAuthorityService 的代码如下：

```java
package com.ch.ch6_2.service;
import java.util.List;
import com.ch.ch6_2.entity.Authority;
import com.ch.ch6_2.entity.User;
public interface UserAndAuthorityService {
    public void saveAll();
    public List<User> findByAuthorityList_id(int id);
    public List<User> findByAuthorityList_name(String name);
    public List<Authority> findByUserList_id(int id);
    public List<Authority> findByUserList_Username(String username);
}
```

UserAndAuthorityServiceImpl 的代码如下：

```java
package com.ch.ch6_2.service;
import java.util.ArrayList;
import java.util.List;
import org.springframework.beans.factory.annotation.Autowired;
import org.springframework.stereotype.Service;
import com.ch.ch6_2.entity.Authority;
import com.ch.ch6_2.entity.User;
import com.ch.ch6_2.repository.AuthorityRepository;
import com.ch.ch6_2.repository.UserRepository;
@Service
public class UserAndAuthorityServiceImpl implements UserAndAuthorityService{
    @Autowired
    private AuthorityRepository authorityRepository;
```

```java
@Autowired
private UserRepository userRepository;
@Override
public void saveAll() {
    //添加权限 1
    Authority at1 = new Authority();
    at1.setName("增加");
    authorityRepository.save(at1);
    //添加权限 2
    Authority at2 = new Authority();
    at2.setName("修改");
    authorityRepository.save(at2);
    //添加权限 3
    Authority at3 = new Authority();
    at3.setName("删除");
    authorityRepository.save(at3);
    //添加权限 4
    Authority at4 = new Authority();
    at4.setName("查询");
    authorityRepository.save(at4);
    //添加用户 1
    User u1 = new User();
    u1.setUsername("陈恒 1");
    u1.setPassword("123");
    ArrayList<Authority> authorityList1 = new ArrayList<Authority>();
    authorityList1.add(at1);
    authorityList1.add(at2);
    authorityList1.add(at3);
    u1.setAuthorityList(authorityList1);
    userRepository.save(u1);
    //添加用户 2
    User u2 = new User();
    u2.setUsername("陈恒 2");
    u2.setPassword("234");
    ArrayList<Authority> authorityList2 = new ArrayList<Authority>();
    authorityList2.add(at2);
    authorityList2.add(at3);
    authorityList2.add(at4);
    u2.setAuthorityList(authorityList2);
    userRepository.save(u2);
}
@Override
public List<User> findByAuthorityList_id(int id) {
    return userRepository.findByAuthorityList_id(id);
}
@Override
public List<User> findByAuthorityList_name(String name) {
    return userRepository.findByAuthorityList_name(name);
}
@Override
public List<Authority> findByUserList_id(int id) {
```

```
        return authorityRepository.findByUserList_id(id);
    }
    @Override
    public List<Authority> findByUserList_Username(String username) {
        return authorityRepository.findByUserList_Username(username);
    }
}
```

4）创建控制器类

在 com.ch.ch6_2.controller 包中，创建名为 TestManyToManyController 的控制器类。TestManyToManyController 的代码如下：

```
package com.ch.ch6_2.controller;
import java.util.List;
import org.springframework.beans.factory.annotation.Autowired;
import org.springframework.web.bind.annotation.RequestMapping;
import org.springframework.web.bind.annotation.RestController;
import com.ch.ch6_2.entity.Authority;
import com.ch.ch6_2.entity.User;
import com.ch.ch6_2.service.UserAndAuthorityService;
@RestController
public class TestManyToManyController {
    @Autowired
    private UserAndAuthorityService userAndAuthorityService;
    @RequestMapping("/saveManyToMany")
    public String save() {
        userAndAuthorityService.saveAll();
        return "权限和用户保存成功!";
    }
    @RequestMapping("/findByAuthorityList_id")
    public List<User> findByAuthorityList_id(int id) {
        return userAndAuthorityService.findByAuthorityList_id(id);
    }
    @RequestMapping("/findByAuthorityList_name")
    public List<User> findByAuthorityList_name(String name) {
        return userAndAuthorityService.findByAuthorityList_name(name);
    }
    @RequestMapping("/findByUserList_id")
    public List<Authority> findByUserList_id(int id) {
        return userAndAuthorityService.findByUserList_id(id);
    }
    @RequestMapping("/findByUserList_Username")
    public List<Authority> findByUserList_Username(String username) {
        return userAndAuthorityService.findByUserList_Username(username);
    }
}
```

5）运行

首先，运行 Ch62Application 主类。然后，访问"http://localhost:8080/ch6_2/saveManyToMany/"，运行效果如图 6.18 所示。

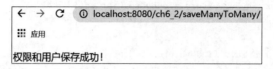

图 6.18　保存数据

通过"http://localhost:8080/ch6_2/findByAuthorityList_id?id=1"查询拥有 id 为 1 的权限的用户列表(关联查询),运行效果如图 6.19 所示。

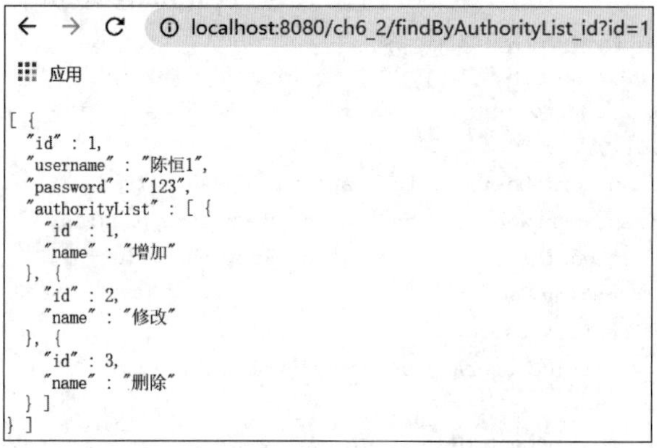

图 6.19　查询拥有 id 为 1 的权限的用户列表

通过"http://localhost:8080/ch6_2/findByAuthorityList_name?name=修改"查询拥有"修改"权限的用户列表(关联查询),运行效果如图 6.20 所示。

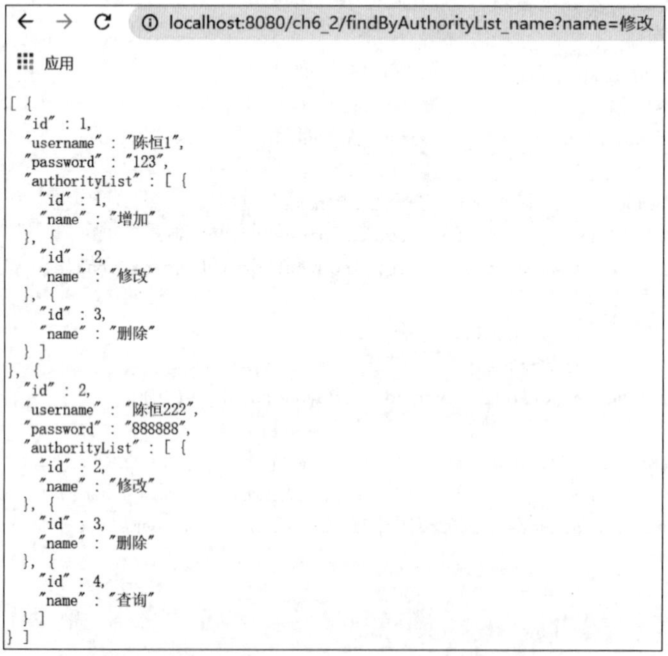

图 6.20　查询拥有"修改"权限的用户列表

第6章 Spring Boot 的数据访问

通过"http://localhost:8080/ch6_2/findByUserList_id?id=2"查询 id 为 2 的用户的权限列表(关联查询),运行效果如图 6.21 所示。

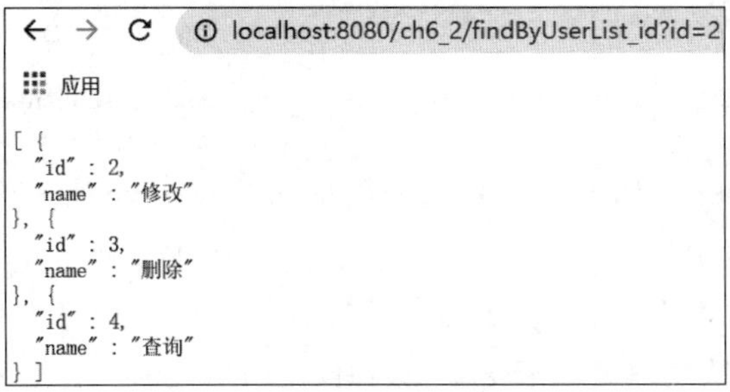

图 6.21 查询 id 为 2 的用户的权限列表

通过"http://localhost:8080/ch6_2/findByUserList_Username?username=陈恒2"查询用户名为"陈恒2"的用户的权限列表(关联查询),运行效果如图 6.22 所示。

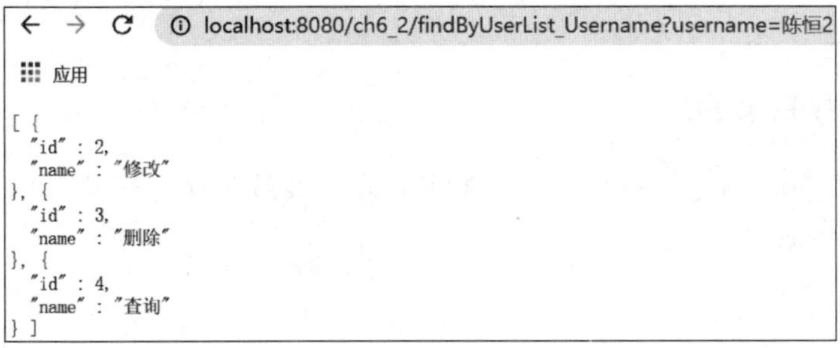

图 6.22 查询用户名为"陈恒2"的用户的权限列表

在 6.1.2 节和 6.1.3 节中的查询方法必须严格按照 Spring Data JPA 的查询关键字命名规范进行查询方法命名。如何摆脱查询关键字和关联查询命名规范约束呢?可以通过 @Query、@NamedQuery 直接定义 JPQL 语句进行数据的访问操作。

6.1.4 @Query 和 @Modifying 注解

视频讲解

1. @Query 注解

使用 @Query 注解可以将 JPQL 语句直接定义在数据访问接口方法上,并且接口方法名不受查询关键字和关联查询命名规范约束。示例代码如下:

```
public interface AuthorityRepository extends JpaRepository<Authority, Integer>{
    /**
     * 根据用户名查询用户所拥有的权限(关联查询)
     */
    @Query("select a from Authority a inner join a.userList u where u.username = ?1")
    public List<Authority> findByUserListUsername(String username);
}
```

使用 @Query 注解定义 JPQL 语句,可以直接返回 List<Map<String, Object>>对

象。示例代码如下：

```
/**
 * 根据作者 id 查询文章信息(标题和内容)
 */
@Query("select new Map(a.title as title, a.content as content) from Article a where a.author.id = ?1 ")
public List<Map<String, Object>> findTitleAndContentByAuthorId(Integer id);
```

使用@Query 注解定义 JPQL 语句，之前的方法是使用参数位置（"?1"指代的是获取方法形参列表中第 1 个参数值，1 代表的是参数位置，以此类推）来获取参数值。除此之外，Spring Data JPA 还支持使用名称来获取参数值，使用格式为":参数名称"。示例代码如下：

```
/**
 * 根据作者名和作者 id 查询文章信息
 */
@Query("select a from Article a where a.author.aname = :aname1 and a.author.id = :id1 ")
public List<Article> findArticleByAuthorAnameAndId(@Param("aname1") String aname, @Param("id1") Integer id);
```

2. @Modifying 注解

可以使用@Modifying 和@Query 注解组合定义在数据访问接口方法上，进行更新查询操作。示例代码如下：

```
/**
 * 根据作者 id 删除作者
 */
@Modifying
@Query("delete from Author a where a.id = ?1")
public int deleteAuthorByAuthorId(int id);
```

下面通过实例讲解@Query 和@Modifying 注解的使用方法。

【例 6-5】 @Query 和@Modifying 注解的使用方法。

首先，为例 6-5 创建基于 Spring Data JPA 的 Spring Boot Web 应用 ch6_3。ch6_3 应用的数据库、pom.xml 以及 application.properties 与 ch6_2 应用基本一样，不再赘述。

其他内容具体实现步骤如下。

1）创建持久化实体类

创建名为 com.ch.ch6_3.entity 的包，并在该包中创建名为 Article 和 Author 的持久化实体类。具体代码分别与 ch6_2 应用的 Article 和 Author 的代码一样，不再赘述。

2）创建数据访问层

创建名为 com.ch.ch6_3.repository 的包，并在该包中创建名为 ArticleRepository 和 AuthorRepository 的接口。

ArticleRepository 的代码如下：

```
package com.ch.ch6_3.repository;
import java.util.List;
import java.util.Map;
```

```java
import org.springframework.data.jpa.repository.JpaRepository;
import org.springframework.data.jpa.repository.Query;
import org.springframework.data.repository.query.Param;
import com.ch.ch6_3.entity.Article;
public interface ArticleRepository extends JpaRepository<Article, Integer>{
    /**
     * 根据作者id查询文章信息(标题和内容)
     */
    @Query("select new Map(a.title as title, a.content as content) from Article a where a.author.id = ?1 ")
    public List<Map<String, Object>> findTitleAndContentByAuthorId(Integer id);
    /**
     * 根据作者名和作者id查询文章信息
     */
    @Query("select a from Article a where a.author.aname = :aname1 and a.author.id = :id1 ")
    public List<Article> findArticleByAuthorAnameAndId(@Param("aname1") String aname, @Param("id1") Integer id);
}
```

AuthorRepository 的代码如下：

```java
package com.ch.ch6_3.repository;
import org.springframework.data.jpa.repository.JpaRepository;
import org.springframework.data.jpa.repository.Modifying;
import org.springframework.data.jpa.repository.Query;
import com.ch.ch6_3.entity.Author;
public interface AuthorRepository extends JpaRepository<Author, Integer>{
    /**
     * 根据文章标题包含的内容查询作者(关联查询)
     */
    @Query("select a from Author a  inner join  a.articleList t where t.title like %?1% ")
    public Author findAuthorByArticleListtitleContaining(String title);
    /**
     * 根据作者id删除作者
     */
    @Modifying
    @Query("delete from Author a where a.id = ?1")
    public int deleteAuthorByAuthorId(int id);
}
```

3）创建业务层

创建名为 com.ch.ch6_3.service 的包，并在该包中创建名为 AuthorAndArticleService 的接口和接口实现类 AuthorAndArticleServiceImpl。

AuthorAndArticleService 的代码如下：

```java
package com.ch.ch6_3.service;
import java.util.List;
import java.util.Map;
import com.ch.ch6_3.entity.Article;
import com.ch.ch6_3.entity.Author;
public interface AuthorAndArticleService {
```

```java
    public List<Map<String, Object>> findTitleAndContentByAuthorId(Integer id);
    public List<Article> findArticleByAuthorAnameAndId(String aname, Integer id);
    public Author findAuthorByArticleListtitleContaining(String title);
    public int deleteAuthorByAuthorId(int id);
}
```

AuthorAndArticleServiceImpl 的代码如下:

```java
package com.ch.ch6_3.service;
import java.util.List;
import java.util.Map;
import org.springframework.beans.factory.annotation.Autowired;
import org.springframework.stereotype.Service;
import com.ch.ch6_3.entity.Article;
import com.ch.ch6_3.entity.Author;
import com.ch.ch6_3.repository.ArticleRepository;
import com.ch.ch6_3.repository.AuthorRepository;
@Service
public class AuthorAndArticleServiceImpl implements AuthorAndArticleService{
    @Autowired
    private AuthorRepository authorRepository;
    @Autowired
    private ArticleRepository articleRepository;
    @Override
    public List<Map<String, Object>> findTitleAndContentByAuthorId(Integer id) {
        return articleRepository.findTitleAndContentByAuthorId(id);
    }
    @Override
    public List<Article> findArticleByAuthorAnameAndId(String aname, Integer id) {
        return articleRepository.findArticleByAuthorAnameAndId(aname, id);
    }
    @Override
    public Author findAuthorByArticleListtitleContaining(String title) {
        return authorRepository.findAuthorByArticleListtitleContaining(title);
    }
    @Override
    public int deleteAuthorByAuthorId(int id) {
        return authorRepository.deleteAuthorByAuthorId(id);
    }
}
```

4) 创建控制器类

创建名为 com.ch.ch6_3.controller 的包,并在该包中创建名为 TestOneToManyController 的控制器类。

TestOneToManyController 的代码如下:

```java
package com.ch.ch6_3.controller;
import java.util.List;
import java.util.Map;
import org.springframework.beans.factory.annotation.Autowired;
import org.springframework.web.bind.annotation.RequestMapping;
import org.springframework.web.bind.annotation.RestController;
import com.ch.ch6_3.entity.Article;
```

```java
import com.ch.ch6_3.entity.Author;
import com.ch.ch6_3.service.AuthorAndArticleService;
@RestController
public class TestOneToManyController {
    @Autowired
    private AuthorAndArticleService authorAndArticleService;
    @RequestMapping("/findTitleAndContentByAuthorId")
    public List<Map<String, Object>> findTitleAndContentByAuthorId(Integer id){
        return authorAndArticleService.findTitleAndContentByAuthorId(id);
    }
    @RequestMapping("/findArticleByAuthorAnameAndId")
    public List<Article> findArticleByAuthorAnameAndId(String aname, Integer id){
        return authorAndArticleService.findArticleByAuthorAnameAndId(aname, id);
    }
    @RequestMapping("/findAuthorByArticleListtitleContaining")
    public Author findAuthorByArticleListtitleContaining(String title) {
        return authorAndArticleService.findAuthorByArticleListtitleContaining(title);
    }
    @RequestMapping("/deleteAuthorByAuthorId")
    public int deleteAuthorByAuthorId(int id) {
        return authorAndArticleService.deleteAuthorByAuthorId(id);
    }
}
```

5) 运行

首先，运行 Ch63Application 主类。然后，通过"http://localhost:8080/ch6_3/findTitleAndContentByAuthorId?id=1"查询作者 id 为 1 的文章标题和内容，运行效果如图 6.23 所示。

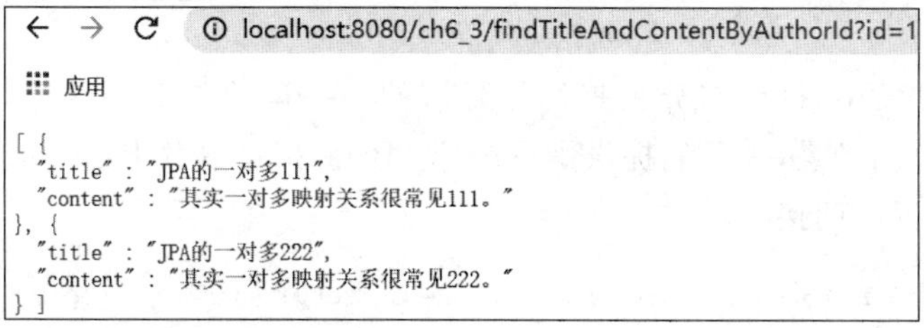

图 6.23 查询作者 id 为 1 的文章标题和内容

通过"http://localhost:8080/ch6_3/findArticleByAuthorAnameAndId?aname=陈恒 2&&id=2"查询作者名为"陈恒 2"且作者 id 为 2 的文件列表，运行效果如图 6.24 所示。

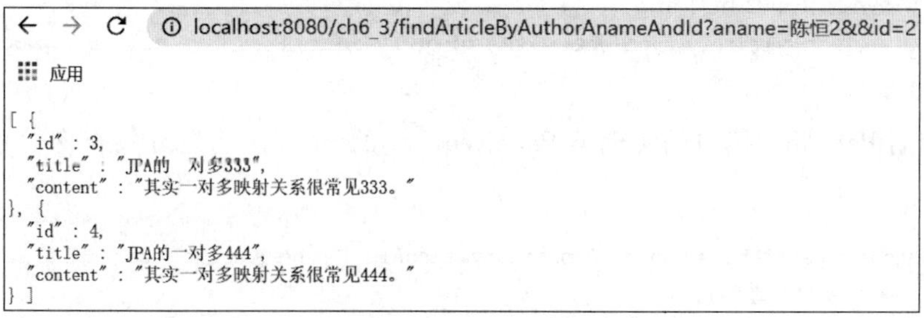

图 6.24 查询作者名为"陈恒 2"且作者 id 为 2 的文件列表

通过"http://localhost:8080/ch6_3/findAuthorByArticleListtitleContaining?title=对多1"查询文章标题包含"对多1"的作者,运行效果如图 6.25 所示。

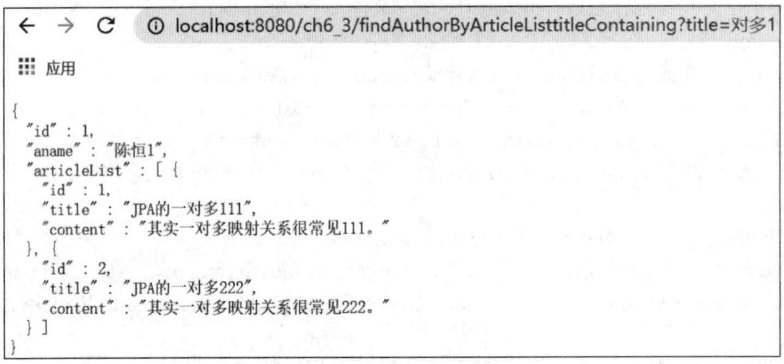

图 6.25　查询文章标题包含"对多1"的作者

通过"http://localhost:8080/ch6_3/deleteAuthorByAuthorId?id=1"删除 id 为 1 的作者,运行效果如图 6.26 所示。

图 6.26　删除 id 为 1 的作者

视频讲解

6.1.5　排序与分页查询

在实际应用开发中,排序与分页查询是必需的。幸运的是 Spring Data JPA 充分考虑了排序与分页查询的场景,为我们提供了 Sort 类、Page 接口以及 Pageable 接口。

例如,如下数据访问接口:

```
public interface AuthorRepository extends JpaRepository<Author, Integer>{
    List<Author> findByAnameContaining(String aname, Sort sort);
}
```

那么,在 Service 层可以这样使用排序:

```
public List<Author> findByAnameContaining(String aname, String sortColumn) {
    //按 sortColumn 降序排序
    return authorRepository.findByAnameContaining(aname, new Sort(Direction.DESC, sortColumn));
}
```

可以使用 Pageable 接口的实现类 PageRequest 的 of 方法构造分页查询对象。示例代码如下:

```
Page<Author> pageData = authorRepository.findAll(PageRequest.of(page - 1, size, new Sort(Direction.DESC, "id")));
```

其中 Page 接口可以获得当前页面的记录、总页数、总记录数等信息。示例代码如下:

```
//获得当前页面的记录
List<Author> allAuthor = pageData.getContent();
model.addAttribute("allAuthor",allAuthor);
//获得总记录数
model.addAttribute("totalCount", pageData.getTotalElements());
//获得总页数
model.addAttribute("totalPage", pageData.getTotalPages());
```

下面通过实例讲解 Spring Data JPA 的排序与分页查询的使用方法。

【例 6-6】 排序与分页查询的使用方法。

首先,为例 6-6 创建基于 Thymeleaf 和 Spring Data JPA 的 Spring Boot Web 应用 ch6_4。ch6_4 应用的数据库、pom.xml、application.properties 以及静态资源等内容与 ch6_1 应用基本一样,不再赘述。

其他内容具体实现步骤如下。

1. 创建持久化实体类

创建名为 com.ch.ch6_4.entity 的包,并在该包中创建名为 Article 和 Author 的持久化实体类。具体代码分别与 ch6_2 应用的 Article 和 Author 的代码一样,不再赘述。

2. 创建数据访问层

创建名为 com.ch.ch6_4.repository 的包,并在该包中创建名为 AuthorRepository 的接口。

AuthorRepository 的代码如下:

```
package com.ch.ch6_4.repository;
import java.util.List;
import org.springframework.data.domain.Sort;
import org.springframework.data.jpa.repository.JpaRepository;
import com.ch.ch6_4.entity.Author;
public interface AuthorRepository extends JpaRepository<Author, Integer>{
    /**
     * 查询作者名含有 name 的作者列表,并排序
     */
    List<Author> findByAnameContaining(String aname, Sort sort);
}
```

3. 创建业务层

创建名为 com.ch.ch6_4.service 的包,并在该包中创建名为 ArticleAndAuthorService 的接口和接口实现类 ArticleAndAuthorServiceImpl。

ArticleAndAuthorService 的代码如下:

```
package com.ch.ch6_4.service;
import java.util.List;
import org.springframework.ui.Model;
import com.ch.ch6_4.entity.Author;
```

```java
public interface ArticleAndAuthorService {
    /**
     * name 代表作者名的一部分(模糊查询),sortColumn 代表排序列
     */
    List<Author> findByAnameContaining(String aname,String sortColumn);
    /**
     * 分页查询作者,page 代表第几页
     */
    public String findAllAuthorByPage(Integer page, Model model);
}
```

ArticleAndAuthorServiceImpl 的代码如下:

```java
package com.ch.ch6_4.service;
import java.util.List;
import org.springframework.beans.factory.annotation.Autowired;
import org.springframework.data.domain.Page;
import org.springframework.data.domain.PageRequest;
import org.springframework.data.domain.Sort;
import org.springframework.data.domain.Sort.Direction;
import org.springframework.stereotype.Service;
import org.springframework.ui.Model;
import com.ch.ch6_4.entity.Author;
import com.ch.ch6_4.repository.AuthorRepository;
@Service
public class ArticleAndAuthorServiceImpl implements ArticleAndAuthorService{
    @Autowired
    private AuthorRepository authorRepository;
    @Override
    public List<Author> findByAnameContaining(String aname, String sortColumn) {
        //按 sortColumn 降序排序
        return authorRepository.findByAnameContaining(aname, new Sort(Direction.DESC, sortColumn));
    }
    @Override
    public String findAllAuthorByPage(Integer page, Model model) {
        if(page == null) {                    //第一次访问 findAllAuthorByPage 方法时
            page = 1;
        }
        int size = 2;                         //每页显示2条
        //分页查询,of 方法的第一个参数代表第几页(比实际小 1),
        //第二个参数代表页面大小,第三个参数代表排序规则
        Page<Author> pageData =
authorRepository.findAll(PageRequest.of(page-1, size, new Sort(Direction.DESC, "id")));
        //获得当前页面数据并转换成 List<Author>,转发到视图页面显示
        List<Author> allAuthor = pageData.getContent();
        model.addAttribute("allAuthor",allAuthor);
        //共多少条记录
        model.addAttribute("totalCount", pageData.getTotalElements());
        //共多少页
        model.addAttribute("totalPage", pageData.getTotalPages());
        //当前页
```

```
            model.addAttribute("page", page);
            return "index";
    }
}
```

4. 创建控制器类

创建名为 com.ch.ch6_4.controller 的包,并在该包中创建名为 TestSortAndPage 的控制器类。

TestSortAndPage 的代码如下:

```
package com.ch.ch6_4.controller;
import java.util.List;
import org.springframework.beans.factory.annotation.Autowired;
import org.springframework.stereotype.Controller;
import org.springframework.ui.Model;
import org.springframework.web.bind.annotation.RequestMapping;
import org.springframework.web.bind.annotation.ResponseBody;
import com.ch.ch6_4.entity.Author;
import com.ch.ch6_4.service.ArticleAndAuthorService;
@Controller
public class TestSortAndPage {
    @Autowired
    private ArticleAndAuthorService articleAndAuthorService;
    @RequestMapping("/findByAnameContaining")
    @ResponseBody
    public List<Author> findByAnameContaining(String aname, String sortColumn){
        return articleAndAuthorService.findByAnameContaining(aname, sortColumn);
    }
    @RequestMapping("/findAllAuthorByPage")
    /**
     * @param page 第几页
     */
    public String findAllAuthorByPage(Integer page, Model model){
        return articleAndAuthorService.findAllAuthorByPage(page, model);
    }
}
```

5. 创建 View 视图页面

在 src/main/resources/templates 目录下,创建视图页面 index.html。

index.html 的代码具体如下:

```
<!DOCTYPE html>
<html xmlns:th="http://www.thymeleaf.org">
<head>
<meta charset="UTF-8">
<title>显示分页查询结果</title>
<link rel="stylesheet" th:href="@{css/bootstrap.min.css}" />
<link rel="stylesheet" th:href="@{css/bootstrap-theme.min.css}" />
```

```html
</head>
<body>
    <div class="panel panel-primary">
        <div class="panel-heading">
            <h3 class="panel-title">Spring Data JPA 分页查询</h3>
        </div>
    </div>
    <div class="container">
        <div class="panel panel-primary">
            <div class="panel-body">
                <div class="table table-responsive">
                    <table class="table table-bordered table-hover">
                        <tbody class="text-center">
                            <tr th:each="author: ${allAuthor}">
                                <td>
                                    <span th:text="${author.id}"></span>
                                </td>
                                <td>
                                    <span th:text="${author.aname}"></span>
                                </td>
                            </tr>
                            <tr>
                                <td colspan="2" align="right">
                                    <ul class="pagination">
                                        <li><a>第<span th:text="${page}"></span>页</a></li>
                                        <li><a>共<span th:text="${totalPage}"></span>页</a></li>
                                        <li><a>共<span th:text="${totalCount}"></span>条</a></li>
                                        <li>
                                            <a th:href="@{findAllAuthorByPage(page=${page-1})}" th:if="${page != 1}">上一页</a></li>
                                        <li>
<a th:href="@{findAllAuthorByPage(page=${page+1})}" th:if="${page != totalPage}">下一页</a></li>
                                    </ul>
                                </td>
                            </tr>
                        </tbody>
                    </table>
                </div>
            </div>
        </div>
    </div>
</body>
</html>
```

6. 运行

首先，运行 Ch64Application 主类。然后，通过"http://localhost:8080/ch6_4/

findByAnameContaining?aname=陈&sortColumn=id"查询作者名含有"陈"的作者列表，并按照 id 降序排序，运行效果如图 6.27 所示。

图 6.27　查询作者名含有"陈"的作者列表，并按照 id 降序排序

通过"http://localhost:8080/ch6_4/findAllAuthorByPage"分页查询作者，并按照 id 降序排序。运行效果如图 6.28 所示。

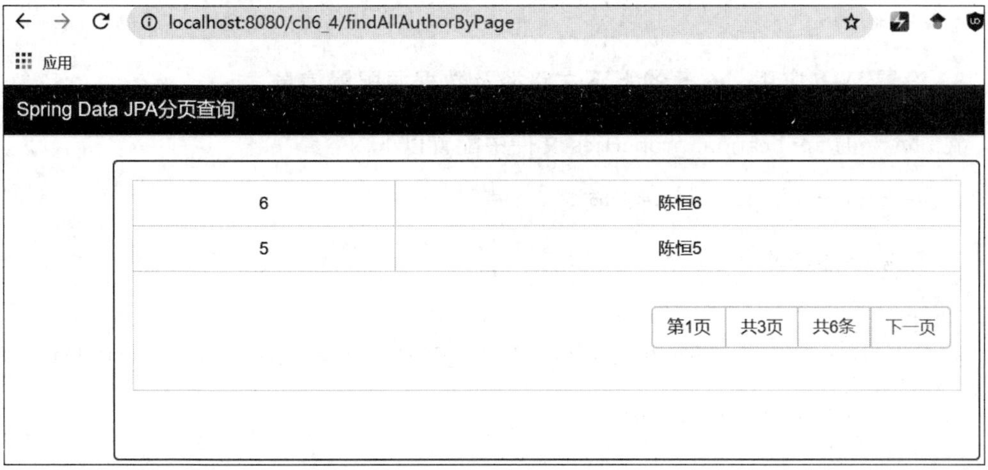

图 6.28　分页查询作者

6.2　Spring Boot 使用 JdbcTemplate

视频讲解

JDBC 模板(JdbcTemplate)是 Spring 对数据库的操作在 JDBC 基础上做了封装，建立了一个 JDBC 存取框架。在 Spring Boot 应用中，如果使用 JdbcTemplate 操作数据库，那么只需在 pom.xml 文件中添加 spring-boot-starter-jdbc 模块，即可通过@Autowired 注解依赖注入 JdbcTemplate 对象，然后调用 JdbcTemplate 提供的方法操作数据库。

下面通过实例讲解如何在 Spring Boot 应用中使用 JdbcTemplate 操作数据库。

【例 6-7】　使用 JdbcTemplate 操作数据库。

具体实现步骤如下。

1. 创建 Spring Boot Web 应用

创建 Spring Boot Web 应用 ch6_5，在该应用中，操作的数据库与 6.1.2 节一样，都是 springbootjpa；操作的数据表是 6.1.3 节创建的 user 表。

2. 修改 pom.xml 文件

在 pom.xml 文件中添加 MySQL 连接器和 spring-boot-starter-jdbc 模块。具体代码如下：

```xml
<dependency>
    <groupId>mysql</groupId>
    <artifactId>mysql-connector-java</artifactId>
    <version>5.1.45</version>
    <!-- MySQL8.x时,请使用8.x的连接器 -->
</dependency>
<dependency>
    <groupId>org.springframework.boot</groupId>
    <artifactId>spring-boot-starter-jdbc</artifactId>
</dependency>
```

3. 设置 Web 应用 ch6_5 的上下文路径及数据源配置信息

在 ch6_5 的 application.properties 文件中配置以下内容：

```
server.servlet.context-path=/ch6_5
###
##数据源信息配置
###
#数据库地址
spring.datasource.url=jdbc:mysql://localhost:3306/springbootjpa?characterEncoding=utf8
#数据库用户名
spring.datasource.username=root
#数据库密码
spring.datasource.password=root
#数据库驱动
spring.datasource.driver-class-name=com.mysql.jdbc.Driver
#数据库MySQL为8.x时,驱动类为com.mysql.cj.jdbc.Driver
spring.jackson.serialization.indent-output=true
#让控制器输出的JSON字符串格式更美观
```

4. 创建实体类

创建名为 com.ch.ch6_5.entity 的包，并在该包中创建 MyUser 实体类。具体代码如下：

```java
package com.ch.ch6_5.entity;
public class MyUser {
    private Integer id;
    private String username;
    private String password;
    //省略 set 和 get 方法
}
```

5. 创建数据访问层

创建名为 com.ch.ch6_5.repository 的包，并在该包中创建 MyUserRepository 接口和

MyUserRepositoryImpl 接口实现类。

MyUserRepository 的代码如下:

```java
package com.ch.ch6_5.repository;
import java.util.List;
import com.ch.ch6_5.entity.MyUser;
public interface MyUserRepository {
    public int saveUser(MyUser myUser);
    public int deleteUser(Integer id);
    public int updateUser(MyUser myUser);
    public List<MyUser> findAll();
    public MyUser findUserById(Integer id);
}
```

MyUserRepositoryImpl 的代码如下:

```java
package com.ch.ch6_5.repository;
import java.util.List;
import org.springframework.beans.factory.annotation.Autowired;
import org.springframework.jdbc.core.BeanPropertyRowMapper;
import org.springframework.jdbc.core.JdbcTemplate;
import org.springframework.jdbc.core.RowMapper;
import org.springframework.stereotype.Repository;
import com.ch.ch6_5.entity.MyUser;
@Repository
public class MyUserRepositoryImpl implements MyUserRepository{
    @Autowired
    private JdbcTemplate jdbcTemplate;
    @Override
    public int saveUser(MyUser myUser) {
        String sql = "insert into user (username, password) values (?,?)";
        Object args[] = {
                myUser.getUsername(),
                myUser.getPassword()
        };
        return jdbcTemplate.update(sql, args);
    }
    @Override
    public int deleteUser(Integer id) {
        String sql = "delete from user where id = ? ";
        Object args[] = {
                id
        };
        return jdbcTemplate.update(sql, args);
    }
    @Override
    public int updateUser(MyUser myUser) {
        String sql = "update user set username = ?, password = ? where id = ? ";
        Object args[] = {
                myUser.getUsername(),
                myUser.getPassword(),
```

```java
                myUser.getId()
        };
        return jdbcTemplate.update(sql, args);
    }
    @Override
    public List<MyUser> findAll() {
        String sql = "select * from user ";
        //定义一个RowMapper
        RowMapper<MyUser> rowMapper = new BeanPropertyRowMapper<MyUser>(MyUser.class);
        return jdbcTemplate.query(sql, rowMapper);
    }
    @Override
    public MyUser findUserById(Integer id) {
        String sql = "select * from user where id = ? ";
        Object args[] = {
                id
        };
        //定义一个RowMapper
        RowMapper<MyUser> rowMapper = new BeanPropertyRowMapper<MyUser>(MyUser.class);
        return jdbcTemplate.queryForObject(sql, args, rowMapper);
    }
}
```

6. 创建业务层

创建名为 com.ch.ch6_5.service 的包，并在该包中创建 MyUserService 接口和 MyUserServiceImpl 实现类。

MyUserService 的代码如下：

```java
package com.ch.ch6_5.service;
import java.util.List;
import com.ch.ch6_5.entity.MyUser;
public interface MyUserService {
    public int saveUser(MyUser myUser);
    public int deleteUser(Integer id);
    public int updateUser(MyUser myUser);
    public List<MyUser> findAll();
    public MyUser findUserById(Integer id);
}
```

MyUserServiceImpl 的代码如下：

```java
package com.ch.ch6_5.service;
import java.util.List;
import org.springframework.beans.factory.annotation.Autowired;
import org.springframework.stereotype.Service;
import com.ch.ch6_5.entity.MyUser;
import com.ch.ch6_5.repository.MyUserRepository;
@Service
public class MyUserServiceImpl implements MyUserService{
    @Autowired
```

```java
    private MyUserRepository myUserRepository;
    @Override
    public int saveUser(MyUser myUser) {
        return myUserRepository.saveUser(myUser);
    }
    @Override
    public int deleteUser(Integer id) {
        return myUserRepository.deleteUser(id);
    }
    @Override
    public int updateUser(MyUser myUser) {
        return myUserRepository.updateUser(myUser);
    }
    @Override
    public List<MyUser> findAll() {
        return myUserRepository.findAll();
    }
    @Override
    public MyUser findUserById(Integer id) {
        return myUserRepository.findUserById(id);
    }
}
```

7. 创建控制器类 MyUserController

创建名为 com.ch.ch6_5.controller 的包,并在该包中创建控制器类 MyUserController。MyUserController 的代码如下:

```java
package com.ch.ch6_5.controller;
import java.util.List;
import org.springframework.beans.factory.annotation.Autowired;
import org.springframework.web.bind.annotation.RequestMapping;
import org.springframework.web.bind.annotation.RestController;
import com.ch.ch6_5.entity.MyUser;
import com.ch.ch6_5.service.MyUserService;
@RestController
public class MyUserController {
    @Autowired
    private MyUserService myUserService;
    @RequestMapping("/saveUser")
    public int saveUser(MyUser myUser) {
        return myUserService.saveUser(myUser);
    }
    @RequestMapping("/deleteUser")
    public int deleteUser(Integer id) {
        return myUserService.deleteUser(id);
    }
    @RequestMapping("/updateUser")
    public int updateUser(MyUser myUser) {
        return myUserService.updateUser(myUser);
    }
    @RequestMapping("/findAll")
```

```
    public List<MyUser> findAll(){
        return myUserService.findAll();
    }
    @RequestMapping("/findUserById")
    public MyUser findUserById(Integer id) {
        return myUserService.findUserById(id);
    }
}
```

8. 运行

首先，运行 Ch65Application 主类。然后，访问"http://localhost:8080/ch6_5/saveUser?username=陈恒3&&password=123456"，运行效果如图 6.29 所示。

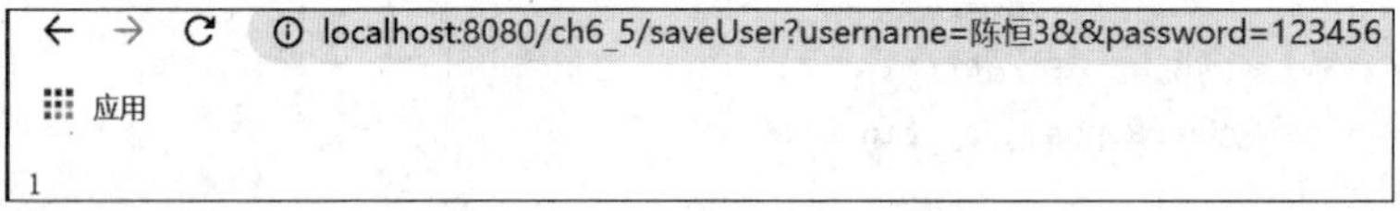

图 6.29　增加用户信息

通过"http://localhost:8080/ch6_5/deleteUser?id=1"删除用户 id 为 1 的用户信息，运行效果如图 6.30 所示。

图 6.30　删除用户信息

通过"http://localhost:8080/ch6_5/updateUser?id=2&&username=陈恒222&&password=888888"修改用户 id 为 2 的用户信息，运行效果如图 6.31 所示。

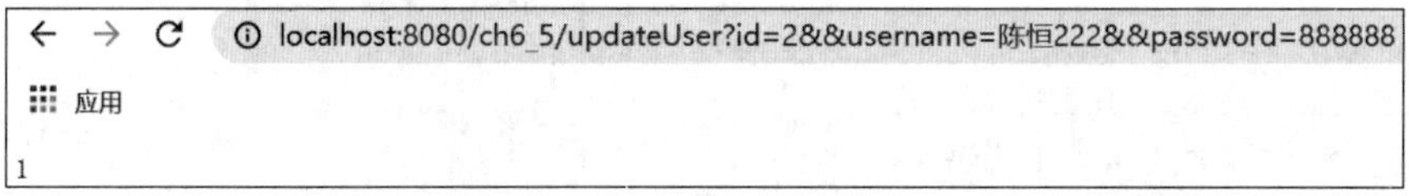

图 6.31　修改用户信息

通过"http://localhost:8080/ch6_5/findAll"查询所有用户信息，运行效果如图 6.32 所示。

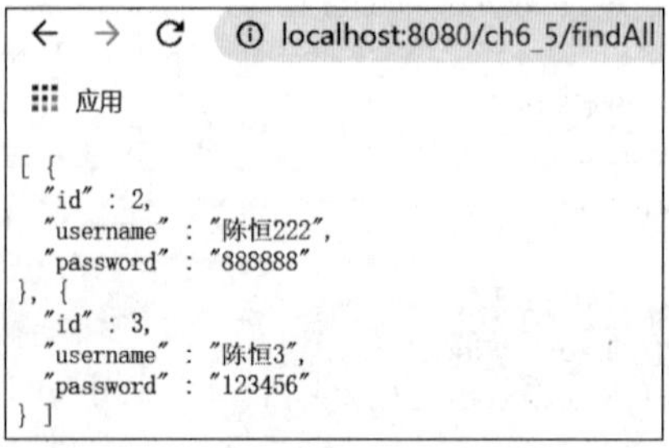

图 6.32　查询所有用户信息

通过"http://localhost:8080/ch6_5/findUserById?id=2"查询用户 id 为 2 的用户信息，运行效果如图 6.33 所示。

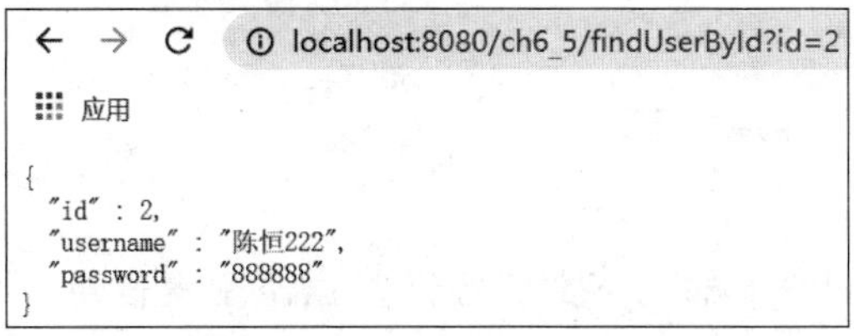

图 6.33 查询用户 id 为 2 的用户信息

6.3 Spring Boot 整合 MyBatis

视频讲解

MyBatis 本是 Apache 的一个开源项目 iBatis，2010 年这个项目由 Apache Software Foundation 迁移到了 Google Code，并且改名为 MyBatis。

MyBatis 是一个基于 Java 的持久层框架。MyBatis 提供的持久层框架包括 SQL Maps 和 Data Access Objects(DAO)，它消除了几乎所有的 JDBC 代码和参数的手工设置以及结果集的检索。MyBatis 使用简单的 XML 或注解用于配置和原始映射，将接口和 Java 的 POJOs(Plain Old Java Objects，普通的 Java 对象)映射成数据库中的记录。

本节仅仅介绍如何在 Spring Boot 应用中使用 MyBatis 框架访问数据库，需要学习 MyBatis 相关知识的读者，请参考作者的另一本教程《Java EE 框架整合开发入门到实战——Spring+Spring MVC+MyBatis(微课版)》。

下面通过实例讲解如何在 Spring Boot 应用中使用 MyBatis 框架操作数据库(基于 XML 的映射配置)。

【例 6-8】 在 Spring Boot 应用中使用 MyBatis 框架操作数据库(基于 XML 的映射配置)。

具体实现步骤如下。

1. 创建 Spring Boot Web 应用

创建 Spring Boot Web 应用 ch6_6，在该应用中，操作的数据库与 6.1.2 节一样，都是 springbootjpa；操作的数据表是 6.1.3 节创建的 user 表。

2. 修改 pom.xml 文件

在 pom.xml 文件中添加 MySQL 连接器和 mybatis-spring-boot-starter 模块。具体代码如下：

```
<dependency>
    <groupId>mysql</groupId>
    <artifactId>mysql-connector-java</artifactId>
    <version>5.1.45</version>
    <!-- MySQL8.x时,请使用 8.x的连接器 -->
```

```xml
</dependency>
<!-- MyBatis-Spring,Spring Boot 应用整合 MyBatis 框架的核心依赖配置 -->
<dependency>
    <groupId>org.mybatis.spring.boot</groupId>
    <artifactId>mybatis-spring-boot-starter</artifactId>
    <version>2.1.0</version>
</dependency>
```

3. 设置 Web 应用 ch6_6 的上下文路径及数据源配置信息

在 ch6_6 的 application.properties 文件中配置如下内容：

```
server.servlet.context-path=/ch6_6
###
## 数据源信息配置
###
# 数据库地址
spring.datasource.url=jdbc:mysql://localhost:3306/springbootjpa?characterEncoding=utf8
# 数据库用户名
spring.datasource.username=root
# 数据库密码
spring.datasource.password=root
# 数据库驱动
spring.datasource.driver-class-name=com.mysql.jdbc.Driver
# 设置包别名(在 Mapper 映射文件中直接使用实体类名)
mybatis.type-aliases-package=com.ch.ch6_6.entity
# 告诉系统到哪里去找 mapper.xml 文件(映射文件)
mybatis.mapperLocations=classpath:mappers/*.xml
# 在控制台输出 SQL 语句日志
logging.level.com.ch.ch6_6.repository=debug
# 让控制器输出的 JSON 字符串格式更美观
spring.jackson.serialization.indent-output=true
```

4. 创建实体类

创建名为 com.ch.ch6_6.entity 的包，并在该包中创建 MyUser 实体类，代码与例 6-7 中的实体类一样，不再赘述。

5. 创建数据访问接口

创建名为 com.ch.ch6_6.repository 的包，并在该包中创建 MyUserRepository 接口。MyUserRepository 的代码如下：

```java
package com.ch.ch6_6.repository;
import java.util.List;
import org.springframework.stereotype.Repository;
import com.ch.ch6_6.entity.MyUser;
@Repository
/*
 * @Repository 可有可无，但有时提示依赖注入找不到(不影响运行)，
 * 加上后可以消去依赖注入的报错信息.
```

```
 * 这里不再需要@Mapper,是因为在启动类中使用@MapperScan注解,
 * 将数据访问层的接口都注解为 Mapper 接口的实现类,
 * @Mapper 与@MapperScan 两者用其一即可
 */
public interface MyUserRepository {
    public List<MyUser> findAll();
}
```

6. 创建 Mapper 映射文件

在 src/main/resources 目录下,创建名为 mappers 的包,并在该包中创建 SQL 映射文件 MyUserMapper.xml。具体代码如下:

```xml
<?xml version="1.0" encoding="UTF-8"?>
<!DOCTYPE mapper
PUBLIC "-//mybatis.org//DTD Mapper 3.0//EN"
"http://mybatis.org/dtd/mybatis-3-mapper.dtd">
<mapper namespace="com.ch.ch6_6.repository.MyUserRepository">
    <select id="findAll" resultType="MyUser">
        select * from user
    </select>
</mapper>
```

7. 创建业务层

创建名为 com.ch.ch6_6.service 的包,并在该包中创建 MyUserService 接口和 MyUserServiceImpl 实现类。

MyUserService 的代码如下:

```java
package com.ch.ch6_6.service;
import java.util.List;
import com.ch.ch6_6.entity.MyUser;
public interface MyUserService {
    public List<MyUser> findAll();
}
```

MyUserServiceImpl 的代码如下:

```java
package com.ch.ch6_6.service;
import java.util.List;
import org.springframework.beans.factory.annotation.Autowired;
import org.springframework.stereotype.Service;
import com.ch.ch6_6.entity.MyUser;
import com.ch.ch6_6.repository.MyUserRepository;
@Service
public class MyUserServiceImpl implements MyUserService{
    @Autowired
    private MyUserRepository myUserRepository;
    @Override
    public List<MyUser> findAll() {
        return myUserRepository.findAll();
    }
}
```

8. 创建控制器类 MyUserController

创建名为 com.ch.ch6_6.controller 的包,并在该包中创建控制器类 MyUserController。MyUserController 的代码如下:

```
package com.ch.ch6_6.controller;
import java.util.List;
import org.springframework.beans.factory.annotation.Autowired;
import org.springframework.web.bind.annotation.RequestMapping;
import org.springframework.web.bind.annotation.RestController;
import com.ch.ch6_6.entity.MyUser;
import com.ch.ch6_6.service.MyUserService;
@RestController
public class MyUserController {
    @Autowired
    private MyUserService myUserService;
    @RequestMapping("/findAll")
    public List<MyUser> findAll(){
        return myUserService.findAll();
    }
}
```

9. 在应用程序的主类中扫描 Mapper 接口

在应用程序的 Ch66Application 主类中,使用@MapperScan 注解扫描 MyBatis 的 Mapper 接口。具体代码如下:

```
package com.ch.ch6_6;
import org.mybatis.spring.annotation.MapperScan;
import org.springframework.boot.SpringApplication;
import org.springframework.boot.autoconfigure.SpringBootApplication;
@SpringBootApplication
//配置扫描 MyBatis 接口的包路径
@MapperScan(basePackages = {"com.ch.ch6_6.repository"})
public class Ch66Application {
    public static void main(String[] args) {
        SpringApplication.run(Ch66Application.class, args);
    }
}
```

10. 运行

首先,运行 Ch66Application 主类。然后,访问"http://localhost:8080/ch6_6/findAll",运行效果如图 6.34 所示。

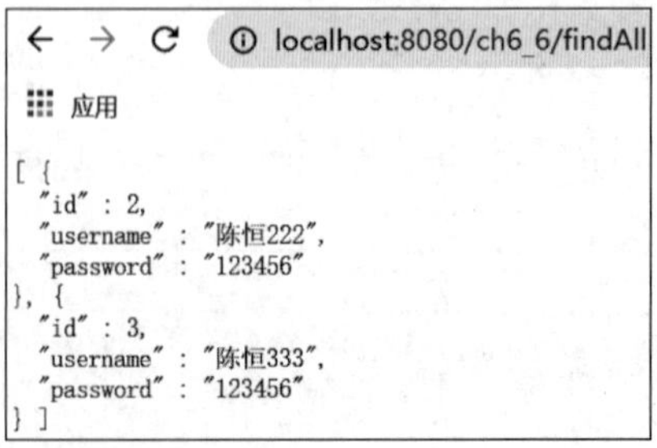

在例 6-8 中,使用基于 XML 的 SQL 映射文件。那么如何使用 SQL 注解的方式操作数据库呢?我们只需要在例 6-8 中做简单修改即可,具体修改如下。

图 6.34 查询所有用户信息

1) 修改 application.properties 文件

删除文件中的如下配置：

```
#设置包别名(在 Mapper 映射文件中直接使用实体类名)
mybatis.type-aliases-package=com.ch.ch6_6.entity
#告诉系统到哪里去找 mapper.xml 文件(映射文件)
mybatis.mapperLocations=classpath:mappers/*.xml
```

2) 删除 Mapper 映射文件

在 src/main/resources 目录下，删除 mappers 包。

3) 修改数据访问接口

将 com.ch.ch6_6.repository 包中的 MyUserRepository 接口修改如下：

```java
package com.ch.ch6_6.repository;
import java.util.List;
import org.apache.ibatis.annotations.Select;
import com.ch.ch6_6.entity.MyUser;
/**
 * MyBatis 映射接口
 */
public interface MyUserRepository {
    @Select("select * from user")
    public List<MyUser> findAll();
}
```

6.4 Spring Boot 的事务管理

6.4.1 Spring Data JPA 的事务支持

视频讲解

我们可以从 org.springframework.data.jpa.repository.support.SimpleJpaRepository 的源代码中看出，Spring Data JPA 对所有的默认接口方法都开启了事务支持，并且对查询类事务默认启用 readOnly。

6.4.2 Spring Boot 的事务支持

在 Spring Boot 中，自动配置了事务管理器，并自动开启了注解事务的支持。

1. 自动配置的事务管理器

在使用 JDBC 访问数据库时，Spring Boot 定义了 DataSourceTransactionManager 的 Bean。具体配置见 org.springframework.boot.autoconfigure.jdbc.DataSourceTransactionManagerAutoConfiguration 类中的定义：

```
@Bean
@ConditionalOnMissingBean(PlatformTransactionManager.class)
public DataSourceTransactionManager transactionManager(DataSourceProperties properties) {
    DataSourceTransactionManager transactionManager = new DataSourceTransactionManager
(this.dataSource);
    if (this.transactionManagerCustomizers != null) {
        this.transactionManagerCustomizers.customize(transactionManager);
    }
    return transactionManager;
}
```

在使用 JPA 访问数据库时，Spring Boot 定义了 PlatformTransactionManager 的实现 JpaTransactionManager 的 Bean。具体配置见 org.springframework.boot.autoconfigure.orm.jpa.JpaBaseConfiguration 类中的定义：

```
@Bean
@ConditionalOnMissingBean
public PlatformTransactionManager transactionManager() {
    JpaTransactionManager transactionManager = new JpaTransactionManager();
    if (this.transactionManagerCustomizers != null) {
        this.transactionManagerCustomizers.customize(transactionManager);
    }
    return transactionManager;
}
```

从上述源码可以看出，如果添加的是 spring-boot-starter-jdbc 依赖，框架会默认注入 DataSourceTransactionManager 实例；如果添加的是 spring-boot-starter-data-jpa 依赖，框架会默认注入 JpaTransactionManager 实例。

2. 自动开启注解事务的支持

Spring Boot 使用 org.springframework.boot.autoconfigure.transaction.TransactionAutoConfiguration 类配置事务的支持，并在该类中自动开启注解事务的支持。具体代码如下：

```
@Configuration
@ConditionalOnBean(PlatformTransactionManager.class)
@ConditionalOnMissingBean(AbstractTransactionManagementConfiguration.class)
public static class EnableTransactionManagementConfiguration {
}
```

不过自动启用该注解有两个前提条件，分别是 @ConditionalOnBean(PlatformTransactionManager.class) 和 @ConditionalOnMissingBean(AbstractTransactionManagementConfiguration.class)，而一般情况，这两个条件都是满足的，所以，在启动类上写不写 @EnableTransactionManagement 注解都可开启事务的支持，然后使用 @Transactional 注解处理事务。有关 @Transactional 注解的使用，请读者参考 1.6.4 节的内容。

6.5 REST

本节将介绍 RESTful 风格接口,并通过 Spring Boot 来实现 RESTful。

6.5.1 REST 简介

REST(Representational State Transfer,表现层状态转化)是 Roy Thomas Fielding 博士在他 2000 年发表的博士论文中提出的一种软件架构风格。它是一种针对网络应用的设计和开发方式,可以降低开发的复杂性,提高系统的可伸缩性。目前在 3 种主流的 Web 服务实现方案中,因为 REST 模式的 Web 服务与复杂的 SOAP 和 XML-RPC 对比来讲更加简洁,越来越多的 Web 服务开始采用 REST 风格设计和实现。

REST 是一组架构约束条件和原则。这些约束有:

(1) 使用客户/服务器模型。客户和服务器之间通过一个统一的接口来互相通信。

(2) 层次化的系统。在一个 REST 系统中,客户端并不会固定地与一个服务器打交道。

(3) 无状态。在一个 REST 系统中,服务端并不会保存有关客户的任何状态。也就是说,客户端自身负责用户状态的维持,并在每次发送请求时都需要提供足够的信息。

(4) 可缓存。REST 系统需要恰当地缓存请求,以尽量减少服务端和客户端之间的信息传输,以提高性能。

(5) 统一的接口。一个 REST 系统需要使用一个统一的接口来完成子系统之间以及服务与用户之间的交互。这使得 REST 系统中的各个子系统可以独自完成演化。

满足这些约束条件和原则的应用程序或设计就是 RESTful。需要注意的是,REST 是设计风格而不是标准。REST 通常基于 HTTP、URI、XML 以及 HTML 这些现有的广泛流行的协议和标准。

理解 RESTful 架构,应该先理解 Representational State Transfer 这个词组到底是什么意思,它的每一个词表达了什么含义。

- 资源(Resources)

"表现层状态转化"中的"表现层"其实指的是"资源"的"表现层"。

"资源"就是网络上的一个实体,或者说是网络上的一个具体信息。"资源"可以是一段文本、一张图片、一段视频,总之就是一个具体的实体。可以使用一个 URI(统一资源定位符)指向资源,每种资源对应一个特定的 URI。我们需要获取资源时,访问它的 URI 即可,因此 URI 是每个资源的地址或独一无二的标识符。REST 风格的 Web 服务,是通过一个简洁清晰的 URI 来提供资源链接,客户端通过对 URI 发送 HTTP 请求获得这些资源,而获取和处理资源的过程让客户端应用的状态发生改变。

- 表现层(Representation)

"资源"是一种信息实体,可以有多种外在的表现形式。将"资源"呈现出来的形式称为它的"表现层"。例如,文本可以使用 txt 格式表现,也可以使用 XML 格式、JSON 格式表现。

- 状态转化(State Transfer)

客户端访问一个网站,就代表了它和服务器的一个互动过程。在这个互动过程中,将涉及数据和状态的变化。我们知道 HTTP 协议是一个无状态的通信协议,这意味着所有状态都保存在服务器端。因此,如果客户端操作服务器,需要通过某种手段(如 HTTP 协议)让服务器端发生"状态变化"。而这种转化是建立在表现层之上的,所以就是"表现层状态转化"。

在流行的各种 Web 框架中,包括 Spring Boot 都支持 REST 开发。REST 并不是一种技术或者规范,而是一种架构风格,包括如何标识资源、如何标识操作接口及操作的版本、如何标识操作的结果等,主要内容如下。

1. 使用"api"作为上下文

在 REST 架构中,建议使用"api"作为上下文,示例如下:

```
http://localhost:8080/api
```

2. 增加一个版本标识

在 REST 架构中,可以通过 URL 标识版本信息,示例如下:

```
http://localhost:8080/api/v1.0
```

3. 标识资源

在 REST 架构中,可以将资源名称放到 URL 中,示例如下:

```
http://localhost:8080/api/v1.0/user
```

4. 确定 HTTP Method

HTTP 协议有 5 个常用的表示操作方式的动词:GET、POST、PUT、DELETE 和 PATCH。它们分别对应 5 种基本操作:GET 用来获取资源;POST 用来增加资源(也可以用于更新资源);PUT 用来更新资源;DELETE 用来删除资源;PATCH 用来更新资源的部分属性。示例如下:

1) 新增用户

```
POST http://localhost:8080/api/v1.0/user
```

2) 查询 id 为 123 的用户

```
GET http://localhost:8080/api/v1.0/user/123
```

3) 更新 id 为 123 的用户

```
PUT http://localhost:8080/api/v1.0/user/123
```

4) 删除 id 为 123 的用户

```
DELETE http://localhost:8080/api/v1.0/user/123
```

5. 确定 HTTP Status

服务器向用户返回的状态码和提示信息,常用的如下。

(1) 200 OK - [GET]：服务器成功返回用户请求的数据。

(2) 201 CREATED - [POST/PUT/PATCH]：用户新建或修改数据成功。

(3) 202 Accepted - [*]：表示一个请求已经进入后台排队(异步任务)。

(4) 204 NO CONTENT - [DELETE]：用户删除数据成功。

(5) 400 INVALID REQUEST - [POST/PUT/PATCH]：用户发出的请求有错误,服务器没有进行新建或修改数据的操作。

(6) 401 Unauthorized - [*]：表示用户没有权限(令牌、用户名、密码错误)。

(7) 403 Forbidden - [*]：表示用户得到授权(与 401 错误相对),但是访问是被禁止的。

(8) 404 NOT FOUND - [*]：用户发出的请求针对的是不存在的记录,服务器没有进行操作。

(9) 406 Not Acceptable - [GET]：用户请求的格式不可得(例如用户请求 JSON 格式,但是只有 XML 格式)。

(10) 410 Gone - [GET]：用户请求的资源被永久删除,且不会再得到。

(11) 422 Unprocesable entity - [POST/PUT/PATCH]：当创建一个对象时,发生一个验证错误。

(12) 500 INTERNAL SERVER ERROR - [*]：服务器发生错误,用户将无法判断发出的请求是否成功。

6.5.2 Spring Boot 整合 REST

在 Spring Boot 的 Web 应用中,自动支持 REST。也就是说,只要 spring-boot-starter-web 依赖在 pom 中,就支持 REST。

【例 6-9】 一个 RESTful 应用示例。

假如在 ch6_2 应用的控制器类中有如下处理方法：

```
@RequestMapping("/findArticleByAuthor_id/{id}")
public List<Article> findByAuthor_id(@PathVariable("id") Integer id) {
    return authorAndArticleService.findByAuthor_id(id);
}
```

那么可以使用如下所示的 REST 风格的 URL 访问上述处理方法：

```
http://localhost:8080/ch6_2/findArticleByAuthor_id/2
```

在例 6-9 中使用了 URL 模板模式映射@RequestMapping("/findArticleByAuthor_id/{id}"),其中{XXX}为占位符,请求的 URL 可以是"/findArticleByAuthor_id/1"或"/findArticleByAuthor_id/2"。通过在处理方法中使用@PathVariable 获取{XXX}中的 XXX 变量值。@PathVariable 用于将请求 URL 中的模板变量映射到功能处理方法的参数

上。如果{XXX}中的变量名 XXX 和形参名称一致,则@PathVariable 不用指定名称。

6.5.3 Spring Data REST

Spring Data JPA 基于 Spring Data 的 repository 之上,可以将 repository 自动输出为 REST 资源。目前,Spring Data REST 支持将 Spring Data JPA、Spring Data MongoDB、Spring Data Neo4j、Spring Data GemFire 以及 Spring Data Cassandra 的 repository 自动转换成 REST 服务。

Spring Boot 对 Spring Data REST 的自动配置存放在 rest 包中,如图 6.35 所示。

通过 SpringBootRepositoryRestConfigurer 类的源码可以得出,Spring Boot 已经自动配置了 RepositoryRestConfiguration,所以在 Spring Boot 应用中使用 Spring Data REST 只需引入 spring-boot-starter-data-rest 的依赖即可。

图 6.35 Spring Data REST 的自动配置

下面通过实例讲解 Spring Data REST 的构建过程。

【例 6-10】 Spring Data REST 的构建过程。

具体实现步骤如下。

1. 创建 Spring Boot 应用 ch6_7

创建 Spring Boot 应用 ch6_7,依赖为 Spring Data JPA 和 Rest Repositories,如图 6.36 所示。

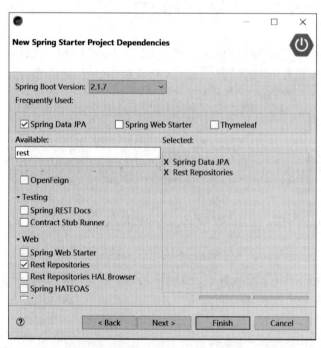

图 6.36 ch6_7 的项目依赖

2. 修改 pom.xml 文件，添加 MySQL 依赖

修改 ch6_7 应用的 pom.xml 文件，添加 MySQL 依赖，具体代码如下：

```xml
<dependency>
    <groupId>mysql</groupId>
    <artifactId>mysql-connector-java</artifactId>
    <version>5.1.45</version>
</dependency>
```

3. 设置 ch6_7 应用的上下文路径及数据源配置信息

在 ch6_7 的 application.properties 文件中配置如下内容：

```
server.servlet.context-path=/api
###
##数据源信息配置
###
#数据库地址
spring.datasource.url=jdbc:mysql://localhost:3306/springbootjpa?characterEncoding=utf8
#数据库用户名
spring.datasource.username=root
#数据库密码
spring.datasource.password=root
#数据库驱动
spring.datasource.driver-class-name=com.mysql.jdbc.Driver
####
#JPA持久化配置
####
#指定数据库类型
spring.jpa.database=MYSQL
#指定是否在日志中显示SQL语句
spring.jpa.show-sql=true
#指定自动创建、更新数据库表等配置
#update表示如果数据库中存在持久化类对应的表就不创建，不存在就创建对应的表
spring.jpa.hibernate.ddl-auto=update
#让控制器输出的JSON字符串格式更美观
spring.jackson.serialization.indent-output=true
```

4. 创建持久化实体类 Student

创建名为 com.ch.ch6_7.entity 的包，并在该包中创建名为 Student 的持久化实体类。具体代码如下：

```java
package com.ch.ch6_7.entity;
import java.io.Serializable;
import javax.persistence.Entity;
import javax.persistence.GeneratedValue;
import javax.persistence.GenerationType;
import javax.persistence.Id;
import javax.persistence.Table;
@Entity
@Table(name = "student_table")
```

```java
public class Student implements Serializable{
    private static final long serialVersionUID = 1L;
    @Id
    @GeneratedValue(strategy = GenerationType.IDENTITY)
    private int id;                                         //主键
    private String sno;
    private String sname;
    private String ssex;
    public Student() {
        super();
    }
    public Student(int id, String sno, String sname, String ssex) {
        super();
        this.id = id;
        this.sno = sno;
        this.sname = sname;
        this.ssex = ssex;
    }
    //省略 set 和 get 方法
}
```

5．创建数据访问层

创建名为 com.ch.ch6_7.repository 的包，并在该包中创建名为 StudentRepository 的接口，该接口继承 JpaRepository 接口，具体代码如下：

```java
package com.ch.ch6_7.repository;
import java.util.List;
import org.springframework.data.jpa.repository.JpaRepository;
import org.springframework.data.repository.query.Param;
import org.springframework.data.rest.core.annotation.RestResource;
import com.ch.ch6_7.entity.Student;
public interface StudentRepository extends JpaRepository<Student, Integer>{
    /**
     * 自定义接口查询方法,暴露为 REST 资源
     */
    @RestResource(path = "snameStartsWith", rel = "snameStartsWith")
    List<Student> findBySnameStartsWith(@Param("sname") String sname);
}
```

在上述数据访问接口中，定义了 findBySnameStartsWith，并使用@RestResource 注解将该方法暴露为 REST 资源。

至此，基于 Spring Data 的 REST 资源服务已经构建完毕，接下来就是使用 REST 客户端测试此服务。

6.5.4　REST 服务测试

在 Web 和移动端开发时，常常会调用服务器端的 RESTful 接口进行数据请求，为了调试，一般会先用工具进行测试，通过测试后才开始在开发中使用。本节将介绍如何使用 Wisdom REST Client 进行 6.5.3 节的 RESTful 接口请求测试。

第6章 Spring Boot 的数据访问

Wisdom REST Client 是用 Java 语言编写的 REST 客户端，是 GitHub 上的开源项目。可以通过 https://github.com/Wisdom-Projects/rest-client 地址下载。该客户端使用方便，解压后双击 tools 下的 restclient-1.2.jar 包，即可运行(前提是已安装 JDK)。

1. 获得列表数据

在 RESTful 架构中，每个网址代表一种资源(resource)，所以网址中不能有动词，只能有名词，而且所用的名词往往与实体名对应。一般来说，数据库中的表都是同种记录的"集合"(collection)，所以 API 中的名词也应该使用复数，如 students。

运行 ch6_7 的主类 Ch67Application，在 student_table 中手工添加几条学生信息后，在 Wisdom REST Client 中，使用 GET 方式访问"http://localhost:8080/api/students"请求路径获得所有学生信息，如图 6.37 和图 6.38 所示。

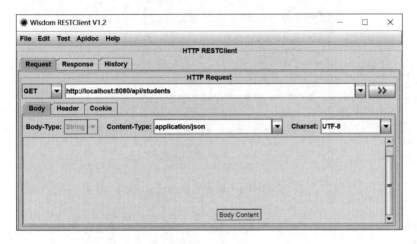

图 6.37　GET 所有学生信息的 Request 界面

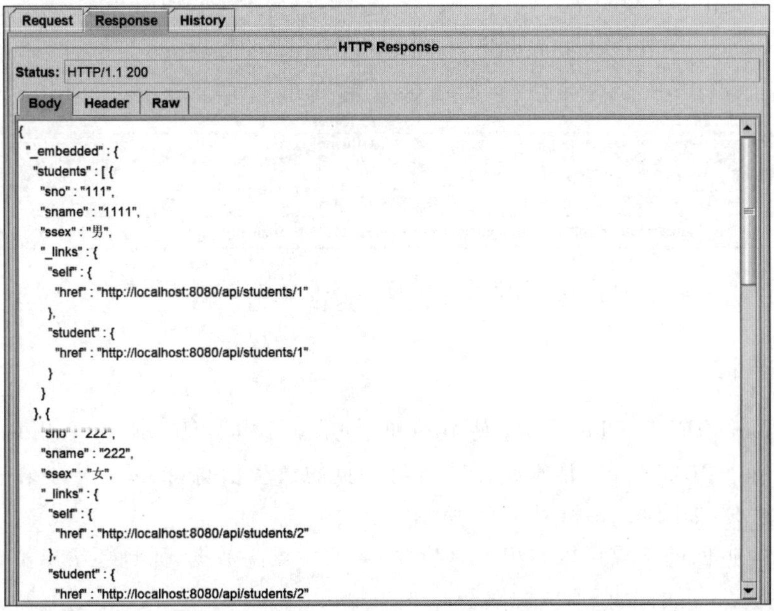

图 6.38　GET 所有学生信息的 Response 界面

2. 获得单一对象

在 Wisdom REST Client 中，使用 GET 方式访问"http://localhost:8080/api/students/1"请求路径获得 id 为 1 的学生信息，如图 6.39 和图 6.40 所示。

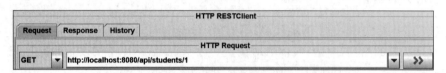

图 6.39　GET id 为 1 的学生信息的 Request 界面

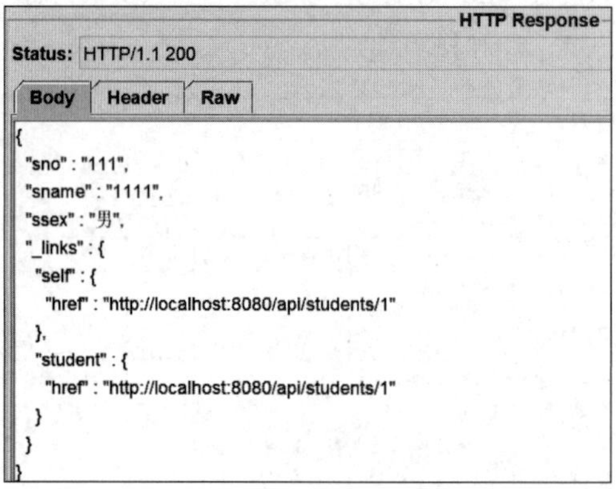

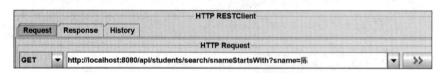

图 6.40　GET id 为 1 的学生信息的 Response 界面

3. 查询

在 Wisdom REST Client 中，search 调用自定义的接口查询方法。因此，可以使用 GET 访问"http://localhost:8080/api/students/search/snameStartsWith?sname=陈"请求路径调用 List< Student > findBySnameStartsWith(@Param("sname") String sname)接口方法，获得姓名前缀为"陈"的学生信息，如图 6.41 和图 6.42 所示。

图 6.41　GET 姓名前缀为"陈"的学生信息的 Request 界面

4. 分页查询

在 Wisdom REST Client 中，使用 GET 方式访问"http://localhost:8080/api/students/?page=0&size=2"请求路径获得第一页的学生信息（page＝0 即第一页，size＝2 即每页数量为 2），如图 6.43 和图 6.44 所示。

从图 6.44 返回的结果可以看出，不仅获得当前分页的数据，而且还给出了上一页、下一页、第一页、最后一页的 REST 资源路径。

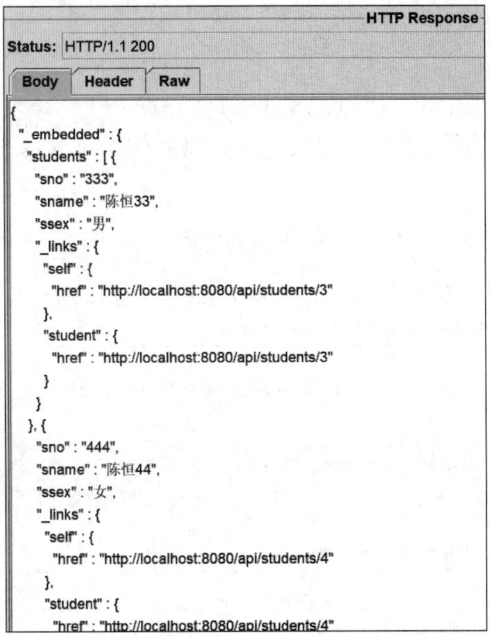

HTTP Response
Status: HTTP/1.1 200
Body Header Raw
{
"_embedded" : {
"students" : [{
"sno" : "333",
"sname" : "陈恒33",
"ssex" : "男",
"_links" : {
"self" : {
"href" : "http://localhost:8080/api/students/3"
},
"student" : {
"href" : "http://localhost:8080/api/students/3"
}
}
}, {
"sno" : "444",
"sname" : "陈恒44",
"ssex" : "女",
"_links" : {
"self" : {
"href" : "http://localhost:8080/api/students/4"
},
"student" : {
"href" : "http://localhost:8080/api/students/4"

图 6.42 GET 姓名前缀为“陈”的学生信息的 Response 界面

HTTP RESTClient
Request Response History
HTTP Request
GET
http://localhost:8080/api/students/?page=0&size=2

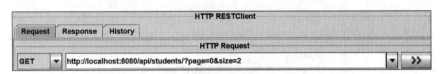

图 6.43 GET 分页查询的 Request 界面

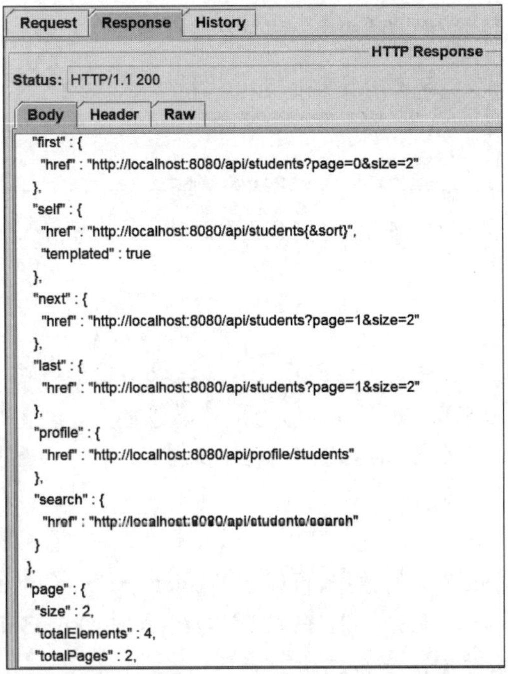

Request Response History
HTTP Response
Status: HTTP/1.1 200
Body Header Raw
"first" : {
"href" : "http://localhost:8080/api/students?page=0&size=2"
},
"self" : {
"href" : "http://localhost:8080/api/students{&sort}",
"templated" : true
},
"next" : {
"href" : "http://localhost:8080/api/students?page=1&size=2"
},
"last" : {
"href" : "http://localhost:8080/api/students?page=1&size=2"
},
"profile" : {
"href" : "http://localhost:8080/api/profile/students"
},
"search" : {
"href" : "http://localhost:8080/api/students/search"
}
},
"page" : {
"size" : 2,
"totalElements" : 4,
"totalPages" : 2,

图 6.44 GET 分页查询的 Response 界面

5. 排序

在 Wisdom REST Client 中，使用 GET 方式访问 "http://localhost:8080/api/students/?sort=sno,desc"请求路径获得按照 sno 属性倒序排序的列表，如图 6.45 和图 6.46 所示。

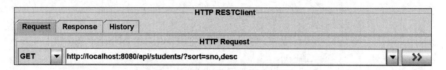

图 6.45　GET 排序查询的 Request 界面

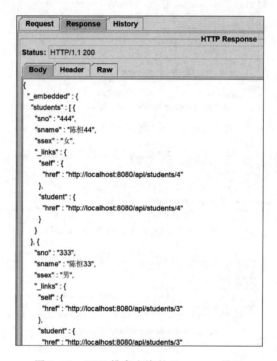

图 6.46　GET 排序查询的 Response 界面

6. 保存

在 Wisdom REST Client 中，发起 POST 方式请求实现新增功能，将要保存的数据放置在请求体中，数据类型为 JSON，如图 6.47 和图 6.48 所示。

从图 6.48 可以看出，保存成功后，新数据的 id 为 5。

7. 更新

假如需要更新新增的 id 为 5 的数据，可以在 Wisdom REST Client 中，使用 PUT 方式访问 "http://localhost:8080/api/students/5"，修改提交的数据，如图 6.49 和图 6.50 所示。

从图 6.50 可以看出，数据已更新成功。

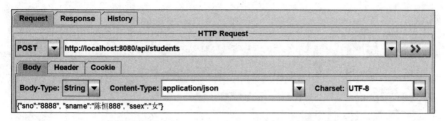

图 6.47　发起 POST 请求实现新增功能

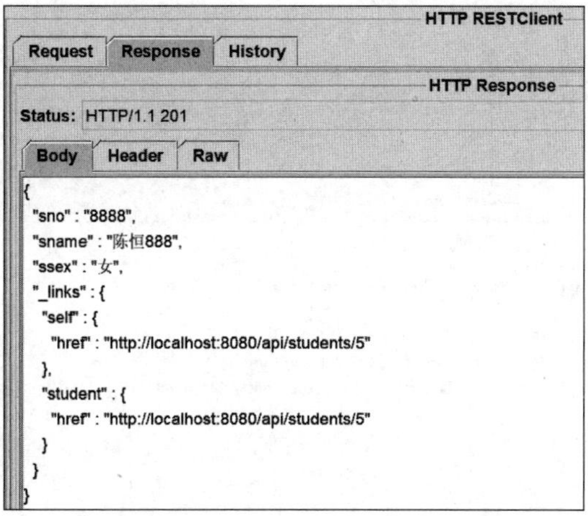

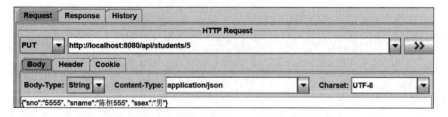

图 6.48　保存成功

图 6.49　发起 PUT 请求实现更新功能

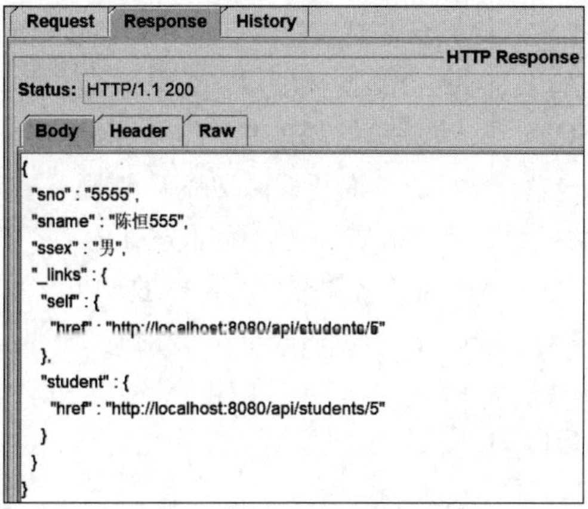

图 6.50　更新成功

8. 删除

假如需要删除新增的 id 为 5 的数据,可以在 Wisdom REST Client 中,使用 DELETE 方式访问"http://localhost:8080/api/students/5",删除数据,如图 6.51 和图 6.52 所示。

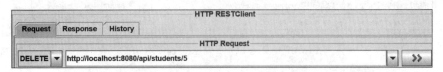

图 6.51　发起 DELETE 请求实现删除功能

图 6.52　删除成功

从图 6.52 返回的状态码 204 可以看出,用户数据删除成功。

6.6　MongoDB

视频讲解

MongoDB 是一个基于分布式文件存储的 NoSQL 数据库,由 C++ 语言编写,旨在为 Web 应用提供可扩展的高性能数据存储解决方案。

MongoDB 是一个介于关系数据库和非关系数据库之间的产品,是非关系数据库中功能最丰富、最像关系数据库的。它支持的数据结构非常松散,是类似 JSON 的 BSON(Binary JSON,二进制 JSON)格式,因此可以存储比较复杂的数据类型。Mongo 最大的特点是它支持的查询语言非常强大,其语法有点类似于面向对象的查询语言,几乎可以实现类似关系数据库单表查询的绝大部分功能,而且还支持对数据建立索引。

本节不会介绍太多关于 MongoDB 数据库本身的知识,主要介绍 Spring Boot 对 MongoDB 的支持,以及基于 Spring Boot 和 MongoDB 的实例。

6.6.1　安装 MongoDB

可以从官方网站 https://www.mongodb.com/download-center/community 下载自己

计算机的操作系统对应版本的 MongoDB，作者编写本书时使用的 MongoDB 是 mongodb-win32-x86_64-2012plus-4.2.0-signed.msi。成功下载后，双击 mongodb-win32-x86_64-2012plus-4.2.0-signed.msi，按照默认安装即可。

可以使用 MongoDB 的图形界面管理工具 MongoDB Compass 可视化操作 MongoDB 数据库。可以使用 mongodb-win32-x86_64-2012plus-4.2.0-signed.msi 自带的 MongoDB Compass，也可以从官方网站 https://www.mongodb.com/download-center/compass 下载。

6.6.2 Spring Boot 整合 MongoDB

1. Spring 对 MongoDB 的支持

Spring 对 MongoDB 的支持主要是通过 Spring Data MongoDB 实现的，Spring Data MongoDB 提供了如下功能。

1）对象/文档映射注解

Spring Data MongoDB 提供了如表 6.2 所示的注解。

表 6.2 Spring Data MongoDB 提供的对象/文档映射注解

注 解	含 义
@Document	映射领域对象与 MongoDB 的一个文档
@Id	映射当前属性是文档对象 ID
@DBRef	当前属性将参考其他文档
@Field	为文档的属性定义名称
@Version	将当前属性作为版本

2）MongoTemplate

与 JdbcTemplate 一样，Spring Data MongoDB 也提供了一个 MongoTemplate，而 MongoTemplate 提供了数据访问的方法。

3）Repository

类似于 Spring Data JPA，Spring Data MongoDB 也提供了 Repository 的支持，使用方式和 Spring Data JPA 一样，示例代码如下：

```
public interface PersonRepository extends MongoRepository<Person, String>{
}
```

2. Spring Boot 对 MongoDB 的支持

Spring Boot 对 MongoDB 的自动配置位于 org.springframework.boot.autoconfigure.mongo 包中，主要配置了数据库连接、MongoTemplate，可以在配置文件中使用以 spring.data.mongodb 为前缀的属性来配置 MongoDB 的相关信息。Spring Boot 对 MongoDB 提供了一些默认属性，如默认端口号为 27017，默认服务器为 localhost，默认数据库为 test，默认无用户名和无密码访问方式，并默认开启了对 Repository 的支持。因此，在 Spring Boot 应用中，只需引入 spring-boot-starter-data-mongodb 依赖即可按照默认配置操作 MongoDB

数据库。

6.6.3 增删改查

本节通过实例，讲解如何在 Spring Boot 应用中，对 MongoDB 数据库进行增删改查。

【例 6-11】 在 Spring Boot 应用中，对 MongoDB 数据库进行增删改查。

具体实现步骤如下。

1. 创建基于 spring-boot-starter-data-mongodb 依赖的 Spring Boot Web 应用 ch6_8

创建 Spring Boot Web 应用 ch6_8，该应用需要引入 spring-boot-starter-data-mongodb 依赖，如图 6.53 所示。

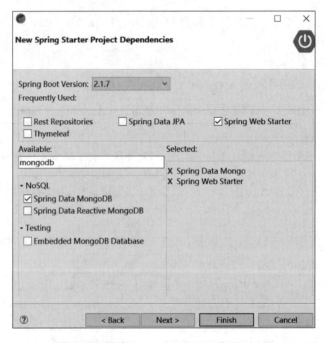

图 6.53　创建 Spring Boot Web 应用 ch6_8

2. 配置 application.properties 文件

在该 Spring Boot 应用 ch6_8 中，使用 MongoDB 的默认数据库连接。所以，不需要在 application.properties 文件中配置数据库连接信息。application.properties 文件的具体内容如下：

```
server.servlet.context-path=/ch6_8
#让控制器输出的JSON字符串格式更美观
spring.jackson.serialization.indent-output=true
```

3. 创建领域模型

创建名为 com.ch.ch6_8.domain 的包，并在该包中创建领域模型 Person（人）以及

Person 去过的 Location（地点）。在 Person 类中，使用@Document 注解对 Person 领域模型和 MongoDB 的文档进行映射。

Person 的代码如下：

```
package com.ch.ch6_8.domain;
import java.util.ArrayList;
import java.util.List;
import org.springframework.data.annotation.Id;
import org.springframework.data.mongodb.core.mapping.Document;
import org.springframework.data.mongodb.core.mapping.Field;
@Document
public class Person {
    @Id
    private String pid;
    private String pname;
    private Integer page;
    private String psex;
    @Field("plocs")
    private List<Location> locations = new ArrayList<Location>();
    public Person() {
        super();
    }
    public Person(String pname, Integer page, String psex) {
        super();
        this.pname = pname;
        this.page = page;
        this.psex = psex;
    }
    //省略 set 和 get 方法
}
```

Location 的代码如下：

```
package com.ch.ch6_8.domain;
public class Location {
    private String locName;
    private String year;
    public Location() {
        super();
    }
    public Location(String locName, String year) {
        super();
        this.locName = locName;
        this.year = year;
    }
}
```

4. 创建数据访问接口

创建名为 com.ch.ch6_8.repository 的包，并在该包中创建数据访问接口 PersonRepository，该接口继承 MongoRepository 接口。PersonRepository 接口的代码如下：

```
package com.ch.ch6_8.repository;
import java.util.List;
import org.springframework.data.mongodb.repository.MongoRepository;
import org.springframework.data.mongodb.repository.Query;
import com.ch.ch6_8.domain.Person;
public interface PersonRepository extends MongoRepository<Person, String>{
    Person findByPname(String pname);           //支持方法名查询,方法名命名规范参照表 6.1
    @Query("{'psex':?0}")                       //JSON 字符串
    List<Person> selectPersonsByPsex(String psex);
}
```

5. 创建控制器层

由于本实例业务简单,我们直接在控制器层调用数据访问层。创建名为 com.ch.ch6_8.controller 的包,并在该包中创建控制器类 TestMongoDBController。

TestMongoDBController 的代码如下:

```
package com.ch.ch6_8.controller;
import java.util.ArrayList;
import java.util.List;
import org.springframework.beans.factory.annotation.Autowired;
import org.springframework.web.bind.annotation.RequestMapping;
import org.springframework.web.bind.annotation.RestController;
import com.ch.ch6_8.domain.Location;
import com.ch.ch6_8.domain.Person;
import com.ch.ch6_8.repository.PersonRepository;
@RestController
public class TestMongoDBController {
    @Autowired
    private PersonRepository personRepository;
    @RequestMapping("/save")
    public List<Person> save() {
        List<Location> locations1 = new ArrayList<Location>();
        Location loc1 = new Location("北京","2019");
        Location loc2 = new Location("上海","2018");
        locations1.add(loc1);
        locations1.add(loc2);
        List<Location> locations2 = new ArrayList<Location>();
        Location loc3 = new Location("广州","2017");
        Location loc4 = new Location("深圳","2016");
        locations2.add(loc3);
        locations2.add(loc4);
        List<Person> persons = new ArrayList<Person>();
        Person p1 = new Person("陈恒 1", 88, "男");
        p1.setLocations(locations1);
        Person p2 = new Person("陈恒 2", 99, "女");
        p2.setLocations(locations2);
        persons.add(p1);
        persons.add(p2);
```

```
        return personRepository.saveAll(persons);
    }
    @RequestMapping("/findByPname")
    public Person findByPname(String pname) {
        return personRepository.findByPname(pname);
    }
    @RequestMapping("/selectPersonsByPsex")
    public List<Person> selectPersonsByPsex(String psex) {
        return personRepository.selectPersonsByPsex(psex);
    }
    @RequestMapping("/updatePerson")
    public Person updatePerson(String oldPname, String newPname) {
        Person p1 = personRepository.findByPname(oldPname);
        if(p1 != null)
            p1.setPname(newPname);
        return personRepository.save(p1);
    }
    @RequestMapping("/deletePerson")
    public void updatePerson(String pname) {
        Person p1 = personRepository.findByPname(pname);
        personRepository.delete(p1);
    }
}
```

6. 运行

首先，运行 Ch68Application 主类。然后，访问"http://localhost:8080/ch6_8/save"测试保存数据，运行效果如图 6.54 所示。

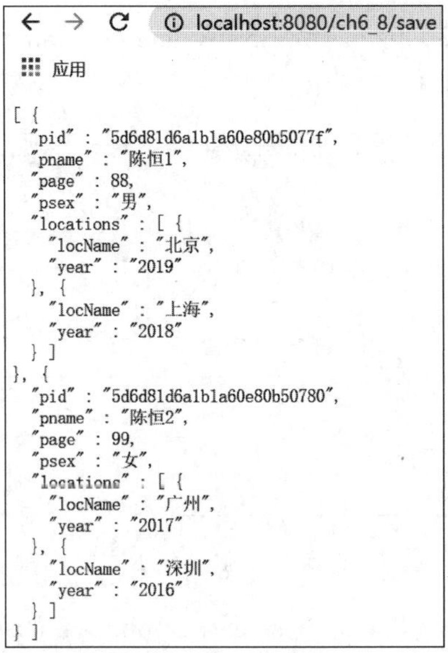

图 6.54 测试保存

保存成功后，使用 MongoDB 的图形界面管理工具 MongoDB Compass 打开查看已保存的数据，如图 6.55 所示。

图 6.55　查看已保存的数据

通过"http://localhost:8080/ch6_8/findByPname?pname＝陈恒 1"查询人名为"陈恒 1"的文档数据，运行效果如图 6.56 所示。

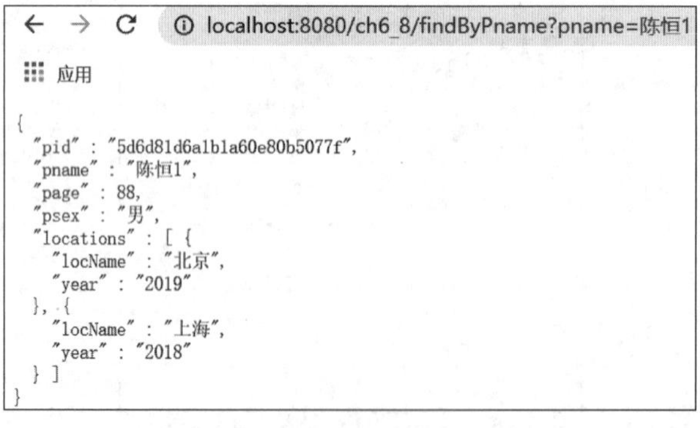

图 6.56　查询人名为"陈恒 1"的文档数据

通过"http://localhost:8080/ch6_8/selectPersonsByPsex?psex＝女"查询性别为"女"的文档数据，运行效果如图 6.57 所示。

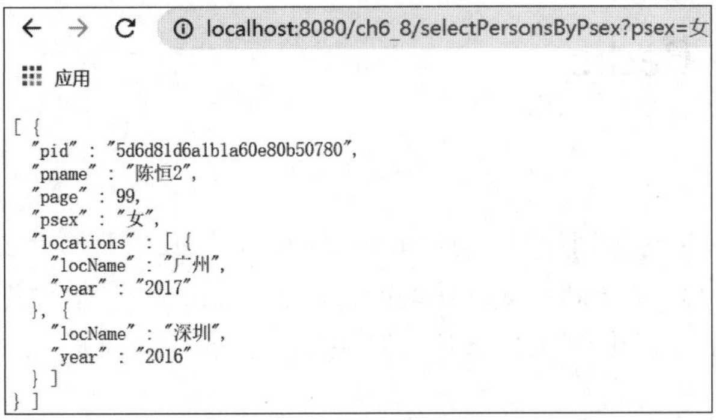

图 6.57 查询性别为"女"的文档数据

通过"http://localhost:8080/ch6_8/updatePerson?oldPname＝陈恒 1&newPname＝陈恒 111"将人名为"陈恒 1"的数据修改成人名为"陈恒 111",运行效果如图 6.58 所示。

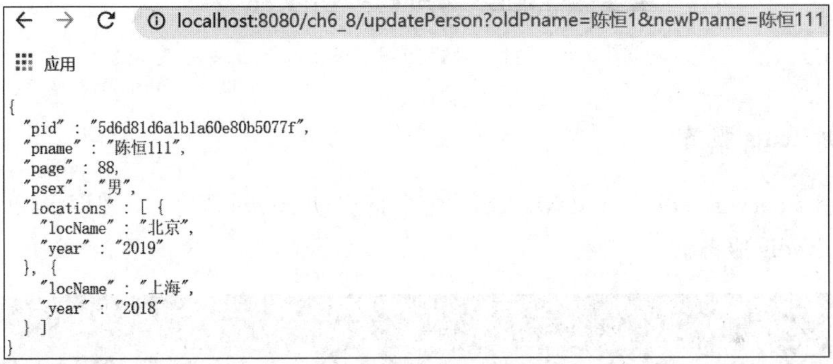

图 6.58 修改数据测试

通过"http://localhost:8080/ch6_8/deletePerson?pname＝陈恒 111"将人名为"陈恒 111"的文档数据删除。

至此,通过 Spring Boot Web 应用对 MongoDB 数据库的操作演示完毕。

6.7 Redis

视频讲解

Redis 是一个开源的使用 ANSI C 语言编写、支持网络、可基于内存亦可持久化的日志型、Key-Value 数据库,并提供多种语言的 API。它支持字符串、哈希表、列表、集合、有序集合、位图、地理空间信息等数据类型,同时也可以作为高速缓存和消息队列代理。但是,Redis 在内存中存储数据,因此,存放在 Redis 中的数据不应该大于内存容量,否则会导致操作系统性能降低。

本节不会介绍太多关于 Redis 数据库本身的知识,主要介绍 Spring Boot 对 Redis 的支持,以及基于 Spring Boot 和 Redis 的实例。

6.7.1 安装 Redis

1. 下载 Redis

编写本书时，Redis 官方网站只提供 Linux 版本的下载。因此，我们只能通过 https://github.com/MSOpenTech/redis/tags 从 GitHub 上下载 Redis，本书下载的版本是 Redis-x64-3.2.100.zip。在运行中输入"cmd"，然后把目录指向解压的 Redis 目录，如图 6.59 所示。

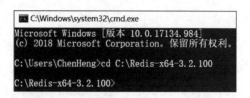

图 6.59 将目录指向解压的 Redis 目录

2. 启动 Redis 服务

使用 redis-server redis.windows.conf 命令行启动 Redis 服务，出现图 6.60 的显示，表示成功启动 Redis 服务。

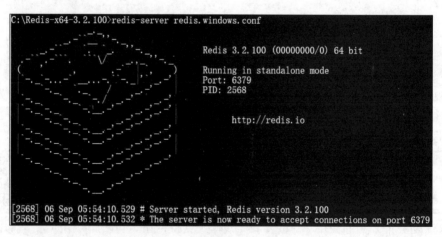

图 6.60 启动 Redis 服务

图 6.60 虽然启动了 Redis 服务，但关闭 cmd 窗口，Redis 服务就消失。所以需要把 Redis 设置成 Windows 下的服务。

关闭 cmd 重新打开 cmd，进入 Redis 解压目录。执行设置服务命令：redis-server --service-install redis.windows-service.conf --loglevel verbose，如图 6.61 所示。

图 6.61 中没有报错，表示成功设置成 Windows 下的服务，刷新服务，会看到 Redis 服务，如图 6.62 所示。

第6章 Spring Boot 的数据访问

图 6.61　将 Redis 服务设置成 Windows 下的服务

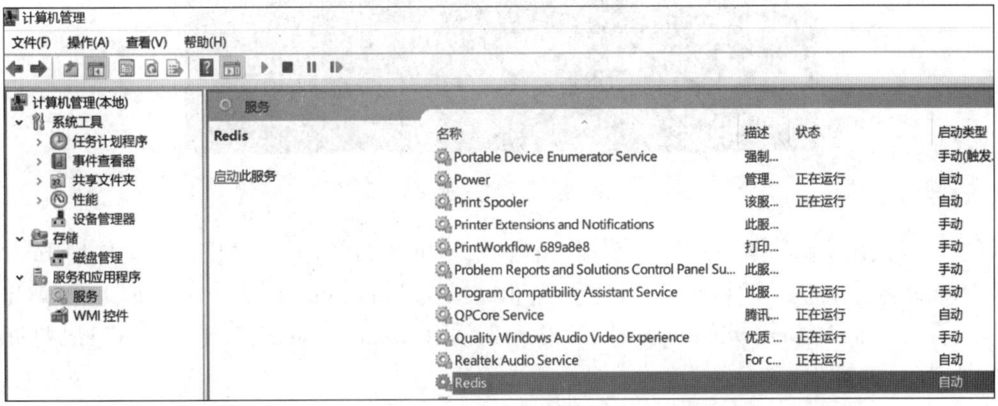

图 6.62　Windows 下的 Redis 服务

3．常用的 Redis 服务命令

卸载服务：redis-server --service-uninstall

开启服务：redis-server --service-start

停止服务：redis-server --service-stop

4．启动 Redis 服务

可以在 cmd 中，使用 redis-server --service-start 命令启动 Redis 服务，如图 6.63 所示。同时也可以在图 6.62 中启动 Redis 服务。

图 6.63　启动 Redis 服务

5. 操作测试 Redis

如图 6.64 所示，启动 Redis 服务后，首先，使用 redis-cli.exe -h 127.0.0.1 -p 6379 命令创建一个地址为 127.0.0.1、端口号为 6379 的 Redis 数据库服务；然后，使用 set key value 和 get key 命令保存和获得数据。

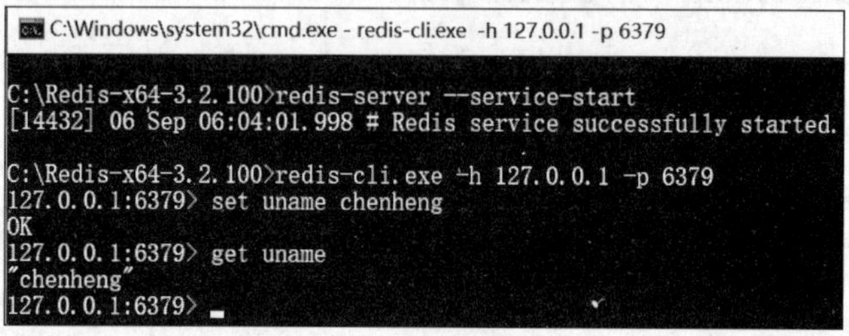

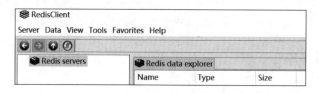

图 6.64 测试 Redis

也可以使用 Redis 客户端查看数据（相当于 get key 命令）。本书提供一个使用 Java 开发的 Redis 客户端 RedisClient，下载地址是 https://github.com/caoxinyu/RedisClient，解压后双击运行 C:\RedisClient-windows\release 目录下的 redisclient-win32.x86_64.2.0.jar，打开如图 6.65 所示的界面。

图 6.65 Redis 客户端 RedisClient 界面

单击图 6.65 的 Server 菜单，添加 Redis 服务，如图 6.66 所示。

选择图 6.66 中的 Add server 菜单项，添加一个地址为 127.0.0.1、端口号为 6379 的 Redis 数据库服务，如图 6.67 所示。

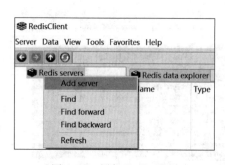

图 6.66 添加 Redis 服务

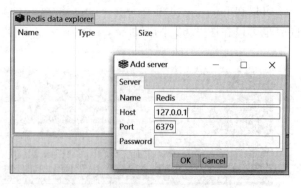

图 6.67 输入 Redis 服务信息界面

单击图 6.67 中的 OK 按钮后，打开如图 6.68 所示的界面。从图 6.68 可以看出 Redis 服务中共有 16 个数据库。其中，db0 是默认的数据库名，也就是说，我们前面存进去的 uname 就在该数据库中。因此，展开 db0 数据库，即可看到 uname 数据，如图 6.69 所示。

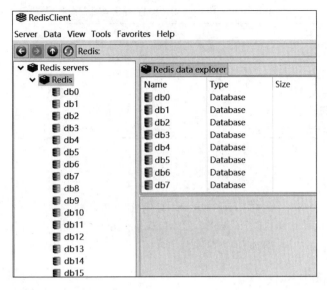

图 6.68 Redis 服务中的数据库

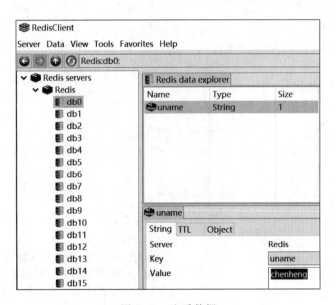

图 6.69 查看数据

6.7.2 Spring Boot 整合 Redis

1. Spring Data Redis

Spring 对 Redis 的支持是通过 Spring Data Redis 来实现的。Spring Data Redis 提供了 RedisTemplate 和 StringRedisTemplate 两个模板来进行数据操作,其中,StringRedisTemplate 只针对键值都是字符串类型的数据进行操作。

RedisTemplate 和 StringRedisTemplate 模板提供的主要数据访问方法如表 6.3 所示。

表 6.3　RedisTemplate 和 StringRedisTemplate 的主要数据访问方法

方　　法	说　　明
opsForValue()	操作只有简单属性的数据
opsForList()	操作含有 List 的数据
opsForSet()	操作含有 Set 的数据
opsForZSet()	操作含有 Zset(有序的 Set)的数据
opsForHash()	操作含有 Hash 的数据

2. Serializer

当数据存储到 Redis 时，键和值都是通过 Spring 提供的 Serializer 序列化到数据的。RedisTemplate 默认使用 JdkSerializationRedisSerializer 序列化，StringRedisTemplate 默认使用 StringRedisSerializer 序列化。

3. Spring Boot 的支持

Spring Boot 对 Redis 的支持位于 org.springframework.boot.autoconfigure.data.redis 包下，如图 6.70 所示。

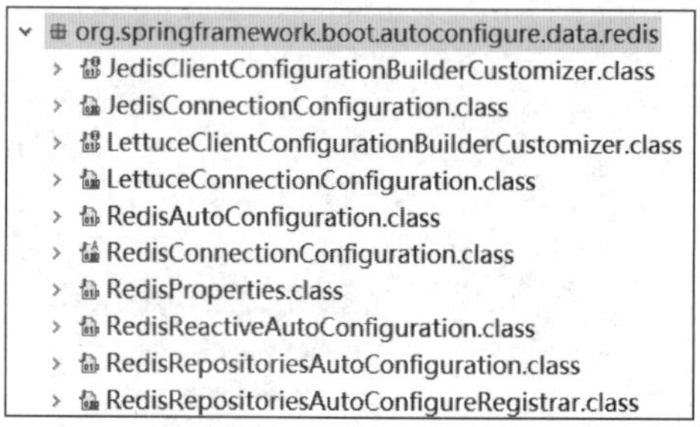

在 RedisAutoConfiguration 配置类中，默认配置了 RedisTemplate 和 StringRedisTemplate，可以直接使用 Redis 存储数据。

在 RedisProperties 类中，可以使用以 spring.redis 为前缀的属性在 application.properties 中配置 Redis，主要属性默认配置如下：

图 6.70　Redis 包

```
spring.redis.database = 0           #数据库名 db0
spring.redis.host = localhost       #服务器地址
spring.redis.port = 6379            #连接端口号
spring.redis.max-idle = 8           #连接池的最大连接数
spring.redis.min-idle = 0           #连接池的最小连接数
spring.redis.max-active = 8         #在给定时间连接池可以分配的最大连接数
spring.redis.max-wait = -1          #当池被耗尽时，抛出异常之前连接分配应阻塞的最大时间量
                                    #(以毫秒为单位)。使用负值表示无限期地阻止
```

从上述默认属性值可以看出，默认配置了数据库名为 db0、服务器地址为 localhost、端口号为 6379 的 Redis。

因此，在 Spring Boot 应用中，只要引入 spring-boot-starter-data-redis 依赖就可以使用默认配置的 Redis 进行数据操作。

6.7.3　使用 StringRedisTemplate 和 RedisTemplate

本节通过实例讲解如何在 Spring Boot 应用中使用 StringRedisTemplate 和 RedisTemplate 模板访问操作 Redis 数据库。

【例 6-12】　在 Spring Boot 应用中使用 StringRedisTemplate 和 RedisTemplate 模板访

问操作 Redis 数据库。

具体实现步骤如下。

1. 创建基于 spring-boot-starter-data-redis 依赖的 Spring Boot Web 应用 ch6_9

创建 Spring Boot Web 应用 ch6_9，该应用需要引入 spring-boot-starter-data-redis 依赖，如图 6.71 所示。

图 6.71　创建 Spring Boot Web 应用 ch6_9

2. 配置 application.properties 文件

在该 Spring Boot 应用 ch6_9 中，使用 Redis 的默认数据库连接。所以，不需要在 application.properties 文件中配置数据库连接信息。

3. 创建实体类

创建名为 com.ch.ch6_9.entity 的包，并在该包中创建名为 Student 的实体类。该类必须实现序列化接口，这是因为使用 Jackson 做序列化需要一个空构造。

Student 的代码如下：

```
package com.ch.ch6_9.entity;
import java.io.Serializable;
public class Student implements Serializable{
    private static final long serialVersionUID = 1L;
    private String sno;
    private String sname;
    private Integer sage;
    public Student() {
```

```java
        super();
    }
    public Student(String sno, String sname, Integer sage) {
        super();
        this.sno = sno;
        this.sname = sname;
        this.sage = sage;
    }
    //省略 get 和 set 方法
}
```

4. 创建数据访问层

创建名为 com.ch.ch6_9.repository 的包,并在该包中创建名为 StudentRepository 的类,该类使用@Repository 注解标注为数据访问层。

StudentRepository 的代码如下:

```java
package com.ch.ch6_9.repository;
import java.util.List;
import javax.annotation.Resource;
import org.springframework.beans.factory.annotation.Autowired;
import org.springframework.data.redis.core.RedisTemplate;
import org.springframework.data.redis.core.StringRedisTemplate;
import org.springframework.data.redis.core.ValueOperations;
import org.springframework.stereotype.Repository;
import com.ch.ch6_9.entity.Student;
@Repository
public class StudentRepository{
    @SuppressWarnings("unused")
    @Autowired
    private StringRedisTemplate stringRedisTemplate;
    @SuppressWarnings("unused")
    @Autowired
    private RedisTemplate<Object, Object> redisTemplate;
    /**
     * 使用@Resource 注解指定 stringRedisTemplate,可注入基于字符串的简单属性操作方法
     * ValueOperations<String, String> valueOpsStr = stringRedisTemplate.opsForValue();
     */
    @Resource(name = "stringRedisTemplate")
    ValueOperations<String, String> valueOpsStr;
    /**
     * 使用@Resource 注解指定 redisTemplate,可注入基于对象的简单属性操作方法
     * ValueOperations<Object, Object> valueOpsObject = redisTemplate.opsForValue();
     */
    @Resource(name = "redisTemplate")
    ValueOperations<Object, Object> valueOpsObject;
    /**
     * 保存字符串到 redis
     */
    public void saveString(String key, String value) {
```

```java
        valueOpsStr.set(key, value);
    }
    /**
     * 保存对象到redis
     */
    public void saveStudent(Student stu) {
        valueOpsObject.set(stu.getSno(), stu);
    }
    /**
     * 保存List数据到redis
     */
    public void saveMultiStudents(Object key, List<Student> stus) {
        valueOpsObject.set(key, stus);
    }
    /**
     * 从redis中获得字符串数据
     */
    public String getString(String key) {
        return valueOpsStr.get(key);
    }
    /**
     * 从redis中获得对象数据
     */
    public Object getObject(Object key) {
        return valueOpsObject.get(key);
    }
}
```

5．创建控制器层

由于本实例业务简单，我们直接在控制器层调用数据访问层。创建名为 com.ch.ch6_9.controller 的包，并在该包中创建控制器类 TestRedisController。

TestRedisController 的代码如下：

```java
package com.ch.ch6_9.controller;
import java.util.ArrayList;
import java.util.List;
import org.springframework.beans.factory.annotation.Autowired;
import org.springframework.web.bind.annotation.RequestMapping;
import org.springframework.web.bind.annotation.RestController;
import com.ch.ch6_9.entity.Student;
import com.ch.ch6_9.repository.StudentRepository;
@RestController
public class TestRedisController {
    @Autowired
    private StudentRepository studentRepository;
    @RequestMapping("/save")
    public void save() {
        studentRepository.saveString("uname", "陈恒");
        Student s1 = new Student("111", "陈恒1", 77);
```

```java
        studentRepository.saveStudent(s1);
        Student s2 = new Student("222","陈恒 2",88);
        Student s3 = new Student("333","陈恒 3",99);
        List<Student> stus = new ArrayList<Student>();
        stus.add(s2);
        stus.add(s3);
        studentRepository.saveMultiStudents("mutilStus",stus);
    }
    @RequestMapping("/getUname")
    public String getUname(String key) {
        return studentRepository.getString(key);
    }
    @RequestMapping("/getStudent")
    public Student getStudent(String key) {
        return(Student)studentRepository.getObject(key);
    }
    @SuppressWarnings("unchecked")
    @RequestMapping("/getMultiStus")
    public List<Student> getMultiStus(String key) {
        return(List<Student>)studentRepository.getObject(key);
    }
}
```

6. 修改配置类 Ch69Application

我们知道 RedisTemplate 默认使用 JdkSerializationRedisSerializer 序列化数据,这对使用 Redis Client 查看数据很不直观,因为 JdkSerializationRedisSerializer 使用二进制形式存储数据。所以,在此我们将自己配置 RedisTemplate,并定义 Serializer。

修改后的配置类 Ch69Application 的代码如下:

```java
package com.ch.ch6_9;
import org.springframework.boot.SpringApplication;
import org.springframework.boot.autoconfigure.SpringBootApplication;
import org.springframework.context.annotation.Bean;
import org.springframework.data.redis.connection.RedisConnectionFactory;
import org.springframework.data.redis.core.RedisTemplate;
import org.springframework.data.redis.serializer.Jackson2JsonRedisSerializer;
import org.springframework.data.redis.serializer.StringRedisSerializer;
import com.fasterxml.jackson.annotation.JsonAutoDetect;
import com.fasterxml.jackson.annotation.PropertyAccessor;
import com.fasterxml.jackson.databind.ObjectMapper;
@SpringBootApplication
public class Ch69Application {
    public static void main(String[] args) {
        SpringApplication.run(Ch69Application.class, args);
    }
    @Bean
    public RedisTemplate<Object, Object> redisTemplate(RedisConnectionFactory
                                                    redisConnectionFactory){
```

```
        RedisTemplate<Object, Object> rTemplate = new RedisTemplate<Object, Object>();
        rTemplate.setConnectionFactory(redisConnectionFactory);
        @SuppressWarnings({ "unchecked", "rawtypes" })
        Jackson2JsonRedisSerializer<Object> jackson2JsonRedisSerializer =
    new Jackson2JsonRedisSerializer(Object.class);
        ObjectMapper om = new ObjectMapper();
        om.setVisibility(PropertyAccessor.ALL, JsonAutoDetect.Visibility.ANY);
        om.enableDefaultTyping(ObjectMapper.DefaultTyping.NON_FINAL);
        jackson2JsonRedisSerializer.setObjectMapper(om);
        //设置值的序列化采用 jackson2JsonRedisSerializer
        rTemplate.setValueSerializer(jackson2JsonRedisSerializer);
        //设置键的序列化采用 StringRedisSerializer
        rTemplate.setKeySerializer(new StringRedisSerializer());
        return rTemplate;
    }
}
```

7. 运行测试

首先，运行 Ch69Application 主类。然后通过"http://localhost:8080/save"测试存储数据。成功运行后，通过 Redis Client 查看数据，如图 6.72 所示。

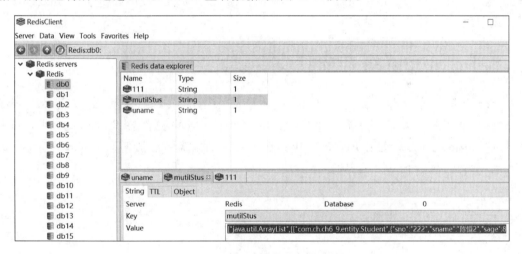

图 6.72 查看保存的数据

通过"http://localhost:8080/getUname?key=uname"查询 key 为 uname 的字符串值，如图 6.73 所示。

通过"http://localhost:8080/getStudent?key=111"查询 key 为 111 的 Student 对象值，如图 6.74 所示。

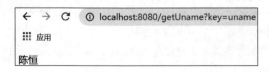

图 6.73 根据 key 查询字符串值　　　　　　图 6.74 根据 key 查询简单对象

通过"http://localhost:8080/getMultiStus?key＝mutilStus"查询 key 为 mutilStus 的 List 集合，如图 6.75 所示。

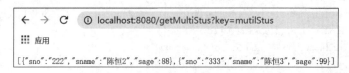

图 6.75　根据 key 查询 List 集合

6.8　数据缓存 Cache

视频讲解

我们知道内存的读取速度远大于硬盘的读取速度。当需要重复地获取相同数据时，一次一次地请求数据库或者远程服务，导致在数据库查询或者远程方法调用上消耗大量的时间，最终导致程序性能降低，这就是数据缓存要解决的问题。

本节将介绍 Spring Boot 应用中 Cache 的一般概念，Spring Cache 对 Cache 进行了抽象，提供了 CacheManager 和 Cache 接口，并提供了 @Cacheable、@CachePut、@CacheEvict、@Caching、@CacheConfig 等注解。Spring Boot 应用基于 Spring Cache，即提供了基于内存实现的缓存管理器用于单体应用系统，也集成了 Redis、EhCache 等缓存服务器用于大型系统或分布式系统。

6.8.1　Spring 缓存支持

Spring 框架定义了 org.springframework.cache.CacheManager 和 org.springframework.cache.Cache 接口来统一不同的缓存技术。针对不同的缓存技术，需要实现不同的 CacheManager。例如，使用 EhCache 作为缓存技术时，需要注册实现 CacheManager 的 Bean。示例代码如下：

```
@Bean
public EhCacheCacheManager cacheManager(CacheManager ehCacheCacheManager) {
    return new EhCacheCacheManager(ehCacheCacheManager);
}
```

CacheManager 的常用实现如表 6.4 所示。

表 6.4　CacheManager 常用实现

CacheManager	描述
SimpleCacheManager	使用简单的 Collection 来存储缓存，主要用于测试
NoOpCacheManager	仅用于测试，不会实际存储缓存
ConcurrentMapCacheManager	使用 ConcurrentMap 存储缓存，Spring 默认采用此技术存储缓存
EhCacheCacheManager	使用 EhCache 作为缓存技术
JCacheCacheManager	支持 JCache(JSR-107)标准实现作为缓存技术，如 Apache Commons JCS
RedisCacheManager	使用 Redis 作为缓存技术
HazelcastCacheManager	使用 Hazelcast 作为缓存技术

一旦配置好 Spring 缓存支持，就可以在 Spring 容器管理的 Bean 中使用缓存注解（基于 AOP 原理），一般情况下，都是在业务层（Service 类）使用这些注解。

1. @Cacheable

@Cacheable 可以标记在一个方法上，也可以标记在一个类上。当标记在一个方法上时表示该方法是支持缓存的；当标记在一个类上时则表示该类所有的方法都是支持缓存的。对于一个支持缓存的方法，在方法执行前，Spring 先检查缓存中是否存在方法返回的数据，如果存在，则直接返回缓存数据；如果不存在，则调用方法并将方法返回值存入缓存。

@Cacheable 注解经常使用 value、key、condition 等属性。

value：缓存的名称，指定一个或多个缓存名称。如 @Cacheable(value="mycache") 或者 @Cacheable(value={"cache1","cache2"})。该属性与 cacheNames 属性意义相同。

key：缓存的 key，可以为空。如果指定，需要按照 SpEL 表达式编写；如果不指定，则默认按照方法的所有参数进行组合。如 @Cacheable(value="testcache", key="#student.id")。

condition：缓存的条件，可以为空。如果指定，需要按照 SpEL 编写，返回 true 或者 false，只有为 true 才进行缓存。如 @Cacheable(value="testcache", condition="#student.id>2")。该属性与 unless 相反，条件成立时，不进行缓存。

2. @CacheEvict

@CacheEvict 是用来标注在需要清除缓存元素的方法或类上的。当标记在一个类上时，表示其中所有方法的执行都会触发缓存的清除操作。@CacheEvict 可以指定的属性有 value、key、condition、allEntries 和 beforeInvocation。其中，value、key 和 condition 的语义与 @Cacheable 对应的属性类似。

allEntries：是否清空所有缓存内容，默认为 false，如果指定为 true，则方法调用后将立即清空所有缓存。如 @CacheEvict(value="testcache", allEntries=true)。

beforeInvocation：是否在方法执行前就清空，默认为 false，如果指定为 true，则在方法还没有执行时就清空缓存。默认情况下，如果方法执行抛出异常，则不会清空缓存。

3. @CachePut

@CachePut 也可以声明一个方法支持缓存功能，与 @Cacheable 不同的是使用 @CachePut 标注的方法在执行前不会去检查缓存中是否存在之前执行过的结果，而是每次都会执行该方法，并将执行结果以键值对的形式存入指定的缓存中。

@CachePut 也可以标注在类上和方法上。@CachePut 的属性与 @Cacheable 的属性一样。

4. @Caching

@Caching 注解可以在一个方法或者类上同时指定多个 Spring Cache 相关的注解。其拥有 3 个属性：cacheable、put 和 evict，分别用于指定 @Cacheable、@CachePut 和 @CacheEvict。示例代码如下：

```
@Caching(
cacheable = @Cacheable("cache1"),
evict = { @CacheEvict("cache2"),@CacheEvict(value = "cache3", allEntries = true) }
)
```

5. @CacheConfig

所有的 Cache 注解都需要提供 Cache 名称,如果每个 Service 方法上都包含相同的 Cache 名称,可能写起来重复。此时,可以使用@CacheConfig 注解作用在类上,设置当前缓存的一些公共设置。

6.8.2　Spring Boot 缓存支持

在 Spring 中使用缓存技术的关键是配置缓存管理器 CacheManager,而 Spring Boot 为我们自动配置了多个 CacheManager 的实现。Spring Boot 的 CacheManager 的自动配置位于 org.springframework.boot.autoconfigure.cache 包中,如图 6.76 所示。

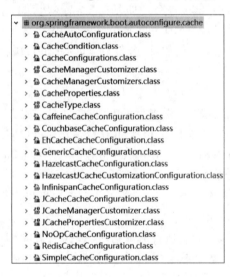

图 6.76　CacheManager 的自动配置

从图 6.76 可以看出,Spring Boot 自动配置了 EhCacheCacheConfiguration、GenericCacheConfiguration、HazelcastCacheConfiguration、HazelcastJCacheCustomizationConfiguration、InfinispanCacheConfiguration、JCacheCacheConfiguration、NoOpCacheConfiguration、RedisCacheConfiguration 和 SimpleCacheConfiguration。默认情况下,Spring Boot 使用的是 SimpleCacheConfiguration,即使用 ConcurrentMapCacheManager。Spring Boot 支持以 spring.cache 为前缀的属性来进行缓存的相关配置。

在 Spring Boot 应用中,使用缓存技术只需在应用中引入相关缓存技术的依赖,并在配置类中使用@EnableCaching 注解开启缓存支持即可。

下面通过实例讲解如何在 Spring Boot 应用中使用默认的缓存技术 ConcurrentMapCacheManager。

第6章 Spring Boot 的数据访问

【例 6-13】 在 Spring Boot 应用中使用默认的缓存技术 ConcurrentMapCacheManager。
具体实现步骤如下。

1. 创建基于 spring-boot-starter-cache 和 spring-boot-starter-data-jpa 依赖的 Spring Boot Web 应用 ch6_10

创建 Spring Boot Web 应用 ch6_10，该应用需要引入 spring-boot-starter-cache 和 spring-boot-starter-data-jpa 依赖，如图 6.77 所示。

图 6.77 创建 Spring Boot Web 应用 ch6_10

2. 配置 application.properties 文件

在该 Spring Boot 应用 ch6_10 中，使用 Spring Data JPA 访问 MySQL 数据库。所以，在 application.properties 文件中配置数据库连接信息，但因为使用默认的缓存技术 ConcurrentMapCacheManager，所以不需要缓存的相关配置。配置信息如下：

```
server.servlet.context-path=/ch6_10
###
## 数据源信息配置
###
# 数据库地址
spring.datasource.url=jdbc:mysql://localhost:3306/springbootjpa?characterEncoding=utf8
# 数据库用户名
spring.datasource.username=root
# 数据库密码
spring.datasource.password=root
# 数据库驱动
spring.datasource.driver-class-name=com.mysql.jdbc.Driver
```

```
####
# JPA 持久化配置
####
# 指定数据库类型
spring.jpa.database = MYSQL
# 指定是否在日志中显示 SQL 语句
spring.jpa.show-sql = true
# 指定自动创建、更新数据库表等配置，update 表示如果数据库中存在持久化类对应的表就不创建，
不存在就创建对应的表
spring.jpa.hibernate.ddl-auto = update
# 让控制器输出的 JSON 字符串格式更美观
spring.jackson.serialization.indent-output = true
```

3. 修改 pom.xml 文件，添加 MySQL 连接依赖

因为在该应用中，访问的数据库是 MySQL，所以需要将 MySQL 连接器依赖添加到 pom.xml 文件。具体代码如下：

```
<dependency>
    <groupId>mysql</groupId>
    <artifactId>mysql-connector-java</artifactId>
    <version>5.1.45</version>
</dependency>
```

4. 创建持久化实体类

创建名为 com.ch.ch6_10.entity 的包，并在该包中创建持久化实体类 Student。Student 的代码如下：

```
package com.ch.ch6_10.entity;
import java.io.Serializable;
import javax.persistence.Entity;
import javax.persistence.GeneratedValue;
import javax.persistence.GenerationType;
import javax.persistence.Id;
import javax.persistence.Table;
import com.fasterxml.jackson.annotation.JsonIgnoreProperties;
@Entity
@Table(name = "student_table")
@JsonIgnoreProperties(value = {"hibernateLazyInitializer"})
public class Student implements Serializable{
    private static final long serialVersionUID = 1L;
    @Id
    @GeneratedValue(strategy = GenerationType.IDENTITY)
    private int id;                                    //主键
    private String sno;
    private String sname;
    private String ssex;
    public Student() {
        super();
    }
```

```
    public Student(int id, String sno, String sname, String ssex) {
        super();
        this.id = id;
        this.sno = sno;
        this.sname = sname;
        this.ssex = ssex;
    }
    //省略 set 和 get 方法
}
```

5. 创建数据访问接口

创建名为 com.ch.ch6_10.repository 的包,并在该包中创建名为 StudentRepository 的数据访问接口。

StudentRepository 的代码如下:

```
package com.ch.ch6_10.repository;
import org.springframework.data.jpa.repository.JpaRepository;
import com.ch.ch6_10.entity.Student;
public interface StudentRepository extends JpaRepository<Student, Integer>{

}
```

6. 创建业务层

创建名为 com.ch.ch6_10.service 的包,并在该包中创建 StudentService 接口和该接口的 StudentServiceImpl 实现类。

StudentService 的代码如下:

```
package com.ch.ch6_10.service;
import com.ch.ch6_10.entity.Student;
public interface StudentService {
    public Student saveStudent(Student student);
    public void deleteCache(Student student);
    public Student selectOneStudent(Integer id);
}
```

StudentServiceImpl 的代码如下:

```
package com.ch.ch6_10.service;
import org.springframework.beans.factory.annotation.Autowired;
import org.springframework.cache.annotation.CacheEvict;
import org.springframework.cache.annotation.CachePut;
import org.springframework.cache.annotation.Cacheable;
import org.springframework.stereotype.Service;
import com.ch.ch6_10.entity.Student;
import com.ch.ch6_10.repository.StudentRepository;
@Service
public class StudentServiceImpl implements StudentService{
    @Autowired
    private StudentRepository studentRepository;
```

```java
@Override
@CachePut(value = "student", key = "#student.id")
public Student saveStudent(Student student) {
    Student s = studentRepository.save(student);
    System.out.println("为key=" + student.getId() + "数据做了缓存");
    return s;
}
@Override
@CacheEvict(value = "student", key = "#student.id")
public void deleteCache(Student student) {
    System.out.println("删除了key=" + student.getId() + "的数据缓存");
}
@Override
@Cacheable(value = "student")
public Student selectOneStudent(Integer id) {
    Student s = studentRepository.getOne(id);
    System.out.println("为key=" + id + "数据做了缓存");
    return s;
}
}
```

在上述 Service 的实现类中，使用@CachePut 注解将新增的或更新的数据保存到缓存，其中缓存名为 student，数据的 key 是 student 的 id；使用@CacheEvict 注解从缓存 student 中删除 key 为 student 的 id 的数据；使用@Cacheable 注解将 key 为 student 的 id 的数据缓存到名为 student 的缓存中。

7. 创建控制器层

创建名为 com.ch.ch6_10.controller 的包，并在该包中创建名为 TestCacheController 的控制器类。

TestCacheController 的代码如下：

```java
package com.ch.ch6_10.controller;
import org.springframework.beans.factory.annotation.Autowired;
import org.springframework.web.bind.annotation.RequestMapping;
import org.springframework.web.bind.annotation.RestController;
import com.ch.ch6_10.entity.Student;
import com.ch.ch6_10.service.StudentService;
@RestController
public class TestCacheController {
    @Autowired
    private StudentService studentService;
    @RequestMapping("/savePut")
    public Student save(Student student) {
        return studentService.saveStudent(student);
    }
    @RequestMapping("/selectAble")
    public Student select(Integer id) {
        return studentService.selectOneStudent(id);
    }
}
```

```
    @RequestMapping("/deleteEvict")
    public String deleteCache(Student student) {
        studentService.deleteCache(student);
        return "ok";
    }
}
```

8. 开启缓存支持

在应用的配置类 Ch610Application 中,使用@EnableCaching 注解开启缓存支持。具体代码如下:

```
package com.ch.ch6_10;
import org.springframework.boot.SpringApplication;
import org.springframework.boot.autoconfigure.SpringBootApplication;
import org.springframework.cache.annotation.EnableCaching;
@EnableCaching
@SpringBootApplication
public class Ch610Application {
    public static void main(String[] args) {
        SpringApplication.run(Ch610Application.class, args);
    }
}
```

9. 运行测试

1) 测试@Cacheable

启动应用程序的主类后,第一次访问 http://localhost:8080/ch6_10/selectAble?id=4,将调用方法查询数据库,并将查询到的数据存储到缓存 student 中。此时控制台输出如图 6.78 所示,页面显示数据如图 6.79 所示。

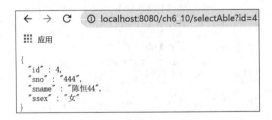

图 6.78 第一次访问查询控制台输出结果　　图 6.79 第一次访问查询页面数据

再次访问"http://localhost:8080/ch6_10/selectAble?id=4",此时控制台没有输出"为 key=4 数据做了缓存"以及 Hibernate 的查询语句,这表明没有调用查询方法,页面数据直接从数据缓存中获得。

2) 测试@CachePut

重启应用程序的主类,访问"http://localhost:8080/ch6_10/savePut?sname=陈恒5&sno=555&ssex=男",此时控制台输出结果如图 6.80 所示,页面数据如图 6.81 所示。

这时访问"http://localhost:8080/ch6_10/selectAble?id=6",控制台无输出,从缓存直接获得数据,页面数据如图 6.81 所示。

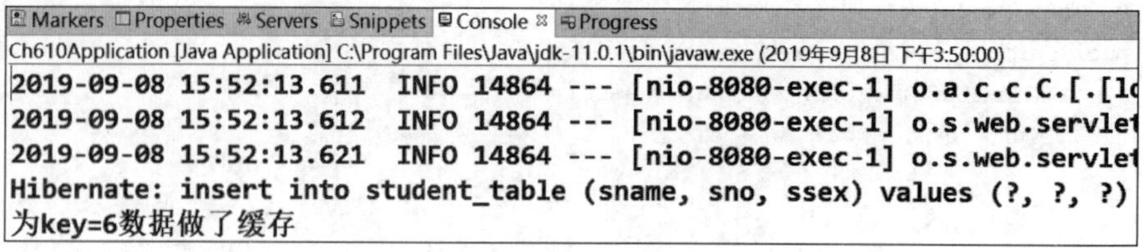

图 6.80　测试@CachePut 控制台输出结果

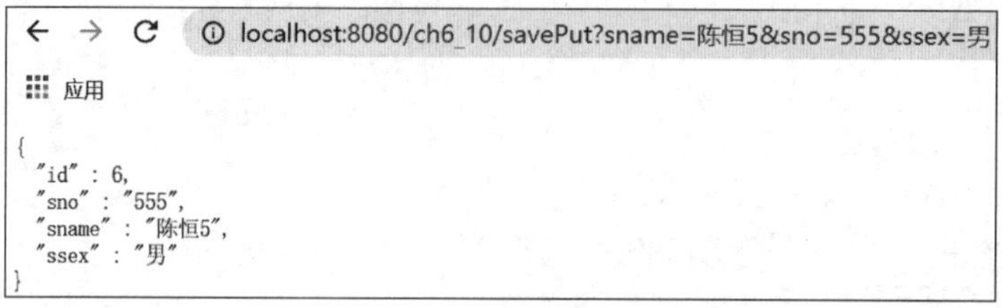

图 6.81　测试@CachePut 页面数据

3）测试@CacheEvict

重启应用程序的主类，首先，访问"http://localhost:8080/ch6_10/selectAble?id＝1"，为 key 为 1 的数据做缓存，再次访问"http://localhost:8080/ch6_10/selectAble?id＝1"，确认数据已从缓存中获取。然后，访问"http://localhost:8080/ch6_10/deleteEvict?id＝1"，从缓存 student 中删除 key 为 1 的数据，此时控制台输出结果如图 6.82 所示。

最后，再次访问"http://localhost:8080/ch6_10/selectAble?id＝1"，此时重新做了缓存，控制台输出结果如图 6.83 所示。

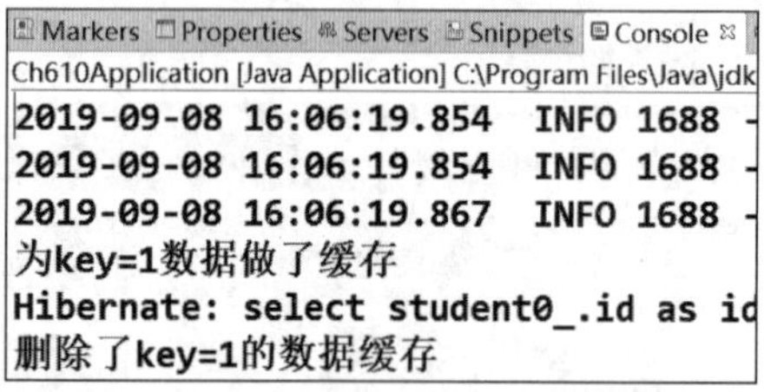

图 6.82　测试@CacheEvict 删除缓存数据 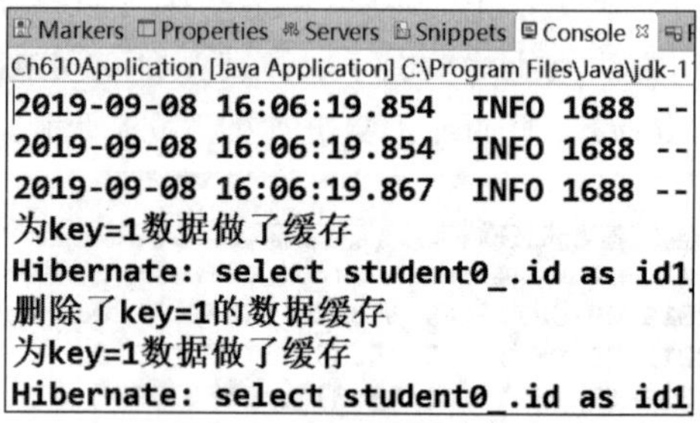　　图 6.83　测试@CacheEvict 重做缓存数据

6.8.3　使用 Redis Cache

在 Spring Boot 中使用 Redis Cache，只需添加 spring-boot-starter-data-redis 依赖即可。下面通过实例测试 Redis Cache。

【例 6-14】　在 6.7.3 节的例 6-12 的基础上，测试 Redis Cache。

具体实现步骤如下。

1. 使用@Cacheable 注解修改控制器方法

将控制器中的方法 getUname 修改如下：

```
@RequestMapping("/getUname")
@Cacheable(value = "myuname")
public String getUname(String key) {
    System.out.println("测试缓存");
    return personRepository.getString(key);
}
```

2. 使用@EnableCaching 注解开启缓存支持

在应用程序的主类 Ch69Application 上使用@EnableCaching 注解开启缓存支持。

3. 测试 Redis Cache

启动应用程序的主类后,多次访问"http://localhost:8080/getUname?key=uname",但"测试缓存"字样在控制台仅打印一次,页面查询结果不变。这说明只有第一次访问时调用了查询方法,后面多次访问都是从缓存直接获得数据。

6.9 本章小结

本章是本书的重点章节,重点讲解了 Spring Data JPA、Spring Boot 整合 MyBatis、Spring Data REST、Spring Boot 整合 MongoDB、Spring Boot 整合 Redis 以及数据缓存 Cache。通过本章的学习,读者不仅掌握了 Spring Boot 访问关系型数据库的解决方案,同时还掌握了 Spring Boot 访问非关系型数据库的解决方案。

习题 6

1. 在 Spring Boot 应用中,数据缓存技术解决了什么问题?
2. 什么是 RESTful 架构?什么是 REST 服务?
3. Spring 框架提供了哪些缓存注解?这些注解如何使用?
4. 在 Spring Data JPA 中,如何实现一对一、一对多、多对多关联查询?请举例说明。

第 7 章

Spring Boot 的安全控制

学习目的与要求

本章重点讲解 Spring Security 安全控制。通过本章的学习，掌握如何使用 Spring Security 安全控制机制解决企业应用程序的安全问题。

本章主要内容

- Spring Security 快速入门。
- Spring Boot Security 操作实例。

在 Web 应用开发中，安全毋庸置疑是十分重要的，选择 Spring Security 来保护 Web 应用是一个非常好的选择。Spring Security 是 Spring 框架的一个安全模块，可以非常方便地与 Spring Boot 应用无缝集成。

7.1　Spring Security 快速入门

7.1.1　什么是 Spring Security

视频讲解

Spring Security 是一个专门针对 Spring 应用系统的安全框架，充分利用了 Spring 框架的依赖注入和 AOP 功能，为 Spring 应用系统提供安全访问控制解决方案。

在 Spring Security 安全框架中，有两个重要概念，即授权（Authorization）

和认证(Authentication)。授权即确定用户在当前应用系统下所拥有的功能权限；认证即确认用户访问当前系统的身份。

7.1.2 Spring Security 的适配器

Spring Security 为 Web 应用提供了一个适配器类 WebSecurityConfigurerAdapter，该类实现了 WebSecurityConfigurer＜WebSecurity＞接口，并提供了两个 configure 方法用于认证和授权操作。

开发者创建自己的 Spring Security 适配器类是非常简单的，只需定义一个继承 WebSecurityConfigurerAdapter 的类，并在该类中使用@Configuration 注解，就可以通过重写两个 configure 方法来配置所需要的安全配置。自定义适配器类的示例代码如下：

```
@Configuration
public class MySecurityConfigurerAdapter extends WebSecurityConfigurerAdapter{
/**
    * 用户认证
    */
    @Override
    protected void configure(AuthenticationManagerBuilder auth) throws Exception {
    }
/**
    * 请求授权
    */
    @Override
    protected void configure(HttpSecurity http) throws Exception {
    }
}
```

7.1.3 Spring Security 的用户认证

在 Spring Security 的适配器类中，通过重写 configure(AuthenticationManagerBuilder auth)方法完成用户认证。

1. 内存中的用户认证

使用 AuthenticationManagerBuilder 的 inMemoryAuthentication()方法可以添加在内存中的用户，并给用户指定角色权限。示例代码如下：

```
@Override
protected void configure(AuthenticationManagerBuilder auth) throws Exception {
    auth.inMemoryAuthentication().withUser("chenheng").password("123456").roles("ADMIN","DBA");
    auth.inMemoryAuthentication().withUser("zhangsan").password("123456").roles("USER");
}
```

上述示例代码中添加了两个用户，一个用户的用户名为"chenheng"，密码为"123456"，用户权限为"ROLE_ADMIN"和"ROLE_DBA"；另一个用户的用户名为"zhangsan"，密码为"123456"，用户权限为"ROLE_USER"。"ROLE_"是 Spring Security 保存用户权限时，默认加上。

2. 通用的用户认证

在实际应用中，可以查询数据库获取用户和权限，这时需要自定义实现 org.springframework.security.core.userdetails.UserDetailsService 接口的类，并重写 public UserDetails loadUserByUsername(String username)方法查询对应的用户和权限。示例代码如下：

```
@Service
public class MyUserSecurityService implements UserDetailsService{
    @Autowired
    private MyUserRepository myUserRepository;
    /**
     * 通过重写 loadUserByUsername 方法查询对应的用户
     * UserDetails 是 Spring Security 的一个核心接口
     * UserDetails 定义了可以获取用户名、密码、权限等与认证相关的信息的方法
     */
    @Override
    public UserDetails loadUserByUsername(String username) throws UsernameNotFoundException {
        //根据用户名(页面接收的用户名)查询当前用户
        MyUser myUser = myUserRepository.findByUsername(username);
        if(myUser == null) {
            throw new UsernameNotFoundException("用户名不存在");
        }
        //GrantedAuthority 代表赋予当前用户的权限(认证权限)
        List<GrantedAuthority> authorities = new ArrayList<GrantedAuthority>();
        //获得当前用户权限集合
        List<Authority> roles = myUser.getAuthorityList();
        //将当前用户的权限保存为用户的认证权限
        for (Authority authority : roles) {
            GrantedAuthority sg = new SimpleGrantedAuthority(authority.getName());
            authorities.add(sg);
        }
        //org.springframework.security.core.userdetails.User 是 Spring Security 内部的实现，
        //专门用于保存用户名、密码、权限等与认证相关的信息
        User su = new User(myUser.getUsername(), myUser.getPassword(), authorities);
        return su;
    }
}
```

除此之外，还需要注册 MyUserSecurityService 完成用户认证。示例代码如下：

```
@Configuration
public class MySecurityConfigurerAdapter extends WebSecurityConfigurerAdapter{
    //依赖注入通用的用户服务类
    @Autowired
    private MyUserSecurityService myUserSecurityService;
    //依赖注入用户认证接口
    @Autowired
```

```
    private AuthenticationProvider authenticationProvider;
    /**
     * DaoAuthenticationProvider 是 AuthenticationProvider 的实现
     */
    @Bean
    public AuthenticationProvider authenticationProvider() {
        DaoAuthenticationProvider provide = new DaoAuthenticationProvider();
        //设置自定义认证方式,用户登录认证
        provide.setUserDetailsService(myUserSecurityService);
        return provide;
    }
    /**
     * 用户认证
     */
    @Override
    protected void configure(AuthenticationManagerBuilder auth) throws Exception {
        //设置认证方式
        auth.authenticationProvider(authenticationProvider);
    }
}
```

7.1.4　Spring Security 的请求授权

在 Spring Security 的适配器类中,通过重写 configure(HttpSecurity http)方法完成用户授权。在 configure(HttpSecurity http)方法中,使用 HttpSecurity 的 authorizeRequests()方法的子节点给指定用户授权访问 URL 模式。可以通过 antMatchers 方法使用 Ant 风格匹配 URL 路径。匹配请求路径后,可以针对当前用户对请求进行安全处理。Spring Security 提供了许多安全处理方法,具体如表 7.1 所示。

表 7.1　Spring Security 的安全处理方法

方　　法	用　　途
anyRequest()	匹配所有请求路径
access(String attribute)	Spring EL 表达式结果为 true 时可以访问
anonymous()	匿名可以访问
authenticated()	用户登录后可访问
denyAll()	用户不能访问
fullyAuthenticated()	用户完全认证可以访问(非 remember-me 下自动登录)
hasAnyAuthority(String...)	参数表示权限,用户权限与其中任一权限相同就可以访问
hasAnyRole(String...)	参数表示角色,用户角色与其中任一角色相同就可以访问
hasAuthority(String authority)	参数表示权限,用户权限与参数相同才可以访问
hasIpAddress(String ipaddressExpression)	参数表示 IP 地址,用户 IP 和参数匹配才可以访问
hasRole(String role)	参数表示角色,用户角色与参数相同才可以访问
permitAll()	任何用户都可以访问
rememberMe()	允许通过 remember-me 登录的用户访问

Spring Security 的请求授权示例代码如下：

```
@Override
protected void configure(HttpSecurity http) throws Exception {
    http.authorizeRequests()
        //首页、登录、注册页面、登录注册功能以及静态资源过滤掉,即可任意访问
        .antMatchers("/toLogin", "/toRegister", "/", "/login", "/register", "/css/**", "/fonts/**", "/js/**").permitAll()
        //这里默认追加 ROLE_,/user/** 是控制器的请求匹配路径
        .antMatchers("/user/**").hasRole("USER")
        .antMatchers("/admin/**").hasAnyRole("ADMIN", "DBA")
        //其他所有请求登录后才能访问
        .anyRequest().authenticated()
        .and()
        //将输入的用户名与密码和授权的进行比较
        .formLogin()
            .loginPage("/login").successHandler(myAuthenticationSuccessHandler)
            .usernameParameter("username").passwordParameter("password")
            //登录失败
            .failureUrl("/login?error")
        .and()
        //注销行为可任意访问
        .logout().permitAll()
        .and()
        //指定异常处理页面
        .exceptionHandling().accessDeniedPage("/deniedAccess");
}
```

上述示例代码解释如下。

http.authorizeRequests()：开始进行请求权限设置。

antMatchers("/toLogin", "/toRegister", "/", "/login", "/register", "/css/**", "/fonts/**", "/js/**").permitAll()："/toLogin""/toRegister""/""/login""/register"等请求以及静态资源过滤掉,即任意用户可访问。

antMatchers("/user/**").hasRole("USER")：请求匹配"/user/**",拥有"ROLE_USER"角色的用户可访问。

antMatchers("/admin/**").hasAnyRole("ADMIN", "DBA")：请求匹配"/admin/**",拥有"ROLE_ADMIN"或"ROLE_DBA"角色的用户可访问。

anyRequest().authenticated()：其余所有请求都需要认证(用户登录后)才可访问。

formLogin()：开始设置登录操作。

loginPage("/login").successHandler(myAuthenticationSuccessHandler)：设置登录的访问地址以及登录成功处理操作。

usernameParameter("username").passwordParameter("password")：登录时接收参数 username 的值作为用户名,接收参数 password 的值作为密码。

failureUrl("/login?error")：指定登录失败后转向的页面和传递的参数。

logout().permitAll()：注销操作,所有用户均可访问。

exceptionHandling().accessDeniedPage("/deniedAccess")：指定异常处理页面。

7.1.5　Spring Security 的核心类

Spring Security 的核心类包括 Authentication、SecurityContextHolder、UserDetails、UserDetailsService、GrantedAuthority、DaoAuthenticationProvider 和 PasswordEncoder。

1. Authentication

Authentication 用来封装用户认证信息的接口，在用户登录认证之前，Spring Security 将相关信息封装为一个 Authentication 具体实现类的对象，在登录认证成功后将生成一个信息更全面、包含用户权限等信息的 Authentication 对象，然后将该对象保存在 SecurityContextHolder 所持有的 SecurityContext 中，方便后续程序进行调用，如当前用户名、访问权限等。

2. SecurityContextHolder

SecurityContextHolder 顾名思义是用来持有 SecurityContext 的类。SecurityContext 中包含当前认证用户的详细信息。Spring Security 使用一个 Authentication 对象描述当前用户的相关信息。例如，最常见的是获得当前登录用户的用户名和权限。示例代码如下：

```
/**
 * 获得当前用户名称
 */
private String getUname() {
    return SecurityContextHolder.getContext().getAuthentication().getName();
}
/**
 * 获得当前用户权限
 */
private String getAuthorities() {
    Authentication authentication = SecurityContextHolder.getContext().getAuthentication();
    List<String> roles = new ArrayList<String>();
    for (GrantedAuthority ga : authentication.getAuthorities()) {
        roles.add(ga.getAuthority());
    }
    return roles.toString();
}
```

3. UserDetails

UserDetails 是 Spring Security 的一个核心接口。该接口定义了一些可以获取用户名、密码、权限等与认证相关的信息的方法。通常需要在应用中获取当前用户的其他信息，如 E-mail、电话等。这时只包含认证相关的 UserDetails 对象可能就不能满足我们的需要了。我们可以实现自己的 UserDetails，在该实现类中定义一些获取用户其他信息的方法，这样就可以直接从当前 SecurityContext 的 Authentication 的 principal 中获取用户的其他信息。Authentication.getPrincipal() 的返回类型是 Object，但通常返回的其实是一个

UserDetails 的实例,通过强制类型转换可以将 Object 转换为 UserDetails 类型。

4. UserDetailsService

UserDetails 是通过 UserDetailsService 的 loadUserByUsername(String username)方法加载的。UserDetailsService 也是一个接口,也需要实现自己的 UserDetailsService 来加载自定义的 UserDetails 信息。

登录认证时,Spring Security 将通过 UserDetailsService 的 loadUserByUsername (String username)方法获取对应的 UserDetails 进行认证,认证通过后将该 UserDetails 赋给认证通过的 Authentication 的 principal,然后再将该 Authentication 保存在 SecurityContext 中。在应用中,如果需要使用用户信息,可以通过 SecurityContextHolder 获取存放在 SecurityContext 中的 Authentication 的 principal,即 UserDetails 实例。

5. GrantedAuthority

Authentication 的 getAuthorities()方法可以返回当前 Authentication 对象拥有的权限(一个 GrantedAuthority 类型的数组),即当前用户拥有的权限。GrantedAuthority 是一个接口,通常是通过 UserDetailsService 进行加载,然后赋给 UserDetails。

6. DaoAuthenticationProvider

在 Spring Security 安全框架中,默认使用 DaoAuthenticationProvider 实现 AuthenticationProvider 接口进行用户认证的处理。DaoAuthenticationProvider 进行认证时,需要一个 UserDetailsService 来获取用户信息 UserDetails。当然我们可以实现自己的 AuthenticationProvider,进而改变认证方式。

7. PasswordEncoder

在 Spring Security 安全框架中,通过 PasswordEncoder 接口完成对密码的加密。Spring Security 对 PasswordEncoder 有多种实现,包括 MD5 加密、SHA-256 加密等,开发者只需直接使用即可。在 Spring Boot 应用中,使用 BCryptPasswordEncoder 加密是较好的选择。BCryptPasswordEncoder 使用 BCrypt 的强散列哈希加密实现,并可以由客户端指定加密强度,强度越高安全性越高。

7.1.6 Spring Security 的验证机制

Spring Security 的验证机制是由许多 Filter 实现的,Filter 将在 Spring MVC 前拦截请求,主要包括注销 Filter(LogoutFilter)、用户名密码验证 Filter(UsernamePasswordAuthenticationFilter)等内容。Filter 再交由其他组件完成细分的功能,最常用的 UsernamePasswordAuthenticationFilter 会持有一个 AuthenticationManager 引用,AuthenticationManager 是一个验证管理器,专门负责验证。AuthenticationManager 持有一个 AuthenticationProvider 集合,AuthenticationProvider 是做验证工作的组件,验证成功或失败之后调用对应的 Hanlder(处理)。

7.2　Spring Boot 的支持

在 Spring Boot 应用中，只需引入 spring-boot-starter-security 依赖即可使用 Spring Security 安全框架，这是因为 Spring Boot 对 Spring Security 提供了自动配置功能。从 org.springframework.boot.autoconfigure.security.SecurityProperties 类中，可以看到使用以 spring.security 为前缀的属性配置了 Spring Security 的相关默认配置。因此，在实际应用开发中只需自定义一个类继承 WebSecurityConfigurerAdapter，无须使用 @EnableWebSecurity 注解，即可自己扩展 Spring Security 的相关配置。

7.3　实际开发中的 Spring Security 操作实例

本节将讲解两个 Spring Boot Security 操作实例，一个是基于 Spring Data JPA 的 Spring Boot Security 操作实例，一个是基于 MyBatis 的 Spring Boot Security 操作实例。

7.3.1　基于 Spring Data JPA 的 Spring Boot Security 操作实例

下面通过一个简单实例演示在 Spring Boot 应用中如何使用基于 Spring Data JPA 的 Spring Security 安全框架。

【例 7-1】　在 Spring Boot 应用中，使用基于 Spring Data JPA 的 Spring Security 安全框架。

具体实现步骤如下。

1. 创建 Spring Boot Web 应用 ch7_1

创建基于 Spring Data JPA、Thymeleaf 及 Spring Security 的 Web 应用 ch7_1，如图 7.1 所示。

2. 修改 pom.xml 文件，添加 MySQL 依赖

在 pom.xml 文件中添加以下依赖：

```xml
<dependency>
    <groupId>mysql</groupId>
    <artifactId>mysql-connector-java</artifactId>
    <version>5.1.45</version>
</dependency>
```

3. 设置 Web 应用 ch7_1 的上下文路径及数据源配置信息

在 ch7_1 应用的 application.properties 文件中配置以下内容：

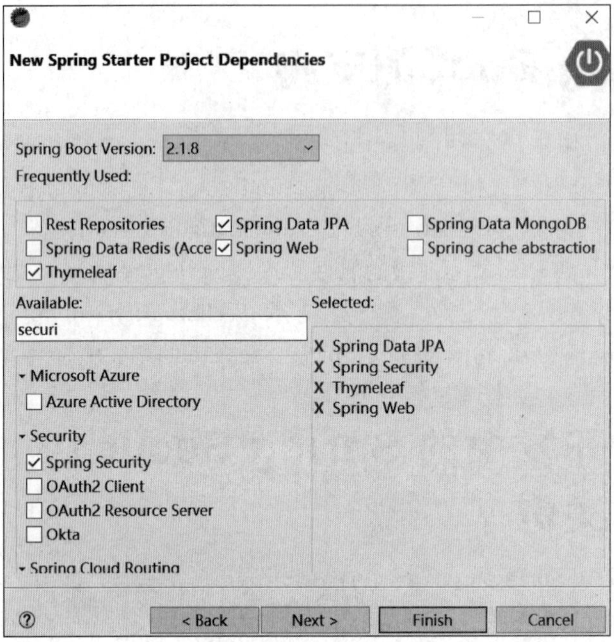

图 7.1 创建 Web 应用 ch7_1

```
server.servlet.context-path=/ch7_1
###
##数据源信息配置
###
#数据库地址
spring.datasource.url=jdbc:mysql://localhost:3306/springbootjpa?characterEncoding=utf8
#数据库用户名
spring.datasource.username=root
#数据库密码
spring.datasource.password=root
#数据库驱动
spring.datasource.driver-class-name=com.mysql.jdbc.Driver
####
#JPA持久化配置
####
#指定数据库类型
spring.jpa.database=MYSQL
#指定是否在日志中显示SQL语句
spring.jpa.show-sql=true
#指定自动创建、更新数据库表等配置,update表示如果数据库中存在持久化类对应的表就不创建,
#不存在就创建对应的表
spring.jpa.hibernate.ddl-auto=update
#让控制器输出的JSON字符串格式更美观
spring.jackson.serialization.indent-output=true
spring.thymeleaf.cache=false
logging.level.org.springframework.security=trace
```

第7章 Spring Boot 的安全控制

4. 整理脚本样式静态文件

JS 脚本、CSS 样式、图片等静态文件默认放置在 src/main/resources/static 目录下，ch7_1 应用引入的 BootStrap 和 jQuery 与例 5-5 中的一样，不再赘述。

5. 创建用户和权限持久化实体类

创建名为 com.ch.ch7_1.entity 的包，并在该包中创建持久化实体类 MyUser 和 Authority。MyUser 类用来保存用户数据，用户名唯一。Authority 类用来保存权限信息。用户和权限是多对多的关系。

MyUser 的代码如下：

```
package com.ch.ch7_1.entity;
import java.io.Serializable;
import java.util.List;
import javax.persistence.CascadeType;
import javax.persistence.Entity;
import javax.persistence.FetchType;
import javax.persistence.GeneratedValue;
import javax.persistence.GenerationType;
import javax.persistence.Id;
import javax.persistence.JoinColumn;
import javax.persistence.JoinTable;
import javax.persistence.ManyToMany;
import javax.persistence.Table;
import javax.persistence.Transient;
import com.fasterxml.jackson.annotation.JsonIgnoreProperties;
@Entity
@Table(name = "user")
@JsonIgnoreProperties(value = { "hibernateLazyInitializer"})
public class MyUser implements Serializable{
    private static final long serialVersionUID = 1L;
    @Id
    @GeneratedValue(strategy = GenerationType.IDENTITY)
    private int id;
    private String username;
    private String password;
    //这里不能是懒加载 lazy,否则在 MyUserSecurityService 的 loadUserByUsername 方法中无法获
    //得权限
    @ManyToMany(cascade = {CascadeType.REFRESH}, fetch = FetchType.EAGER)
    @JoinTable(name = "user_authority",joinColumns = @JoinColumn(name = "user_id"),
        inverseJoinColumns = @JoinColumn(name = "authority_id"))
    private List<Authority> authorityList;
    //repassword 不映射到数据表
    @Transient
    private String repassword;
    //省略 set 和 get 方法
}
```

需要注意的是，在实际开发中 MyUser 还可以实现 org.springframework.security.core.userdetails.UserDetails 接口，实现该接口后即可成为 Spring Security 所使用的用户。本例为了区分 Spring Data JPA 的 pojo 和 Spring Security 的用户对象，并没有实现 UserDetails 接口，而是在实现 UserDetailsService 接口的类中进行绑定。

Authority 的代码如下：

```java
package com.ch.ch7_1.entity;
import java.io.Serializable;
import java.util.List;
import javax.persistence.Column;
import javax.persistence.Entity;
import javax.persistence.GeneratedValue;
import javax.persistence.GenerationType;
import javax.persistence.Id;
import javax.persistence.ManyToMany;
import javax.persistence.Table;
import com.fasterxml.jackson.annotation.JsonIgnore;
import com.fasterxml.jackson.annotation.JsonIgnoreProperties;
@Entity
@Table(name = "authority")
@JsonIgnoreProperties(value = { "hibernateLazyInitializer"})
public class Authority implements Serializable{
    private static final long serialVersionUID = 1L;
    @Id
    @GeneratedValue(strategy = GenerationType.IDENTITY)
    private int id;
    @Column(nullable = false)
    private String name;
    @ManyToMany(mappedBy = "authorityList")
    @JsonIgnore
    private List<MyUser> userList;
    //省略 set 和 get 方法
}
```

6. 创建数据访问层接口

创建名为 com.ch.ch7_1.repository 的包，并在该包中创建名为 MyUserRepository 的接口，该接口继承了 JpaRepository。具体代码如下：

```java
package com.ch.ch7_1.repository;
import org.springframework.data.jpa.repository.JpaRepository;
import com.ch.ch7_1.entity.MyUser;
public interface MyUserRepository extends JpaRepository<MyUser, Integer>{
    //根据用户名查询用户,方法名命名符合 Spring Data JPA 规范
    MyUser findByUsername(String username);
}
```

在 com.ch.ch7_1.repository 包中，创建名为 AuthorityRepository 的接口，该接口继承了 JpaRepository,具体代码如下：

```
package com.ch.ch7_1.repository;
import org.springframework.data.jpa.repository.JpaRepository;
import com.ch.ch7_1.entity.Authority;
public interface AuthorityRepository extends JpaRepository<Authority, Integer>{
}
```

7. 创建业务层

创建名为 com.ch.ch7_1.service 的包，并在该包中创建 UserService 接口和 UserServiceImpl 实现类。

UserService 的代码如下：

```
package com.ch.ch7_1.service;
import javax.servlet.http.HttpServletRequest;
import javax.servlet.http.HttpServletResponse;
import org.springframework.ui.Model;
import com.ch.ch7_1.entity.MyUser;
public interface UserService {
    public String register(MyUser userDomain);
    public String loginSuccess(Model model);
    public String main(Model model);
    public String deniedAccess(Model model);
    public String logout(HttpServletRequest request, HttpServletResponse response);
}
```

UserServiceImpl 的代码如下：

```
package com.ch.ch7_1.service;
import java.util.ArrayList;
import java.util.List;
import javax.servlet.http.HttpServletRequest;
import javax.servlet.http.HttpServletResponse;
import org.springframework.beans.factory.annotation.Autowired;
import org.springframework.security.core.Authentication;
import org.springframework.security.core.GrantedAuthority;
import org.springframework.security.core.context.SecurityContextHolder;
import org.springframework.security.crypto.bcrypt.BCryptPasswordEncoder;
import org.springframework.security.web.authentication.logout.SecurityContextLogoutHandler;
import org.springframework.stereotype.Service;
import org.springframework.ui.Model;
import com.ch.ch7_1.entity.Authority;
import com.ch.ch7_1.entity.MyUser;
import com.ch.ch7_1.repository.MyUserRepository;
import com.ch.ch7_1.repository.AuthorityRepository;
@Service
public class UserServiceImpl implements UserService{
    @Autowired
    private MyUserRepository myUserRepository;
    @Autowired
    private AuthorityRepository authorityRepository;
    /**
```

```java
 * 实现注册
 */
@Override
public String register(MyUser userDomain) {
    String username = userDomain.getUsername();
    List<Authority> authorityList = new ArrayList<Authority>();
    //管理员权限
    if("admin".equals(username)) {
        Authority a1 = new Authority();
        Authority a2 = new Authority();
        a1.setId(1);
        a1.setName("ROLE_ADMIN");
        a2.setId(2);
        a2.setName("ROLE_DBA");
        authorityList.add(a1);
        authorityList.add(a2);
    }else {                                     //用户权限
        Authority a1 = new Authority();
        a1.setId(3);
        a1.setName("ROLE_USER");
        authorityList.add(a1);
    }
    //注册权限
    authorityRepository.saveAll(authorityList);
    userDomain.setAuthorityList(authorityList);
    //加密密码
    String secret = new BCryptPasswordEncoder().encode(userDomain.getPassword());
    userDomain.setPassword(secret);
    MyUser mu = myUserRepository.save(userDomain);
    if(mu != null)                              //注册成功
        return "/login";
    return "/register";                         //注册失败
}
/**
 * 用户登录成功
 */
@Override
public String loginSuccess(Model model) {
    model.addAttribute("user", getUname());
    model.addAttribute("role", getAuthorities());
    return "/user/loginSuccess";
}
/**
 * 管理员登录成功
 */
@Override
public String main(Model model) {
    model.addAttribute("user", getUname());
    model.addAttribute("role", getAuthorities());
    return "/admin/main";
}
```

```java
/**
 * 注销用户
 */
@Override
public String logout(HttpServletRequest request, HttpServletResponse response) {
    //获得用户认证信息
    Authentication authentication = SecurityContextHolder.getContext().getAuthentication();
    if(authentication != null) {
        //注销
        new SecurityContextLogoutHandler().logout(request, response, authentication);
    }
    return "redirect:/login?logout";
}
/**
 * 没有权限拒绝访问
 */
@Override
public String deniedAccess(Model model) {
    model.addAttribute("user", getUname());
    model.addAttribute("role", getAuthorities());
    return "deniedAccess";
}
/**
 * 获得当前用户名称
 */
private String getUname() {
    return SecurityContextHolder.getContext().getAuthentication().getName();
}
/**
 * 获得当前用户权限
 */
private String getAuthorities() {
    Authentication authentication = SecurityContextHolder.getContext().getAuthentication();
    List<String> roles = new ArrayList<String>();
    for (GrantedAuthority ga : authentication.getAuthorities()) {
        roles.add(ga.getAuthority());
    }
    return roles.toString();
}
}
```

8. 创建控制器类

创建名为 com.ch.ch7_1.controller 的包，并在该包中创建控制器类 TestSecurityController。具体代码如下：

```java
package com.ch.ch7_1.controller;
import javax.servlet.http.HttpServletRequest;
import javax.servlet.http.HttpServletResponse;
import org.springframework.beans.factory.annotation.Autowired;
import org.springframework.stereotype.Controller;
```

```java
import org.springframework.ui.Model;
import org.springframework.web.bind.annotation.ModelAttribute;
import org.springframework.web.bind.annotation.RequestMapping;
import com.ch.ch7_1.entity.MyUser;
import com.ch.ch7_1.service.UserService;
@Controller
public class TestSecurityController {
    @Autowired
    private UserService userService;
    @RequestMapping("/")
    public String index() {
        return "/index";
    }
    @RequestMapping("/toLogin")
    public String toLogin() {
        return "/login";
    }
    @RequestMapping("/toRegister")
    public String toRegister(@ModelAttribute("userDomain") MyUser userDomain) {
        return "/register";
    }
    @RequestMapping("/register")
    public String register(@ModelAttribute("userDomain") MyUser userDomain) {
        return userService.register(userDomain);
    }
    @RequestMapping("/login")
    public String login() {
        //这里什么都不错,由Spring Security负责登录验证
        return "/login";
    }
    @RequestMapping("/user/loginSuccess")
    public String loginSuccess(Model model) {
        return userService.loginSuccess(model);
    }
    @RequestMapping("/admin/main")
    public String main(Model model) {
        return userService.main(model);
    }
    @RequestMapping("/logout")
    public String logout(HttpServletRequest request, HttpServletResponse response) {
        return userService.logout(request, response);
    }
    @RequestMapping("/deniedAccess")
    public String deniedAccess(Model model) {
        return userService.deniedAccess(model);
    }
}
```

9. 创建应用的安全控制相关实现

创建名为 com.ch.ch7_1.security 的包,并在该包中创建 MyUserSecurityService、MyAuthenticationSuccessHandler 和 MySecurityConfigurerAdapter 类。

MyUserSecurityService 实现了 UserDetailsService 接口,并通过重写 loadUserByUsername(String username)方法查询对应的用户,并将用户名、密码、权限等与认证相关的信息封装在 UserDetails 对象中。

MyUserSecurityService 的代码如下:

```java
package com.ch.ch7_1.security;
import java.util.ArrayList;
import java.util.List;
import org.springframework.beans.factory.annotation.Autowired;
import org.springframework.security.core.GrantedAuthority;
import org.springframework.security.core.authority.SimpleGrantedAuthority;
import org.springframework.security.core.userdetails.User;
import org.springframework.security.core.userdetails.UserDetails;
import org.springframework.security.core.userdetails.UserDetailsService;
import org.springframework.security.core.userdetails.UsernameNotFoundException;
import org.springframework.stereotype.Service;
import com.ch.ch7_1.entity.Authority;
import com.ch.ch7_1.entity.MyUser;
import com.ch.ch7_1.repository.MyUserRepository;
/**
 * 获得对应的 UserDetails,保存与认证相关的信息
 */
@Service
public class MyUserSecurityService implements UserDetailsService{
    @Autowired
    private MyUserRepository myUserRepository;
    /**
     * 通过重写 loadUserByUsername 方法查询对应的用户
     * UserDetails 是 Spring Security 的一个核心接口
     * UserDetails 定义了可以获取用户名、密码、权限等与认证相关的信息的方法
     */
    @Override
    public UserDetails loadUserByUsername(String username) throws UsernameNotFoundException {
        //根据用户名(页面接收的用户名)查询当前用户
        MyUser myUser = myUserRepository.findByUsername(username);
        if(myUser == null) {
            throw new UsernameNotFoundException("用户名不存在");
        }
        //GrantedAuthority 代表赋予当前用户的权限(认证权限)
        List<GrantedAuthority> authorities = new ArrayList<GrantedAuthority>();
        //获得当前用户权限集合
        List<Authority> roles = myUser.getAuthorityList();

        //将当前用户的权限保存为用户的认证权限
        for (Authority authority : roles) {
            GrantedAuthority sg = new SimpleGrantedAuthority(authority.getName());
            authorities.add(sg);
        }
        //org.springframework.security.core.userdetails.User 是 Spring Security 的内部实现,
        //专门用于保存用户名、密码、权限等与认证相关的信息
        User su = new User(myUser.getUsername(), myUser.getPassword(), authorities);
        return su;
    }
}
```

MyAuthenticationSuccessHandler 继承了 SimpleUrlAuthenticationSuccessHandler 类，并重写了 handle（HttpServletRequest request，HttpServletResponse response，Authentication authentication）方法，根据当前认证用户的角色指定对应的 URL。

MyAuthenticationSuccessHandler 的代码如下：

```java
package com.ch.ch7_1.security;
import java.io.IOException;
import java.util.ArrayList;
import java.util.Collection;
import java.util.List;
import javax.servlet.ServletException;
import javax.servlet.http.HttpServletRequest;
import javax.servlet.http.HttpServletResponse;
import org.springframework.security.core.Authentication;
import org.springframework.security.core.GrantedAuthority;
import org.springframework.security.web.DefaultRedirectStrategy;
import org.springframework.security.web.RedirectStrategy;
import org.springframework.security.web.authentication.SimpleUrlAuthenticationSuccessHandler;
import org.springframework.stereotype.Component;
@Component
/**
 * 用户授权、认证成功处理类
 */
public class MyAuthenticationSuccessHandler extends SimpleUrlAuthenticationSuccessHandler{
    //Spring Security 的重定向策略
    private RedirectStrategy redirectStrategy = new DefaultRedirectStrategy();
    /**
     * 重写 handle 方法,通过 RedirectStrategy 重定向到指定的 URL
     */
    @Override
    protected void handle(HttpServletRequest request, HttpServletResponse response, Authentication authentication)throws IOException, ServletException {
        //根据当前认证用户的角色返回适当的 URL
        String tagetURL = getTargetURL(authentication);
        //重定向到指定的 URL
        redirectStrategy.sendRedirect(request, response, tagetURL);
    }
    /**
     * 从 Authentication 对象中提取当前登录用户的角色,并根据其角色返回适当的 URL
     */
    protected String getTargetURL(Authentication authentication) {
        String url = "";
        //获得当前登录用户的权限(角色)集合
        Collection<? extends GrantedAuthority> authorities = authentication.getAuthorities();
        List<String> roles = new ArrayList<String>();
        //将权限(角色)名称添加到 List 集合
        for (GrantedAuthority au : authorities) {
            roles.add(au.getAuthority());
        }
        //判断不同角色的用户跳转到不同的 URL
        //这里的 URL 是控制器的请求匹配路径
        if(roles.contains("ROLE_USER")) {
            url = "/user/loginSuccess";
```

```
            }else if(roles.contains("ROLE_ADMIN")) {
                url = "/admin/main";
            }else {
                url = "/deniedAccess";
            }
            return url;
        }
    }
```

MySecurityConfigurerAdapter 类继承了 WebSecurityConfigurerAdapter 类, 并通过重写 configure(AuthenticationManagerBuilder auth)方法实现用户认证, 重写 configure(HttpSecurity http)方法实现用户授权操作。

MySecurityConfigurerAdapter 的代码如下:

```
package com.ch.ch7_1.security;
import org.springframework.beans.factory.annotation.Autowired;
import org.springframework.context.annotation.Bean;
import org.springframework.context.annotation.Configuration;
import org.springframework.security.authentication.AuthenticationProvider;
import org.springframework.security.authentication.dao.DaoAuthenticationProvider;
import org.springframework.security.config.annotation.authentication.builders
        .AuthenticationManagerBuilder;
import org.springframework.security.config.annotation.web.builders.HttpSecurity;
import org.springframework.security.config.annotation.web.configuration
        .WebSecurityConfigurerAdapter;
import org.springframework.security.crypto.bcrypt.BCryptPasswordEncoder;
import org.springframework.security.crypto.password.PasswordEncoder;
/**
 * 认证和授权处理类
 */
@Configuration
public class MySecurityConfigurerAdapter extends WebSecurityConfigurerAdapter{
    //依赖注入通用的用户服务类
    @Autowired
    private MyUserSecurityService myUserSecurityService;
    //依赖注入加密接口
    @Autowired
    private PasswordEncoder passwordEncoder;
    //依赖注入用户认证接口
    @Autowired
    private AuthenticationProvider authenticationProvider;
    //依赖注入认证处理成功类,验证用户成功后处理不同用户跳转到不同的页面
    @Autowired
    private MyAuthenticationSuccessHandler myAuthenticationSuccessHandler;
    /**
     * BCryptPasswordEncoder 是 PasswordEncoder 的接口实现
     * 实现加密功能
     */
    @Bean
    public PasswordEncoder passwordEncoder() {
        return new BCryptPasswordEncoder();
```

```java
    }
    /**
     * DaoAuthenticationProvider 是 AuthenticationProvider 的实现
     */
    @Bean
    public AuthenticationProvider authenticationProvider() {
        DaoAuthenticationProvider provide = new DaoAuthenticationProvider();
        //不隐藏用户未找到异常
        provide.setHideUserNotFoundExceptions(false);
        //设置自定义认证方式,用户登录认证
        provide.setUserDetailsService(myUserSecurityService);
        //设置密码加密程序认证
        provide.setPasswordEncoder(passwordEncoder);
        return provide;
    }
    /**
     * 用户认证
     */
    @Override
    protected void configure(AuthenticationManagerBuilder auth) throws Exception {
        System.out.println("configure(AuthenticationManagerBuilder auth) ");
        //设置认证方式
        auth.authenticationProvider(authenticationProvider);
    }
    /**
     * 请求授权
     * 用户授权操作
     */
    @Override
    protected void configure(HttpSecurity http) throws Exception {
        System.out.println("configure(HttpSecurity http)");
        http.authorizeRequests()
            //首页、登录、注册页面、登录注册功能以及静态资源过滤掉,即可任意访问
            .antMatchers("/toLogin", "/toRegister", "/", "/login", "/register", "/css/**", "/fonts/**", "/js/**").permitAll()
            //这里默认追加 ROLE_,/user/** 是控制器的请求匹配路径
            .antMatchers("/user/**").hasRole("USER")
            .antMatchers("/admin/**").hasAnyRole("ADMIN", "DBA")
            //其他所有请求登录后才能访问
            .anyRequest().authenticated()
            .and()
            //将输入的用户名与密码和授权的进行比较
            .formLogin()
                .loginPage("/login").successHandler(myAuthenticationSuccessHandler)
                .usernameParameter("username").passwordParameter("password")
                //登录失败
                .failureUrl("/login?error")
            .and()
```

```
            //注销行为可任意访问
            .logout().permitAll()
            .and()
            //指定异常处理页面
            .exceptionHandling().accessDeniedPage("/deniedAccess");
    }
}
```

10. 创建用于测试的视图页面

在 src/main/resources/templates 目录下创建应用首页、注册、登录以及拒绝访问页面；在 src/main/resources/templates/admin 目录下创建管理员用户认证成功后访问的页面；在 src/main/resources/templates/user 目录下创建普通用户认证成功后访问的页面。

首页页面 index.html 的代码如下：

```html
<!DOCTYPE html>
<html xmlns:th="http://www.thymeleaf.org">
<head>
<meta charset="UTF-8">
<title>首页</title>
<link rel="stylesheet" th:href="@{css/bootstrap.min.css}" />
</head>
<body>
    <div class="panel panel-primary">
        <div class="panel-heading">
            <h3 class="panel-title">Spring Security测试首页</h3>
        </div>
    </div>
    <div class="container">
        <div>
            <a th:href="@{/toLogin}">去登录</a><br><br>
            <a th:href="@{/toRegister}">去注册</a>
        </div>
    </div>
</body>
</html>
```

注册页面 register.html 的代码如下：

```html
<!DOCTYPE html>
<html xmlns:th="http://www.thymeleaf.org">
<head>
<meta charset="UTF-8">
<title>注册页面</title>
<link rel="stylesheet" th:href="@{/css/bootstrap.min.css}" />
<script th:src="@{/js/jquery.min.js}"></script>
<script type="text/javascript" th:inline="javascript">
    function checkBpwd(){
        if($("#username").val() == ""){
            alert("用户名必须输入!");
```

```html
                $("#username").focus();
                return false;
            }else if($("#password").val() == ""){
                alert("密码必须输入!");
                $("#password").focus();
                return false;
            }else if($("#password").val() != $("#repassword").val()){
                alert("两次密码不一致!");
                $("#password").focus();
                return false;
            }else{
                document.myform.submit();
            }
        }
    }
</script>
<body>
    <div class="container">
        <div class="bg-primary" style="width:100%;height:70px;padding-top:10px;">
            <h2 align="center">用户注册</h2>
        </div>
        <br>
        <br>
        <form th:action="@{/register}" name="myform" method="post" th:object="${userDomain}" class="form-horizontal" role="form">
            <div class="form-group has-success">
                <label class="col-sm-2 col-md-2 control-label">用户名</label>
                <div class="col-sm-4 col-md-4">
                    <input type="text" class="form-control"
                        placeholder="请输入您的用户名"
                        th:field="*{username}"/>
                </div>
            </div>
            <div class="form-group has-success">
                <label class="col-sm-2 col-md-2 control-label">密码</label>
                <div class="col-sm-4 col-md-4">
                    <input type="password" class="form-control"
                        placeholder="请输入您的密码" th:field="*{password}" />
                </div>
            </div>
            <div class="form-group has-success">
                <label class="col-sm-2 col-md-2 control-label">确认密码</label>
                <div class="col-sm-4 col-md-4">
                    <input type="password" class="form-control"
                        placeholder="请输入您的密码" th:field="*{repassword}"/>
                </div>
            </div>
            <div class="form-group">
                <div class="col-sm-offset-2 col-sm-10">
                    <button type="button" onclick="checkBpwd()" class="btn btn-success">注册</button>
                    <button type="reset" class="btn btn-primary">重置</button>
```

```html
            </div>
        </div>
    </form>
</div>
</body>
</html>
```

登录页面 login.html 的代码如下:

```html
<!DOCTYPE html>
<html xmlns:th="http://www.thymeleaf.org">
<head>
<meta charset="UTF-8">
<title>登录页面</title>
<link rel="stylesheet" th:href="@{css/bootstrap.min.css}" />
<script type="text/javascript" th:src="@{js/jquery.min.js}"></script>
<script type="text/javascript">
    $(function(){
        $("#loginBtn").click(function(){
            var username = $("#username");
            var password = $("#password");
            var msg = "";
            if(username.val() == ""){
                msg = "用户名不能为空!";
                username.focus();
            }else if(password.val() == ""){
                msg = "密码不能为空!";
                password.focus();
            }
            if(msg != ""){
                alert(msg);
                return false;
            }
            $("#myform").submit();
        });
    });
</script>
<body>
    <div class="container">
    <div class="bg-primary"  style="width:100%; height: 70px;padding-top: 10px;">
            <h2 align="center">用户登录</h2>
    </div>
      <br>
      <br>
    <form th:action="@{/login}"  id="myform" method="post"  class="form-horizontal" role="form">
            <!-- 用户名或密码错误 -->
            <div th:if="${param.error != null}">
                <div class="alert alert-danger">
                    <p><font color="red">用户名或密码错误!</font></p>
                </div>
```

```html
            </div>
            <!-- 注销 -->
            <div th:if="${param.logout != null}">
                <div class="alert alert-success">
                    <p><font color="red">用户已注销成功!</font></p>
                </div>
            </div>
            <div class="form-group has-success">
                <label class="col-sm-2 col-md-2 control-label">用户名</label>
                <div class="col-sm-4 col-md-4">
                    <input type="text" class="form-control"
                     placeholder="请输入您的用户名"
                     name="username" id="username"/>
                </div>
            </div>
            <div class="form-group has-success">
                <label class="col-sm-2 col-md-2 control-label">密码</label>
                <div class="col-sm-4 col-md-4">
                    <input type="password" class="form-control"
                     placeholder="请输入您的密码" name="password" id="password"/>
                </div>
            </div>
            <div class="form-group">
                <div class="col-sm-offset-2 col-sm-10">
                    <button type="button" id="loginBtn" class="btn btn-success">登录</button>
                    <button type="reset" class="btn btn-primary">重置</button>
                </div>
            </div>
        </form>
    </div>
</body>
</html>
```

拒绝访问页面 deniedAccess.html 的代码如下：

```html
<!DOCTYPE html>
<html xmlns:th="http://www.thymeleaf.org">
<head>
<meta charset="UTF-8">
<title>首页</title>
<link rel="stylesheet" th:href="@{/css/bootstrap.min.css}"/>
</head>
<body>
    <div class="panel panel-primary">
        <div class="panel-heading">
            <h3 class="panel-title">拒绝访问页面</h3>
        </div>
    </div>
    <div class="container">
        <div>
```

```html
        <h3><span th:text="${user}"></span>您没有权限访问该页面!您的权限是<span th:text="${role}"></span>。</h3><br><br>
            <a th:href="@{/logout}">安全退出</a>
        </div>
    </div>
</body>
</html>
```

管理员用户成功登录后访问的页面main.html的代码如下:

```html
<!DOCTYPE html>
<html xmlns:th="http://www.thymeleaf.org">
<head>
<meta charset="UTF-8">
<title>首页</title>
<link rel="stylesheet" th:href="@{/css/bootstrap.min.css}" />
</head>
<body>
    <div class="panel panel-primary">
        <div class="panel-heading">
            <h3 class="panel-title">管理员页面</h3>
        </div>
    </div>
    <div class="container">
        <div>
            <h3>欢迎<span th:text="${user}"></span>访问管理员页面!您的权限是<span th:text="${role}"></span>。</h3><br><br>
            <a th:href="@{/user/loginSuccess}">去访问用户登录成功页面</a><br><br>
            <a th:href="@{/logout}">安全退出</a>
        </div>
    </div>
</body>
</html>
```

普通用户成功登录后访问的页面loginSuccess.html的代码如下:

```html
<!DOCTYPE html>
<html xmlns:th="http://www.thymeleaf.org">
<head>
<meta charset="UTF-8">
<title>首页</title>
<link rel="stylesheet" th:href="@{/css/bootstrap.min.css}" />
</head>
<body>
    <div class="panel panel-primary">
        <div class="panel-heading">
            <h3 class="panel-title">登录成功页面</h3>
        </div>
    </div>
    <div class="container">
        <div>
            <h3>欢迎<span th:text="${user}"></span>登录成功!您的权限是
            <span th:text="${role}"></span>。</h3><br><br>
```

```
                <a th:href = "@{/admin/main}">去访问管理员页面</a><br><br>
                <a th:href = "@{/logout}">安全退出</a>
            </div>
        </div>
    </body>
</html>
```

11. 测试应用

运行 Ch71Application 的主方法启动项目。Spring Boot 应用启动后,观察控制台,发现 MySecurityConfigurerAdapter 的两个 configure 方法都已经被执行,说明自定义的用户认证和用户授权工作已经生效。控制台的输出结果如图 7.2 所示。

图 7.2　启动 ch7_1 应用时控制台的输出结果

在浏览器地址栏中输入"http://localhost:8080/ch7_1/""/toLogin""/toRegister""/""/login""/register"等其中任何一个请求都将正常访问,其他请求都将被重定向到"http://localhost:8080/ch7_1/login"登录页面,因为没有登录,用户没有访问权限,如图 7.3 所示。

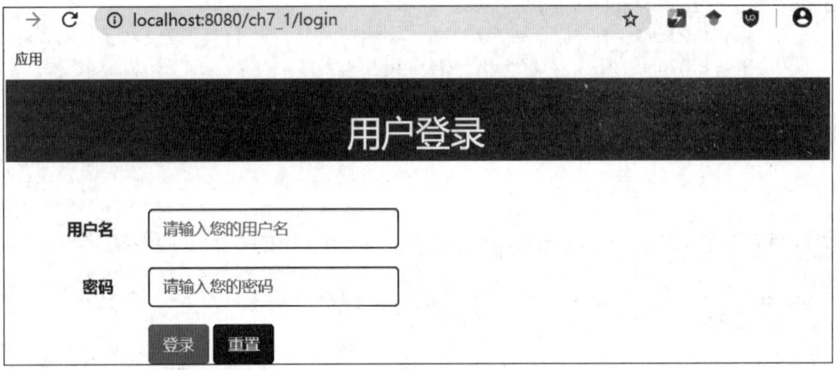

图 7.3　登录页面

可以通过"http://localhost:8080/ch7_1/"访问首页面,如图 7.4 所示;然后,单击"去注册"超链接打开注册页面,如图 7.5 所示。用户成功注册后,打开登录页面进行用户登录。

如果在图 7.5 中输入的用户名不是 admin,那么就是注册了一个普通用户,其权限为 ROLE_USER;如果在图 7.5 中输入的用户名是 admin,那么就是注册了一个管理员用户,其权限为 ROLE_ADMIN 和 ROLE_DBA。

在图 7.3 登录页面中任意输入用户名和密码,单击"登录"按钮,提示用户名或密码错误,如图 7.6 所示。

图 7.4　首页面

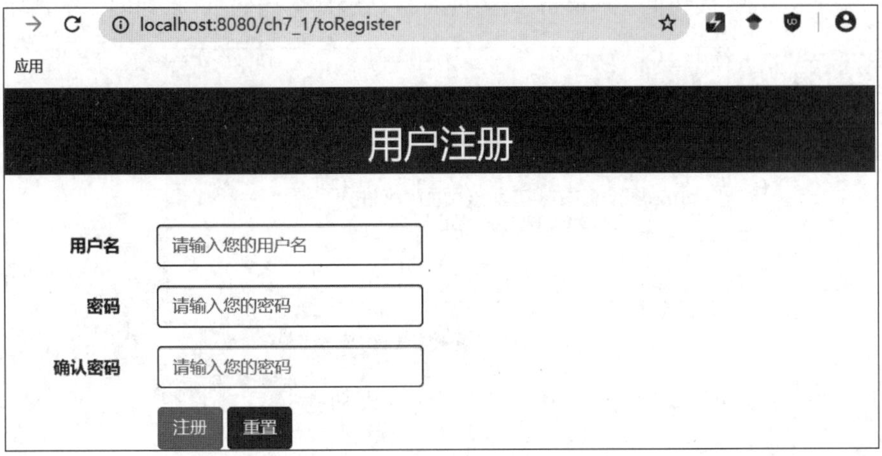

图 7.5 注册页面

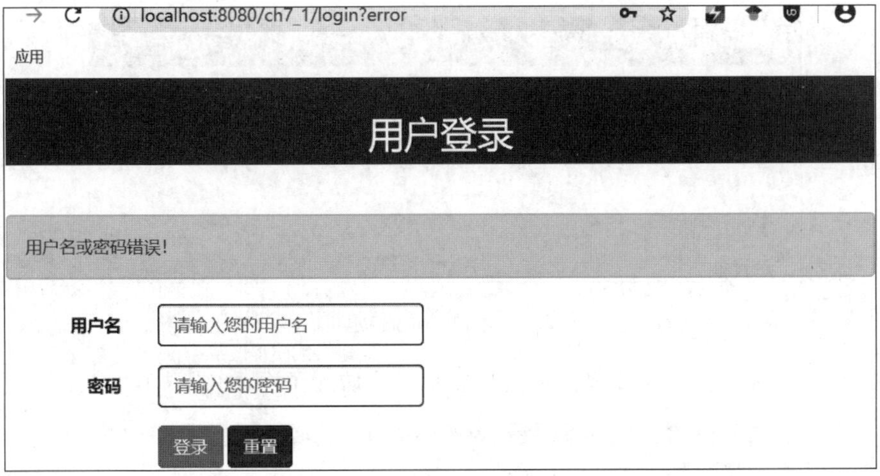

图 7.6 用户名或密码错误

在图 7.3 登录页面中输入管理员用户名和密码,成功登录后打开管理员主页面,如图 7.7 所示。

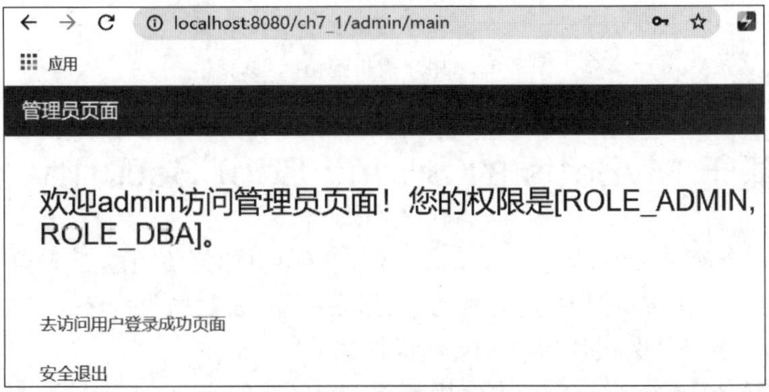

图 7.7 管理员成功登录页面

单击图 7.7 中的"去访问用户登录成功页面",显示拒绝访问页面,如图 7.8 所示。

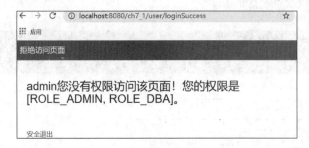

图 7.8　管理员被拒绝访问页面

在图 7.3 登录页面中输入普通用户名和密码,成功登录后打开用户登录成功页面,如图 7.9 所示。

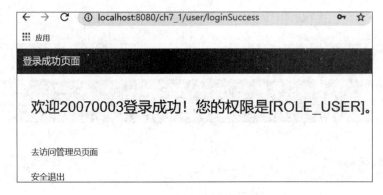

图 7.9　用户登录成功页面

单击图 7.9 中的"去访问管理员页面",显示拒绝访问页面,如图 7.10 所示。

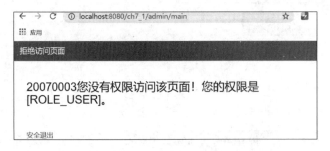

图 7.10　用户被拒绝访问页面

7.3.2　基于 MyBatis 的 Spring Boot Security 操作实例

下面通过一个简单实例,演示在 Spring Boot 应用中如何使用基于 MyBatis 的 Spring Security 安全框架。在该实例中,所有页面、数据表结构、表数据与例 7-1 完全一致,只是修改了几处代码,主要是持久层使用的 MyBatis 框架。

【例 7-2】　在 Spring Boot 应用中使用基于 MyBatis 的 Spring Security 安全框架。

具体实现步骤如下。

1. 创建 Spring Boot Web 应用

创建基于 Thymeleaf 及 Spring Security 的 Web 应用 ch7_2，如图 7.11 所示。

图 7.11　创建 Web 应用 ch7_2

2. 修改 pom.xml 文件

在 pom.xml 文件中添加 MySQL 连接器和 mybatis-spring-boot-starter 模块。具体代码如下：

```
<dependency>
    <groupId>mysql</groupId>
    <artifactId>mysql-connector-java</artifactId>
    <version>5.1.45</version>
</dependency>
<!-- MyBatis-Spring,Spring Boot 应用整合 MyBatis 框架的核心依赖配置 -->
<dependency>
    <groupId>org.mybatis.spring.boot</groupId>
    <artifactId>mybatis-spring-boot-starter</artifactId>
    <version>2.1.0</version>
</dependency>
```

3. 设置 Web 应用 ch7_2 的上下文路径及数据源配置信息

在 ch7_2 的 application.properties 文件中配置以下内容：

```
server.servlet.context-path=/ch7_2
```

```
###
## 数据源信息配置
###
# 数据库地址
spring.datasource.url=jdbc:mysql://localhost:3306/springbootjpa?characterEncoding=utf8
# 数据库用户名
spring.datasource.username=root
# 数据库密码
spring.datasource.password=root
# 数据库驱动
spring.datasource.driver-class-name=com.mysql.jdbc.Driver
# 设置包别名(在 Mapper 映射文件中直接使用实体类名)
mybatis.type-aliases-package=com.ch.ch7_2.entity
# 告诉系统到哪里去找 mapper.xml 文件(映射文件)
mybatis.mapperLocations=classpath:mappers/*.xml
logging.level.org.springframework.security=trace
# 在控制台输出 SQL 语句日志
logging.level.com.ch.ch7_2.repository=debug
```

4. 创建实体类

创建名为 com.ch.ch7_2.entity 的包,并在该包中创建 MyUser 和 Authority 实体类,代码与例 7-1 中的实体类基本一样。

MyUser 的代码如下:

```
package com.ch.ch7_2.entity;
import java.io.Serializable;
import java.util.List;
public class MyUser implements Serializable{
    private static final long serialVersionUID = 1L;
    private int id;
    private String username;
    private String password;
    private List<Authority> authorityList;
    private String repassword;
    //省略 set 和 get 方法
}
```

Authority 的代码如下:

```
package com.ch.ch7_2.entity;
import java.io.Serializable;
import java.util.List;
public class Authority implements Serializable{
    private static final long serialVersionUID = 1L;
    private int id;
    private String name;
    private List<MyUser> userList;
    //省略 set 和 get 方法
}
```

5. 创建数据访问接口

创建名为 com.ch.ch7_2.repository 的包，并在该包中创建 MyUserRepository 接口。具体代码如下：

```java
package com.ch.ch7_2.repository;
import java.util.List;
import org.apache.ibatis.annotations.Mapper;
import org.apache.ibatis.annotations.Param;
import com.ch.ch7_2.entity.Authority;
import com.ch.ch7_2.entity.MyUser;
@Mapper
public interface MyUserRepository{
    /**
     * 根据用户名查询用户信息
     */
    MyUser findByUsername(String username);
    /**
     * 根据用户id查询用户的所有权限
     */
    List<Authority> findRoleByUser(Integer id);
    /**
     * 注册用户
     */
    int save(MyUser mu);
    /**
     * 保存用户权限
     */
    int saveUserAuthority(@Param("user_id") Integer user_id, @Param("authority_id") Integer authority_id);
}
```

6. 创建 Mapper 映射文件

在 src/main/resources 目录下，创建名为 mappers 的包，并在该包中创建 SQL 映射文件 MyUserMapper.xml。具体代码如下：

```xml
<?xml version="1.0" encoding="UTF-8"?>
<!DOCTYPE mapper
PUBLIC "-//mybatis.org//DTD Mapper 3.0//EN"
"http://mybatis.org/dtd/mybatis-3-mapper.dtd">
<mapper namespace="com.ch.ch7_2.repository.MyUserRepository">
    <!-- 根据用户名级联查询用户权限 -->
    <resultMap type="MyUser" id="myResult">
        <id property="id" column="id"/>
        <result property="username" column="username"/>
        <result property="password" column="password"/>
        <!-- 级联查询 -->
        <collection
            property="authorityList"
```

```xml
            ofType = "Authority"
            column = "id"
            fetchType = "eager"
            select = "com.ch.ch7_2.repository.MyUserRepository.findRoleByUser"
        />
    </resultMap>
    <!-- 根据用户名查询用户信息 -->
    <select id = "findByUsername" parameterType = "string" resultMap = "myResult">
        select * from user where username = #{username}
    </select>
    <!-- 根据用户id查询用户权限 -->
    <select id = "findRoleByUser" parameterType = "integer" resultType = "Authority">
        select id, name from authority a, user_authority ua where a.id = ua.authority_id and ua.user_id = #{id}
    </select>
    <!-- 注册用户,并将主键保存到MyUser对象的id属性中 -->
    <insert id = "save" parameterType = "MyUser" keyProperty = "id" useGeneratedKeys = "true">
        insert into user(id, username, password) values(null, #{username}, #{password})
    </insert>
    <!-- 添加用户权限详情 -->
    <insert id = "saveUserAuthority">
        insert into user_authority(user_id, authority_id) values(#{user_id}, #{authority_id})
    </insert>
</mapper>
```

7. 修改业务层

创建名为 com.ch.ch7_2.service 的包,将 ch7_1 的业务层接口和实现类复制到本包中,并将实现类的注册方法修改如下:

```java
/**
 * 实现注册
 */
@Override
public String register(MyUser userDomain) {
    String username = userDomain.getUsername();
    //加密密码
    String secret = new BCryptPasswordEncoder().encode(userDomain.getPassword());
    userDomain.setPassword(secret);
    int n = myUserRepository.save(userDomain);
    //管理员权限
    if("admin".equals(username)) {
        myUserRepository.saveUserAuthority(userDomain.getId(), 1);
        myUserRepository.saveUserAuthority(userDomain.getId(), 2);
    }else {                                    //用户权限
        myUserRepository.saveUserAuthority(userDomain.getId(), 3);
    }
    if(n != 0)                                 //注册成功
        return "/login";
    return "/register";                        //注册失败
}
```

8. 修改 Ch72Application 类

在 Ch72Application 类头部使用 @MapperScan 注解扫描 Mapper 接口。具体代码如下：

```
package com.ch.ch7_2;
import org.mybatis.spring.annotation.MapperScan;
import org.springframework.boot.SpringApplication;
import org.springframework.boot.autoconfigure.SpringBootApplication;
@SpringBootApplication
@MapperScan("com.ch.ch7_2.repository")
public class Ch72Application {
    public static void main(String[] args) {
        SpringApplication.run(Ch72Application.class, args);
    }
}
```

其他的源程序及运行测试与 ch7_1 完全一样，不再赘述。

从 7.3.1 节和 7.3.2 节可以看出，这两个 Spring Boot Security 操作实例的实现原理相同，只是数据库操作不同而已。

7.4 本章小结

本章首先介绍了 Spring Security 快速入门，然后详细介绍了实际开发中的 Spring Security 操作实例。通过本章的学习，读者应该了解 Spring Security 安全机制的基本原理，掌握如何在实际应用开发中，使用 Spring Security 安全机制提供系统安全解决方案。

习题 7

1. 开发者如何自定义 Spring Security 的适配器？
2. Spring Security 的用户认证和请求授权是如何实现的？请举例说明。

第8章 异步消息

学习目的与要求

本章主要讲解了企业级消息代理 JMS 和 AMQP。通过本章的学习,理解异步消息通信原理,掌握异步消息通信技术。

本章主要内容

- JMS。
- AMQP。

当我们跨越多个微服务进行通信时,异步消息就显得至关重要了。例如在电子商务系统中,订单服务在下单时需要和库存服务进行通信,完成库存的扣减操作,这时就需要基于异步消息和最终一致性的通信方式来进行这样的操作,并且能够在发生故障时正常工作。

8.1 消息模型

视频讲解

异步消息的主要目的是解决跨系统的通信。所谓异步消息,即消息发送者无须等待消息接收者的处理及返回,甚至无须关心消息是否发送与接收成功。在异步消息中有两个极其重要的概念,即消息代理和目的地。当消息发送者发送消息后,消息将由消息代理管理,消息代理保证消息传递到目的地。

异步消息的目的地主要有两种形式,即队列和主题。队列用于点对点式的消息通信,即端到端通信(单接收者);主题用于发布/订阅式的消息通信,即广播通信(多接收者)。

8.1.1 点对点式

在点对点式的消息通信中,消息代理获得发送者发送的消息后,将消息存入一个队列,当有消息接收者接收消息时,将从队列中取出消息传递给接收者,这时队列中清除该消息。

在点对点式的消息通信中,确保的是每一条消息只有唯一的发送者和接收者,但并不能说明只有一个接收者可以从队列中接收消息。这是因为队列中有多个消息,点对点式的消息通信只保证每一条消息只有唯一的发送者和接收者。

8.1.2 发布/订阅式

多接收者是消息通信中一种更加灵活的方式,而点对点式的消息通信只保证每一条消息只有唯一的接收者。这时可以使用发布/订阅式的消息通信解决多接收者的问题。和点对点式不同,发布/订阅式是消息发送者将消息发送到主题,而多个消息接收者监听这个主题。此时的消息发送者称为发布者,接收者称为订阅者。

8.2 企业级消息代理

异步消息传递技术常用的有 JMS 和 AMQP。JMS 是面向基于 Java 的企业应用的异步消息代理。AMQP 是面向所有应用的异步消息代理。

视频讲解

8.2.1 JMS

JMS(Java Messaging Service,Java 消息服务)是 Java 平台上有关面向消息中间件的技术规范,它便于消息系统中的 Java 应用程序进行消息交换,并且通过提供标准的产生、发送、接收消息的接口简化企业应用的开发。

1. JMS 元素

JMS 由以下元素组成。
1) JMS 消息代理实现

连接面向消息中间件的,JMS 消息代理接口的一个实现。JMS 的消息代理实现可以是 Java 平台的 JMS 实现,也可以是非 Java 平台的面向消息中间件的适配器。开源的 JMS 实现有 Apache ActiveMQ、JBoss 社区所研发的 HornetQ、The OpenJMS Group 的 OpenJMS 等实现。

2) JMS 客户
生产或消费基于消息的 Java 应用程序或对象。
3) JMS 生产者
创建并发送消息的 JMS 客户。

4）JMS 消费者

接收消息的 JMS 客户。

5）JMS 消息

JMS 消息包括可以在 JMS 客户之间传递的数据对象。JMS 定义了 5 种不同的消息正文格式，以及调用的消息类型，允许发送并接收一些不同形式的数据，提供现有消息格式的一些级别的兼容性。常见的消息格式有 StreamMessage（指 Java 原始值的数据流消息）、MapMessage（映射消息）、TextMessage（文本消息）、ObjectMessage（一个序列化的 Java 对象消息）和 BytesMessage（字节消息）。

6）JMS 队列

一个容纳那些被发送的等待阅读的消息区域。与队列名字所暗示的意思不同，消息的接收顺序并不一定要与消息的发送顺序相同。一旦一个消息被阅读，该消息将被从队列中移走。

7）JMS 主题

一种支持发送消息给多个订阅者的机制。

2. JMS 的应用接口

JMS 的应用接口包括以下接口类型。

1）ConnectionFactory 接口（连接工厂）

用户用来创建到 JMS 消息代理实现的连接的被管对象。JMS 客户通过可移植的接口访问连接，这样当下层的实现改变时，代码不需要进行修改。管理员在 JNDI 名字空间中配置连接工厂，这样，JMS 客户才能够查找到它们。根据目的地的不同，用户将使用队列连接工厂，或者主题连接工厂。

2）Connection 接口（连接）

连接代表了应用程序和消息服务器之间的通信链路。在获得了连接工厂后，就可以创建一个与 JMS 消息代理实现（提供者）的连接。根据不同的连接类型，连接允许用户创建会话，以发送和接收队列和主题到目的地。

3）Destination 接口（目的地）

目的地是一个包装了消息目的地标识符的被管对象，消息目的地是指消息发布和接收的地点，或者是队列，或者是主题。JMS 管理员创建这些对象，然后用户通过 JNDI 发现它们。和连接工厂一样，管理员可以创建两种类型的目的地：点对点模型的队列和发布者/订阅者模型的主题。

4）Session 接口（会话）

Session 接口（会话）表示一个单线程的上下文，用于发送和接收消息。由于会话是单线程的，所以消息是连续的，就是说消息是按照发送的顺序一个一个接收的。会话的好处是它支持事务。如果用户选择了事务支持，会话上下文将保存一组消息，直到事务被提交才发送这些消息。在提交事务之前，用户可以使用回滚操作取消这些消息。一个会话允许用户创建消息，生产者来发送消息，消费者来接收消息。

5）MessageConsumer 接口（消息消费者）

MessageConsumer 接口（消息消费者）是由会话创建的对象，用于接收发送到目的地的消息。消费者可以同步地（阻塞模式），或非同步地（非阻塞模式）接收队列和主题类型的消息。

6) MessageProducer 接口（消息生产者）

MessageProducer 接口（消息生产者）是由会话创建的对象，用于发送消息到目的地。用户可以创建某个目的地的发送者，也可以创建一个通用的发送者，在发送消息时指定目的地。

7) Message 接口（消息）

Message 接口（消息）是在消费者和生产者之间传送的对象，也就是说从一个应用程序传送到另一个应用程序。一个消息有 3 个主要部分。

（1）消息头（必须）：包含用于识别和为消息寻找路由的操作设置。

（2）一组消息属性（可选）：包含额外的属性，支持其他消息代理实现和用户的兼容。可以创建定制的字段和过滤器（消息选择器）。

（3）一个消息体（可选）：允许用户创建 5 种类型的消息（文本消息、映射消息、字节消息、流消息和对象消息）。

JMS 各接口角色之间的关系如图 8.1 所示。

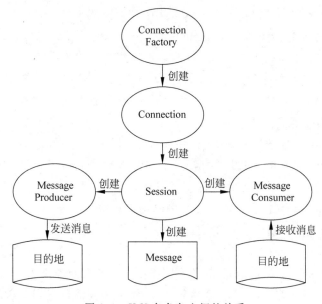

图 8.1 JMS 各角色之间的关系

8.2.2 AMQP

AMQP(Advanced Message Queuing Protocol,高级消息队列协议)是一个提供统一消息服务的应用层标准高级消息队列协议，是应用层协议的一个开放标准，为面向消息的中间件设计。基于此协议的客户端与消息中间件可传递消息，并不受客户端/中间件的不同产品、不同开发语言等条件的限制。AMQP 的技术术语如下。

AMQP 模型（AMQP Model）：一个由关键实体和语义表示的逻辑框架，遵从 AMQP 规范的服务器必须提供这些实体和语义。为了实现本规范中定义的语义，客户端可以发送命令来控制 AMQP 服务器。

连接(Connection)：一个网络连接，例如 TCP/IP 套接字连接。

会话(Session)：端点之间的命名对话。在一个会话上下文中，保证"恰好传递一次"。

信道(Channel)：多路复用连接中的一条独立的双向数据流通道，为会话提供物理传输介质。

客户端(Client)：AMQP 连接或者会话的发起者。AMQP 是非对称的，客户端生产和消费消息，服务器存储和路由这些消息。

服务器(Server)：接受客户端连接，实现 AMQP 消息队列和路由功能的进程，也称为"消息代理"。

端点(Peer)：AMQP 对话的任意一方。一个 AMQP 连接包括两个端点，一个是客户端，一个是服务器。

搭档(Partner)：当描述两个端点之间的交互过程时，使用术语"搭档"来表示"另一个"端点的简记法。例如定义端点 A 和端点 B，当它们进行通信时，端点 B 是端点 A 的搭档，端点 A 是端点 B 的搭档。

片段集(Assembly)：段的有序集合，形成一个逻辑工作单元。

段(Segment)：帧的有序集合，形成片段集中一个完整子单元。

帧(Frame)：AMQP 传输的一个原子单元。一个帧是一个段中的任意分片。

控制(Control)：单向指令，AMQP 规范假设这些指令的传输是不可靠的。

命令(Command)：需要确认的指令，AMQP 规范规定这些指令的传输是可靠的。

异常(Exception)：在执行一个或者多个命令时可能发生的错误状态。

类(Class)：一批用来描述某种特定功能的 AMQP 命令或者控制。

消息头(Header)：描述消息数据属性的一种特殊段。

消息体(Body)：包含应用程序数据的一种特殊段。消息体段对于服务器来说完全透明——服务器不能查看或者修改消息体。

消息内容(Content)：包含在消息体段中的消息数据。

交换器(Exchange)：服务器中的实体，用来接收生产者发送的消息并将这些消息路由给服务器中的队列。

交换器类型(Exchange Type)：基于不同路由语义的交换器类。

消息队列(Message Queue)：一个命名实体，用来保存消息直到发送给消费者。

绑定器(Binding)：消息队列和交换器之间的关联。

绑定器关键字(Binding Key)：绑定的名称。一些交换器类型可能使用这个名称作为定义绑定器路由行为的模式。

路由关键字(Routing Key)：一个消息头，交换器可以用这个消息头决定如何路由某条消息。

持久存储(Durable)：一种服务器资源，当服务器重启时，保存的消息数据不会丢失。

临时存储(Transient)：一种服务器资源，当服务器重启时，保存的消息数据会丢失。

持久化(Persistent)：服务器将消息保存在可靠磁盘存储中，当服务器重启时，消息不会丢失。

非持久化(Non-Persistent)：服务器将消息保存在内存中,当服务器重启时,消息可能丢失。

消费者(Consumer)：一个从消息队列中请求消息的客户端应用程序。

生产者(Producer)：一个向交换器发布消息的客户端应用程序。

虚拟主机(Virtual Host)：一批交换器、消息队列和相关对象。虚拟主机是共享相同的身份认证和加密环境的独立服务器域。客户端应用程序在登录到服务器之后,可以选择一个虚拟主机。

8.3 Spring Boot 的支持

视频讲解

8.3.1 JMS 的自动配置

Spring Boot 对 JMS 的自动配置位于 org.springframework.boot.autoconfigure.jms 包下,支持 JMS 的实现有 ActiveMQ 和 Artemis,如图 8.2 所示。

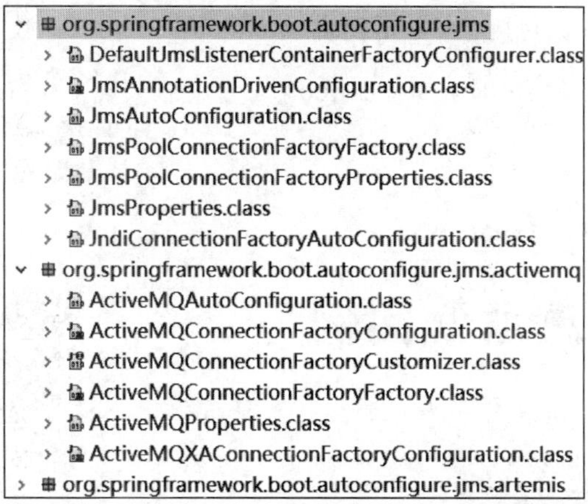

图 8.2 Spring Boot 对 JMS 的自动配置

以 ActiveMQ 为例,Spring Boot 定义了 ActiveMQConnectionFactory 的 Bean 作为连接,并通过以 spring.activemq 为前缀的属性配置 ActiveMQ 的连接属性,主要包含以下内容：

```
spring.activemq.broker-url = tcp://localhost:61616    #消息代理路径
spring.activemq.user =
spring.activemq.password =
spring.activemq.in-memory = true
```

另外,Spring Boot 在 JmsAutoConfiguration 自动配置类中配置了 JmsTemplate；并且在 JmsAnnotationDrivenConfiguration 配置类中开启了注解式消息监听的支持,即自动开启@EnableJms。

8.3.2 AMQP 的自动配置

Spring Boot 对 AMQP 的自动配置位于 org.springframework.boot.autoconfigure.amqp 包下，RabbitMQ 是 AMQP 的主要实现，如图 8.3 所示。

> ⊞ org.springframework.boot.autoconfigure.amqp
> > 🗎 AbstractRabbitListenerContainerFactoryConfigurer.class
> > 🗎 DirectRabbitListenerContainerFactoryConfigurer.class
> > 🗎 RabbitAnnotationDrivenConfiguration.class
> > 🗎 RabbitAutoConfiguration.class
> > 🗎 RabbitProperties.class
> > 🗎 RabbitRetryTemplateCustomizer.class
> > 🗎 RetryTemplateFactory.class
> > 🗎 SimpleRabbitListenerContainerFactoryConfigurer.class

图 8.3 Spring Boot 对 AMQP 的自动配置

在 RabbitAutoConfiguration 自动配置类中，配置了连接的 RabbitConnectionFactoryBean 和 RabbitTemplate，并且在 RabbitAnnotationDrivenConfiguration 配置类中开启了 @EnableRabbit。从 RabbitProperties 类中，可以看出 RabbitMQ 的配置可通过以 spring.rabbitmq 为前缀的属性进行配置，主要包含以下内容：

```
spring.rabbitmq.host = localhost      # RabbitMQ 服务器地址，默认为 localhost
spring.rabbitmq.port = 5672           # RabbitMQ 端口，默认为 5672
spring.rabbitmq.username = guest      # 默认用户名
spring.rabbitmq.password = guest      # 默认密码
```

8.4 异步消息通信实例

本节通过两个实例，讲解异步消息通信的实现过程。

视频讲解

8.4.1 JMS 实例

本节我们使用 JMS 的一种实现 ActiveMQ 讲解 JMS 实例。因此，需要事先安装 ActiveMQ（注意需要安装 jdk）。读者可访问 http://activemq.apache.org/下载符合自己的 ActiveMQ。编写本书时，作者下载了 Windows 版本的 apache-activemq-5.15.10-bin.zip。该版本的 ActiveMQ，解压缩即可完成安装。

解压缩后，双击"apache-activemq-5.15.10\bin\win64"下的"wrapper.exe"或"activemq.bat"启动 ActiveMQ，如图 8.4 所示。然后，通过"http://localhost:8161"运行 ActiveMQ 的管理界面，管理员账号和密码默认为 admin/admin，如图 8.5 所示。

启动 ActiveMQ 服务后，下面通过实例讲解如何使用 JMS 的实现 ActiveMQ 进行两个应用系统之间的点对点式通信。

【例 8-1】 使用 JMS 的实现 ActiveMQ 进行两个应用系统之间的点对点式通信。

具体实现步骤如下：

图 8.4 启动 ActiveMQ

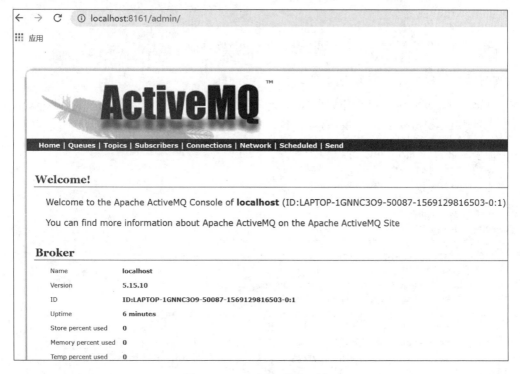

图 8.5 ActiveMQ 的管理界面

1. 创建基于 Apache ActiveMQ5 的 Spring Boot 应用 ch8_1Sender(消息发送者)

创建基于 Apache ActiveMQ5 的 Spring Boot 应用 ch8_1Sender，该应用作为消息发送者，如图 8.6 所示。

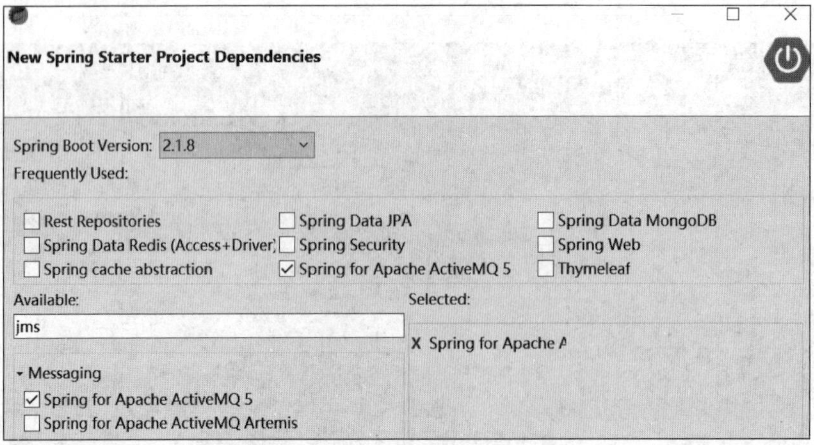

图 8.6　创建基于 Apache ActiveMQ5 的 Spring Boot 应用

2. 配置 ActiveMQ 的消息代理地址

在 ch8_1Sender 应用的配置文件 application.properties 中，配置 ActiveMQ 的消息代理地址。具体代码如下：

```
spring.activemq.broker-url=tcp://localhost:61616
```

3. 定义消息

在 com.ch.ch8_1sender 包下，创建消息定义类 MyMessage，该类需要实现 MessageCreator 接口，并重写接口方法 createMessage 进行消息定义。具体代码如下：

```java
package com.ch.ch8_1sender;
import java.util.ArrayList;
import javax.jms.JMSException;
import javax.jms.MapMessage;
import javax.jms.Message;
import javax.jms.Session;
import org.springframework.jms.core.MessageCreator;
public class MyMessage implements MessageCreator{
    @Override
    public Message createMessage(Session session) throws JMSException {
        MapMessage mapm = session.createMapMessage();
        ArrayList<String> arrayList = new ArrayList<String>();
        arrayList.add("陈恒1");
        arrayList.add("陈恒2");
        mapm.setObject("mesg1", arrayList);          //只能存Java的基本对象
        mapm.setString("mesg2", "测试消息2");
        return mapm;
    }
}
```

4. 发送消息

在 ch8_1Sender 应用的主类 Ch81SenderApplication 中，实现 Spring Boot 的 CommandLineRunner 接口，并重写 run 方法，用于程序启动后执行的代码。在该 run 方法

中，使用 JmsTemplate 的 send 方法向目的地 mydestination 发送 MyMessage 的消息，也相当于在消息代理上定义了一个叫作 mydestination 的目的地。具体代码如下：

```java
package com.ch.ch8_1sender;
import org.springframework.beans.factory.annotation.Autowired;
import org.springframework.boot.CommandLineRunner;
import org.springframework.boot.SpringApplication;
import org.springframework.boot.autoconfigure.SpringBootApplication;
import org.springframework.jms.core.JmsTemplate;
@SpringBootApplication
public class Ch81SenderApplication implements CommandLineRunner{
    @Autowired
    private JmsTemplate jmsTemplate;
    public static void main(String[] args) {
        SpringApplication.run(Ch81SenderApplication.class, args);
    }
    /**
     * 这里为了方便操作使用 run 方法发送消息，
     * 当然完全可以使用控制器通过 Web 访问
     */
    @Override
    public void run(String... args) throws Exception {
        //new MyMessage()回调接口方法 createMessage 产生消息
        jmsTemplate.send("mydestination", new MyMessage());
    }
}
```

5．创建消息接收者

按照步骤 1 创建 Spring Boot 应用 ch8_1Receive，该应用作为消息接收者，并按照步骤 2 配置 ch8_1Receive 的 ActiveMQ 的消息代理地址。

6．定义消息监听器接收消息

在应用 ch8_1Receive 的 com.ch.ch8_1receive 包中，创建消息监听器类 ReceiverMsg。在该类中使用@JmsListener 注解不停地监听目的地 mydestination 是否有消息发送过来，如果有就获取消息。具体代码如下：

```java
package com.ch.ch8_1receive;
import java.util.ArrayList;
import javax.jms.JMSException;
import javax.jms.MapMessage;
import org.springframework.jms.annotation.JmsListener;
import org.springframework.stereotype.Component;
@Component
public class ReceiverMsg {
    @JmsListener(destination = "mydestination")
    public void receiverMessage(MapMessage mapm) throws JMSException {
        @SuppressWarnings("unchecked")
        ArrayList<String> arrayList = (ArrayList<String>)mapm.getObject("mesg1");
        System.out.println(arrayList);
        System.out.println(mapm.getString("mesg2"));
    }
}
```

7. 运行测试

先启动消息接收者 ch8_1Receive 应用,然后单击图 8.5 中的 Queues,可看到如图 8.7 所示的界面。

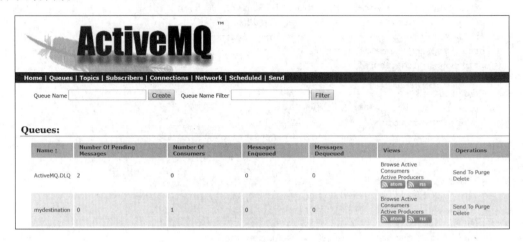

图 8.7　ActiveMQ 的 Queues

从图 8.7 可以看出目的地 mydestination 有一个消费者,正在等待接收消息。此时,启动消息发送者 ch8_1Sender 应用后,可在接收者 ch8_1Receive 应用的控制台上看到有消息打印,如图 8.8 所示。再去刷新图 8.7,可看到如图 8.9 所示的界面。

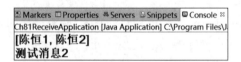

图 8.8　消息接收者接收的消息

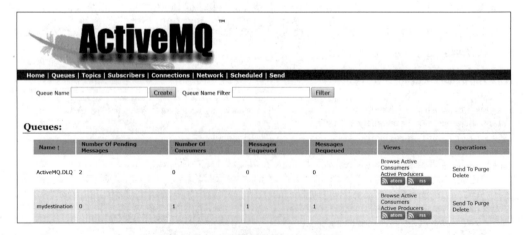

图 8.9　ActiveMQ 的 Queues

从图 8.9 可以看出目的地 mydestination 有一个消息入列(表示发送成功),有一个消息出列(表示接收成功)。

8.4.2　AMQP 实例

本节使用 AMQP 的主要实现 RabbitMQ 讲解 AMQP 实例,因此需要事先安装 RabbitMQ。因为 RabbitMQ 是基于 erlang 语言开发的,所以安装 RabbitMQ 之前,先下载安装 erlang。erlang 语言的下载地址为 https://www.erlang.org/downloads;RabbitMQ 的下载地址为 https://www.rabbitmq.com/download.html。编写本书时,下载的 erlang 语言版本是"otp_win64_22.0.exe",下载的 RabbitMQ 版本是"rabbitmq-server-3.7.18.exe"。

运行 erlang 语言安装包"otp_win64_22.0.exe",一直单击 Next 按钮即可完成 erlang 安装。安装 erlang 后需要配置环境变量 ERLANG_HOME 以及在 path 中新增 %ERLANG_HOME%\bin,如图 8.10 和图 8.11 所示。

图 8.10　ERLANG_HOME

图 8.11　在 path 中新增 %ERLANG_HOME%\bin

运行 RabbitMQ 安装包"rabbitmq-server-3.7.18.exe",一直单击 Next 按钮即可完成 RabbitMQ 安装。安装 RabbitMQ 后需要配置环境变量 RABBITMQ_SERVER=C:\Program Files\RabbitMQ Server\rabbitmq_server-3.7.18 以及在 path 中新增 %RABBITMQ_SERVER%\sbin,操作界面与图 8.10 和图 8.11 类似。

在 cmd 命令行窗口,进入 RabbitMQ 的 sbin 目录下,运行 rabbitmq-plugins.bat enable rabbitmq_management 命令,打开 RabbitMQ 的管理组件,如图 8.12 所示。

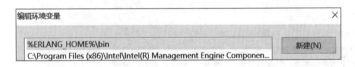

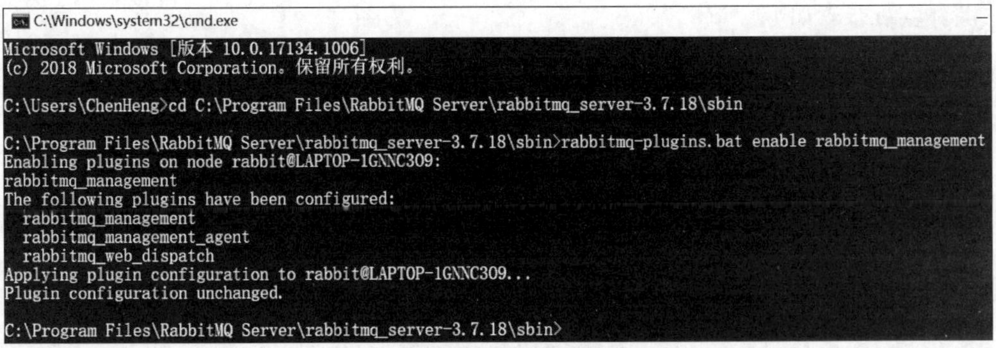

图 8.12　打开 RabbitMQ 的管理组件

以管理员方式打开 cmd 命令，运行 net start RabbitMQ 命令，提示 RabbitMQ 服务已经启动，如图 8.13 所示。

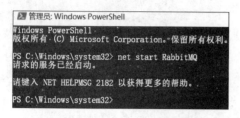

图 8.13　启动 RabbitMQ 服务

在浏览器地址栏中输入"http://localhost:15672"，账号和密码默认为 guest/guest，进入 RabbitMQ 的管理界面，如图 8.14 所示。

至此，完成了 RabbitMQ 服务器的搭建。

在例 8-1 中，不管是消息发送者（生产者）还是消息接收者（消费者），都必须知道一个指定的目的地（队列）才能发送、获取消息。如果同一个消息，要求每个消费者都处理的话，就需要发布/订阅式的消息分发模式。

图 8.14　RabbitMQ 的管理界面

下面通过实例讲解如何使用 RabbitMQ 实现发布/订阅式异步消息通信。在本例中，创建一个发布者应用、两个订阅者应用。该实例中的 3 个应用都是使用 Spring Boot 默认为我们配置的 RabbitMQ，主机为 localhost、端口号为 5672，所以无须在配置文件中配置 RabbitMQ 的连接信息。另外，3 个应用需要使用 Weather 实体类封装消息，并且使用 JSON 数据格式发布和订阅消息。

【例 8-2】　使用 RabbitMQ 实现发布/订阅式异步消息通信。

具体实现步骤如下。

1. 创建发布者应用 ch8_2Sender

创建发布者应用 ch8_2Sender，包括以下步骤。

（1）创建基于 RabbitMQ 的 Spring Boot 应用 ch8_2Sender，如图 8.15 所示。

（2）在 ch8_2Sender 应用的 pom.xml 中添加 spring-boot-starter-json 依赖。具体代码如下：

```
<dependency>
    <groupId>org.springframework.boot</groupId>
    <artifactId>spring-boot-starter-json</artifactId>
</dependency>
```

第8章 异步消息

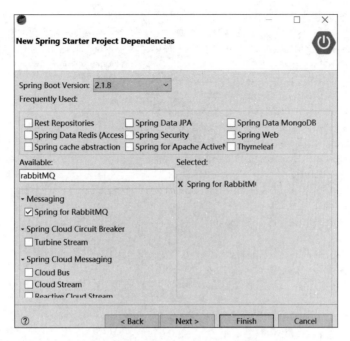

图 8.15　创建基于 RabbitMQ 的 Spring Boot 应用 ch8_2Sender

（3）在 ch8_2Sender 应用中创建名为 com.ch.ch8_2Sender.entity 的包，并在该包中创建 Weather 实体类。具体代码如下：

```
package com.ch.ch8_2Sender.entity;
import java.io.Serializable;
public class Weather implements Serializable{
    private static final long serialVersionUID = -8221467966772683998L;
    private String id;
    private String city;
    private String weatherDetail;
    //省略 set 和 get 方法
    @Override
    public String toString() {
        return "Weather [id = " + id + ", city = " + city + ", weatherDetail = " + weatherDetail + "]";
    }
}
```

（4）在 ch8_2Sender 应用的主类 Ch82SenderApplication 中，实现 Spring Boot 的 CommandLineRunner 接口，并重写 run 方法，用于程序启动后执行的代码。在该 run 方法中，使用 RabbitTemplate 的 convertAndSend 方法将特定的路由 weather.message 发送 Weather 消息对象到指定的交换机 weather-exchange。在发布消息前，需要使用 ObjectMapper 将 Weather 对象转换成 byte[] 类型的 JSON 数据。具体代码如下：

```
package com.ch.ch8_2Sender;
import org.springframework.amqp.core.Message;
import org.springframework.amqp.core.MessageBuilder;
```

```java
import org.springframework.amqp.core.MessageDeliveryMode;
import org.springframework.amqp.rabbit.connection.CorrelationData;
import org.springframework.amqp.rabbit.core.RabbitTemplate;
import org.springframework.amqp.support.converter.Jackson2JsonMessageConverter;
import org.springframework.beans.factory.annotation.Autowired;
import org.springframework.boot.CommandLineRunner;
import org.springframework.boot.SpringApplication;
import org.springframework.boot.autoconfigure.SpringBootApplication;
import com.ch.ch8_2Sender.entity.Weather;
import com.fasterxml.jackson.databind.ObjectMapper;
@SpringBootApplication
public class Ch82SenderApplication implements CommandLineRunner{
    @Autowired
    private ObjectMapper objectMapper;
    @Autowired
    RabbitTemplate rabbitTemplate;
    public static void main(String[] args) {
        SpringApplication.run(Ch82SenderApplication.class, args);
    }
    /**
     * 定义发布者
     */
    @Override
    public void run(String... args) throws Exception {
        //定义消息对象
        Weather weather = new Weather();
        weather.setId("010");
        weather.setCity("北京");
        weather.setWeatherDetail("今天晴到多云,南风 5-6 级,温度 19-26℃");
        //指定 Json 转换器,Jackson2JsonMessageConverter 默认将消息转换成 byte[]类型的消息
        rabbitTemplate.setMessageConverter(new Jackson2JsonMessageConverter());
        //objectMapper 将 weather 对象转换为 JSON 字节数组
        Message msg = MessageBuilder.withBody(objectMapper.writeValueAsBytes(weather))
                .setDeliveryMode(MessageDeliveryMode.NON_PERSISTENT)
                .build();
        //消息唯一 ID
        CorrelationData correlationData = new CorrelationData(weather.getId());
        //使用已封装好的 convertAndSend(String exchange , String routingKey , Object message,
        //CorrelationData correlationData)
        //将特定的路由 key 发送消息到指定的交换机
        rabbitTemplate.convertAndSend(
                "weather-exchange",      //分发消息的交换机名称
                "weather.message",       //用来匹配消息的路由 Key
                msg,                     //消息体
                correlationData);
    }
}
```

2. 创建订阅者应用 ch8_2Receiver-1

创建订阅者应用 ch8_2Receiver-1,具体步骤如下。

（1）创建基于 RabbitMQ 的 Spring Boot 应用 ch8_2Receiver-1。

（2）在 ch8_2Receiver-1 应用的 pom.xml 中添加 spring-boot-starter-json 依赖。

（3）将 ch8_2Sender 中的 Weather 实体类复制到 com.ch.ch8_2Receiver1 包中。

（4）在 com.ch.ch8_2Receiver1 包中创建订阅者类 Receiver1，在该类中使用 @RabbitListener 和 @RabbitHandler 注解监听发布者并接收消息，具体代码如下：

```java
package com.ch.ch8_2Receiver1;
import org.springframework.amqp.rabbit.annotation.Exchange;
import org.springframework.amqp.rabbit.annotation.Queue;
import org.springframework.amqp.rabbit.annotation.QueueBinding;
import org.springframework.amqp.rabbit.annotation.RabbitHandler;
import org.springframework.amqp.rabbit.annotation.RabbitListener;
import org.springframework.beans.factory.annotation.Autowired;
import org.springframework.messaging.handler.annotation.Payload;
import org.springframework.stereotype.Component;
import com.fasterxml.jackson.databind.ObjectMapper;
/**
 * 定义订阅者 Receiver1
 */
@Component
public class Receiver1 {
    @Autowired
    private ObjectMapper objectMapper;
    @RabbitListener(
            bindings =
            @QueueBinding(
                //队列名 weather-queue1 保证和别的订阅者不一样,可以随机起名
                value = @Queue(value = "weather-queue1",durable = "true"),
                //weather-exchange 与发布者的交换机名相同
                exchange = @Exchange(value = "weather-exchange",durable = "true",type = "topic"),
                //weather.message 与发布者的消息的路由 Key 相同
                key = "weather.message"
            )
    )
    @RabbitHandler
    public void receiveWeather(@Payload byte[] weatherMessage)throws Exception{
        System.out.println("---------- 订阅者 Receiver1 接收到消息 --------");
        //将 JSON 字节数组转换为 Weather 对象
        Weather w = objectMapper.readValue(weatherMessage, Weather.class);
        System.out.println("Receiver1 收到的消息内容:" + w);
    }
}
```

3. 创建订阅者应用 ch8_2Receiver-2

与创建订阅者应用 ch8_2Receiver-1 的步骤一样，这里不再赘述。但需要注意的是两个订阅者的队列名不同。

4. 测试运行

首先,运行发布者应用 ch8_2Sender 的主类 Ch82SenderApplication。

其次,运行订阅者应用 ch8_2Receiver-1 的主类 Ch82Receiver1Application,此时接收到的消息如图 8.16 所示。

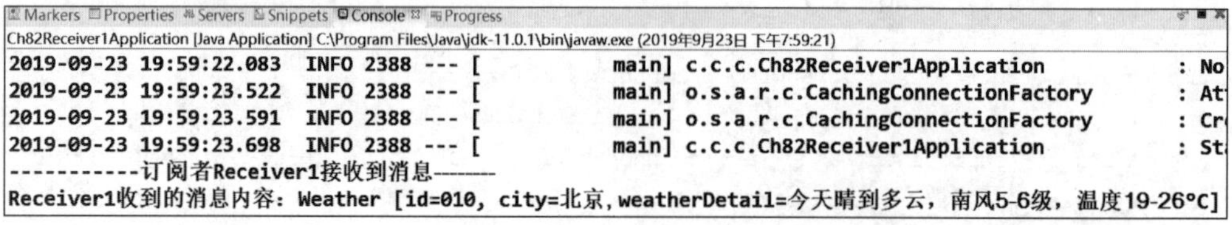

```
Markers Properties Servers Snippets Console Progress
Ch82Receiver1Application [Java Application] C:\Program Files\Java\jdk-11.0.1\bin\javaw.exe (2019年9月23日 下午7:59:21)
2019-09-23 19:59:22.083  INFO 2388 --- [           main] c.c.c.Ch82Receiver1Application         : No
2019-09-23 19:59:23.522  INFO 2388 --- [           main] o.s.a.r.c.CachingConnectionFactory     : At
2019-09-23 19:59:23.591  INFO 2388 --- [           main] o.s.a.r.c.CachingConnectionFactory     : Cr
2019-09-23 19:59:23.698  INFO 2388 --- [           main] c.c.c.Ch82Receiver1Application         : St
-----------订阅者Receiver1接收到消息-------
Receiver1收到的消息内容: Weather [id=010, city=北京,weatherDetail=今天晴到多云, 南风5-6级, 温度19-26°C]
```

图 8.16 订阅者 ch8_2Receiver-1 接收到的消息

最后,运行订阅者应用 ch8_2Receiver-2 的主类 Ch82Receiver2Application,此时接收到的消息如图 8.17 所示。

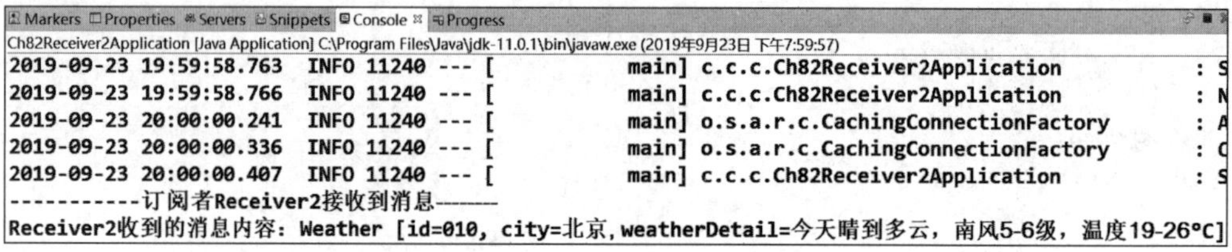

```
Markers Properties Servers Snippets Console Progress
Ch82Receiver2Application [Java Application] C:\Program Files\Java\jdk-11.0.1\bin\javaw.exe (2019年9月23日 下午7:59:57)
2019-09-23 19:59:58.763  INFO 11240 --- [           main] c.c.c.Ch82Receiver2Application         : S
2019-09-23 19:59:58.766  INFO 11240 --- [           main] c.c.c.Ch82Receiver2Application         : N
2019-09-23 20:00:00.241  INFO 11240 --- [           main] o.s.a.r.c.CachingConnectionFactory     : A
2019-09-23 20:00:00.336  INFO 11240 --- [           main] o.s.a.r.c.CachingConnectionFactory     : C
2019-09-23 20:00:00.407  INFO 11240 --- [           main] c.c.c.Ch82Receiver2Application         : S
-----------订阅者Receiver2接收到消息-------
Receiver2收到的消息内容: Weather [id=010, city=北京,weatherDetail=今天晴到多云, 南风5-6级, 温度19-26°C]
```

图 8.17 订阅者 ch8_2Receiver-2 接收到的消息

从例 8-2 可以看出,一个发布者发布的消息,可以被多个订阅者订阅,这就是所谓的发布/订阅式异步消息通信。

8.5 本章小结

本章主要介绍了多个应用系统之间的异步消息。通过本章的学习,读者应该了解 Spring Boot 对 JMS 和 AMQP 的支持,掌握如何在实际应用开发中,使用 JMS 或 AMQP 提供异步通信解决方案。

习题 8

1. 在多个应用系统之间的异步消息中,有哪些消息模型?
2. JMS 和 AMQP 有什么区别?

Spring Boot 的热部署与单元测试

📕 学习目的与要求

本章主要讲解了 Spring Boot 开发的热部署以及 Spring Boot 的单元测试。通过本章的学习，掌握开发热部署的使用技巧，理解单元测试的原理。

📖 本章主要内容

- 开发的热部署。
- Spring Boot 的单元测试。

在实际应用开发过程中，业务变化、代码错误等发生时，难免修改程序。为了正确运行出修改的结果，我们往往需要重启应用，否则将不能看到修改后的效果，这一启动过程是非常浪费时间的，导致开发效率低。因此，我们有必要学习 Spring Boot 开发的热部署，自动实现应用的重启和部署，大大提高开发调试效率。

9.1 开发的热部署

开发热部署的目的是使应用自动重启和部署，提高开发效率。本节将讲解如何实现 Spring Boot 开发的热部署，包括前端模板引擎和后端程序的热部署。

9.1.1 模板引擎的热部署

在 Spring Boot 应用中，使用模板引擎的页面默认是开启缓存的，如果修改了页面内容,则刷新页面是得不到修改后的页面的效果的。因此,可以在配置文件 application .properties 中关闭模板引擎的缓存。示例代码如下。

关闭 Thymeleaf 缓存的配置:

spring.thymeleaf.cache = false

关闭 FreeMarker 缓存的配置:

spring.freemarker.cache = false

关闭 Groovy 缓存的配置:

spring.groovy.template.cache = false

9.1.2 使用 spring-boot-devtools 进行热部署

在 Spring Boot 应用的 pom.xml 文件中添加 spring-boot-devtools 依赖即可实现页面和代码的热部署。

spring-boot-devtools 是一个为开发者服务的模块,最重要的功能就是自动实现将修改的应用代码更新到最新的应用上。其工作原理是使用两个 ClassLoader:一个 ClassLoader 加载那些不会改变的类(如第三方 JAR 包);一个 ClassLoader 加载更新的类。称为 Restart ClassLoader。这样在有代码修改时,原来的 Restart ClassLoader 被丢弃,重新创建一个 Restart ClassLoader 加载更新的类。由于只加载部分修改的类,所以实现了较快的重启。

下面通过实例讲解如何使用 spring-boot-devtools 进行热部署。

【例 9-1】 使用 spring-boot-devtools 进行热部署。

具体实现步骤如下。

1. 创建 Spring Boot Web 应用

创建 Spring Boot Web 应用 ch9_1。

2. 添加 spring-boot-devtools 依赖

在 ch9_1 应用的 pom.xml 文件中,添加 spring-boot-devtools 依赖。具体代码如下:

```
<dependency>
    <groupId>org.springframework.boot</groupId>
    <artifactId>spring-boot-devtools</artifactId>
</dependency>
```

3. 创建控制器类

在 com.ch.ch9_1 包中,创建控制器类 TestDevToolsController。具体代码如下:

```
package com.ch.ch9_1;
import org.springframework.web.bind.annotation.RequestMapping;
import org.springframework.web.bind.annotation.RestController;
```

```
@RestController
public class TestDevToolsController {
    @RequestMapping("/testDevTools")
    public String testDevTools() {
        return "test DevTools 111";
    }
}
```

4．测试运行

首先，运行 Ch91Application 主类，启动 ch9_1 应用。然后，通过"http://localhost:8080/testDevTools"请求 TestDevToolsController 类中的 testDevTools 方法。运行效果如图 9.1 所示。

现在，将 testDevTools 方法中的 return 语句修改如下：

```
return "testDevTools 222";
```

无须重启 ch9_1 应用，直接刷新"http://localhost:8080/testDevTools"，运行效果如图 9.2 所示。

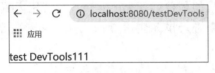

图 9.1　请求 testDevTools 方法

图 9.2　刷新页面效果

从例 9-1 可以看出，spring-boot-devtools 实现了代码修改后的热部署，同样，也可实现新增类、修改配置文件等的热部署。

9.2　Spring Boot 的单元测试

测试是系统开发中非常重要的工作，单元测试在帮助开发人员编写高品质的程序、提升代码质量方面发挥了极大的作用，本节将介绍 Spring Boot 的单元测试。

Spring Boot 为测试提供了一个名为 spring-boot-starter-test 的 Starter。使用 STS 创建 Spring Boot 应用时，将自动添加 spring-boot-starter-test 依赖。这样在测试时，就没有必要再添加额外的 JAR 包。spring-boot-starter-test 主要提供了以下测试库。

（1）JNnit：标准的单元测试 Java 应用程序。

（2）Spring Test & Spring Boot Test：针对 Spring Boot 应用程序的单元测试。

（3）Mockito：Java mocking 框架，用于模拟任何 Spring 管理的 Bean，例如在单元测试中模拟一个第三方系统 Service 接口返回的数据，而不去真正调用第三方系统。

（4）AssertJ：一个流畅的 assertion 库，同时也提供了更多的期望值与测试返回值的比较方式。

（5）JSONassert：对 JSON 对象或 JSON 字符串断言的库。

（6）JsonPath：提供类似于 Xpath（一门在 XML 文档中查找信息的语言）那样的符号来

获取 JSON 数据片段。

9.2.1 Spring Boot 单元测试程序模板

在 Spring Boot 应用中，使用一系列注解增强单元测试以支持 Spring Boot 测试。通常 Spring Boot 单元测试程序类似于如下模板：

```
@RunWith(SpringRunner.class)
@SpringBootTest
public class GoodsServiceTest {
    //注入要测试的 service
    @Autowired
    private GoodsService goodsService;
    @Test
    public void testGoodsService() {
        //调用 GoodsService 的方法进行测试
    }
}
```

@RunWith 注解是 JUnit 标准的一个注解，目的是告诉 JUnit 框架不要使用内置的方式进行单元测试，而应使用@RunWith 指明的类来进行单元测试，所有的 Spring 单元测试总是使用 SpringRunner.class。

@SpringBootTest 用于 Spring Boot 应用测试，它默认根据包名逐级往上找，一直找到 Spring Boot 主程序（包含@SpringBootApplication 注解的类），并在单元测试时启动该主程序来创建 Spring 上下文环境。

9.2.2 测试 Service

单元测试 Service 代码与通过 Controller 调用 Service 代码相比，需要特别考虑该 Service 是否依赖其他还未开发完毕的 Service（第三方接口）。如果依赖其他还未开发完毕的 Service，需要使用 Mockito 来模拟未完成的 Service。

假设在 UserService 中依赖 CreditService（第三方接口）的 getCredit 方法获得用户积分。UserService 的定义如下：

```
@Service
public class UserServiceImpl implements UserService{
    @Autowired
    private CreditService creditService;
    @Autowired
    UserRepository userRepository;
    @Override
    public int getCredit(Integer uid){
        User user = userRepository.getOne(uid);
        if(user != null)
            return creditService.getCredit(uid);
        else
```

```
        return -1;
    }
}
```

那么，如何测试 UserService 呢？单元测试不能实际调用 CreditService（因为 CreditService 是第三方系统），因此，单元测试类需要使用 Mockito 的注解 @MockBean 自动注入 Spring 管理的 Service，用来提供模拟实现。在 Spring 上下文中，CreditService 实现已经被模拟实现代替了。UserService 测试类的代码模板如下：

```
...
import org.mockito.BDDMockito;
import org.springframework.boot.test.mock.mockito.MockBean;
@RunWith(SpringRunner.class)
@SpringBootTest
@Transactional
public class UserServiceTest {
    //注入要测试的 service
    @Autowired
    private UserService userService;
    @MockBean
    private CreditService creditService;
    @Test
    public void testUserService() {
        int uid = 1;
        int expectedCredit = 50;
        /* given 是 BDDMockito 的一个静态方法，用来模拟一个 Service 方法调用返回，anyInt() 表
        示可以传入任何参数，willReturn 方法说明这个调用将返回 50。*/
        BDDMockito.given(creditService.getCredit(anyInt())).willReturn(expectedCredit);
        int credit = userService.getCredit(uid);
        /* assert 定义测试的条件，expectedCredit 与 credit 相等时，assertEquals 方法保持沉
        默，不等时抛出异常 */
        assertEquals(expectedCredit, credit);
    }
}
```

9.2.3 测试 Controller

在 Spring Boot 应用中，可以单独测试 Controller 代码，用来验证与 Controller 相关的 URL 路径映射、文件上传、参数绑定、参数校验等特性。可以通过 @WebMvcTest 注解来完成 Controller 单元测试，当然也可以通过 @SpringBootTest 测试 Controller。通过 @WebMvcTest 注解测试 Controller 的代码模板如下：

```
...
import org.mockito.BDDMockito;
import org.springframework.boot.test.mock.mockito.MockBean;
@RunWith(SpringRunner.class)
//被测试的 Controller
```

```
@WebMvcTest(UserController.class)
public class UserControllerTest{
    //MockMvc 是 Spring 提供的专用于测试 Controller 的类
    @Autowired
    private MockMvc mvc;
    /*用@MockBean 模拟实现 UserService,这是因为在测试 Controller 时,Spring 容器并不会初
    始化@Service 注解的 Service 类。*/
    @MockBean
    private UserService userService;
    @Test
    public void testMvc(){
        int uid = 1;
        int expectedCredit = 50;
        /*given 是 BDDMockito 的一个静态方法,用来模拟一个 Service 方法调用返回。这里模拟
        userService*/
        BDDMockito.given(userService.getCredit(uid)).willReturn(50);
        /*perform 完成一次 Controller 的调用,Controller 测试是一种模拟测试,实际上并未发
        起一次真正的 HTTP 请求;get 方法模拟了一次 Get 请求,请求地址为/getCredit/{id},这里
        的{id}被其后的参数 uid 代替,因此请求路径是/getCredit/1;andExpect 表示请求期望的
        返回结果。*/
        mvc.peform(get("/getCredit/{id}", uid))
            .andExpect(content().string(String.valueOf(expectedCredit)));
    }
}
```

需要注意的是,在使用@WebMvcTest 注解测试 Controller 时,带有@Service 以及别的注解组件类不会自动被扫描注册为 Spring 容器管理的 Bean,而@SpringBootTest 注解告诉 Spring Boot 去寻找一个主配置类(一个带@SpringBootApplication 的类),并使用它来启动 Spring 应用程序上下文,注入所有 Bean。另外,还需要注意的是,MockMvc 用来在 Servlet 容器内对 Controller 进行单元测试,并未真正发起了 HTTP 请求调用 Controller。

@WebMvcTest 用于从服务器端对 Controller 层进行统一测试;如果需要从客户端与应用程序交互时,应该使用@SpringBootTest 做集成测试。

9.2.4 模拟 Controller 请求

MockMvc 的核心方法是:

`public ResultActions perform(RequestBuilder requestBuilder)`

RequestBuilder 类可以通过调用 MockMvcRequestBuilders 的 get、post、multipart 等方法来模拟 Controller 请求。常用示例代码如下。

模拟一个 get 请求:

`mvc.peform(get("/getCredit/{id}", uid));`

模拟一个 post 请求:

`mvc.peform(post("/getCredit/{id}", uid));`

模拟文件上传：

```
mvc.peform(multipart("/upload").file("file", "文件内容".getBytes("UTF-8")));
```

模拟请求参数：

```
//模拟提交 errorMessage 参数
mvc.peform(get("/getCredit/{id}/{uname}", uid, uname).param("errorMessage", "用户名或密码错误"));
//模拟提交 check
mvc.peform(get("/getCredit/{id}/{uname}", uid, uname).param("job", "收银员", "IT"));
```

9.2.5 比较 Controller 请求返回的结果

我们知道，MockMvc 的 perform 方法返回 ResultActions 实例，这个实例代表了请求 Controller 返回的结果。它提供了一系列 andExpect 方法来对请求 Controller 返回的结果进行比较。示例代码如下：

```
mvc.peform(get("/getOneUser/10"))
    .andExpect(status().isOk())    //期望请求成功，即状态码为 200
    //期望返回内容是 application/json
    .andExpect(content().contentType(MediaType.APPLICATION_JSON))
    //使用 JsonPath 比较返回的 JSON 内容
    .andExpect(jsonPath("$.name").value("chenheng"));    //检查返回内容
```

除了上述对请求 Controller 返回的结果进行比较，还有如下的常见结果比较。

1. 比较返回的视图

```
mvc.peform(get("/getOneUser/10"))
    .andExpect(view().name("/userDetail"));
```

2. 比较模型

```
mvc.peform(post("/addOneUser"))
    .andExpect(status().isOk())
    .andExpect(model().size(1))
    .andExpect(model().attributeExists("oneUser"))
    .andExpect(model().attribute("oneUser", "chenheng"))
```

3. 比较转发或重定向

```
mvc.peform(post("/addOneUser"))
    .andExpect(forwardedUrl("/user/selectAll"));    //或者 redirectedUrl("/user/selectAll")
```

4. 比较返回的内容

```
andExpect(content().string("测试很好玩"));    //比较返回的字符串
andExpect(content().xml(xmlContent));        //返回内容是 XML，并且与 xmlContent(变量)一样
andExpect(content().json(jsonContent));      //返回内容是 JSON，并且与 jsonContent(变量)一样
```

9.2.6 测试实例

本节将演示一个简单的测试实例,分别使用@WebMvcTest 和@SpringBootTest 两种方式测试某一个控制器方法是否满足测试用例。

【例 9-2】 使用@WebMvcTest 和@SpringBootTest 两种方式测试某一个控制器方法。

具体实现步骤如下。

1. 创建基于 Spring Data JPA 的 Web 应用 ch9_2

创建基于 Spring Data JPA 的 Web 应用 ch9_2,如图 9.3 所示。

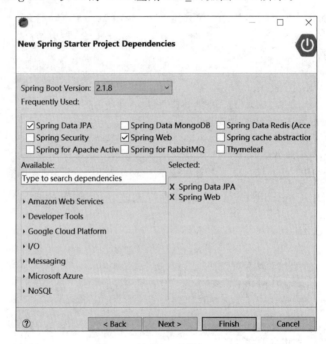

图 9.3 创建 Web 应用 ch9_2

2. 修改 pom.xml 文件,引入 MySQL 依赖

修改 pom.xml 文件,引入 MySQL 依赖。具体代码如下:

```
<dependency>
    <groupId>mysql</groupId>
    <artifactId>mysql-connector-java</artifactId>
    <version>5.1.45</version>
</dependency>
```

3. 配置数据库连接等基本属性

修改配置文件 application.properties 的内容,配置数据库连接等基本属性。具体代码如下:

```
server.servlet.context-path=/ch9_2
###
##数据源信息配置
###
#数据库地址
spring.datasource.url=jdbc:mysql://localhost:3306/springbootjpa?characterEncoding=utf8
#数据库用户名
spring.datasource.username=root
#数据库密码
spring.datasource.password=root
#数据库驱动
spring.datasource.driver-class-name=com.mysql.jdbc.Driver
####
#JPA持久化配置
####
#指定数据库类型
spring.jpa.database=MYSQL
#指定是否在日志中显示SQL语句
spring.jpa.show-sql=true
#指定自动创建、更新数据库表等配置,update表示如果数据库中存在持久化类对应的表就不创建,
#不存在就创建对应的表
spring.jpa.hibernate.ddl-auto=update
#让控制器输出的JSON字符串格式更美观
spring.jackson.serialization.indent-output=true
```

4. 创建持久化实体类

创建名为com.ch.ch9_2.entity的包,并在该包中创建名为Student的持久化实体类。具体代码如下:

```
package com.ch.ch9_2.entity;
import java.io.Serializable;
import javax.persistence.Entity;
import javax.persistence.GeneratedValue;
import javax.persistence.GenerationType;
import javax.persistence.Id;
import javax.persistence.Table;
import com.fasterxml.jackson.annotation.JsonIgnoreProperties;
@Entity
@Table(name = "student_table")
/** 解决 No serializer found for class org.hibernate.proxy.pojo.bytebuddy.ByteBuddyInterceptor
异常*/
@JsonIgnoreProperties(value = {"hibernateLazyInitializer"})
public class Student implements Serializable{
    private static final long serialVersionUID = 1L;
    @Id
    @GeneratedValue(strategy = GenerationType.IDENTITY)
    private int id;                          //主键
    private String sno;
    private String sname;
    private String ssex;
```

```java
    public Student() {
        super();
    }
    public Student(int id, String sno, String sname, String ssex) {
        super();
        this.id = id;
        this.sno = sno;
        this.sname = sname;
        this.ssex = ssex;
    }
    //省略 get 方法和 set 方法
}
```

5. 创建数据访问层

创建名为 com.ch.ch9_2.repository 的包，并在该包中创建数据访问接口 StudentRepository。具体代码如下：

```java
package com.ch.ch9_2.repository;
import org.springframework.data.jpa.repository.JpaRepository;
import com.ch.ch9_2.entity.Student;
public interface StudentRepository extends JpaRepository<Student, Integer>{
}
```

6. 创建控制器层

创建名为 com.ch.ch9_2.controller 的包，并在该包中创建控制器类 StudentController。具体代码如下：

```java
package com.ch.ch9_2.controller;
import org.springframework.beans.factory.annotation.Autowired;
import org.springframework.web.bind.annotation.GetMapping;
import org.springframework.web.bind.annotation.PathVariable;
import org.springframework.web.bind.annotation.PostMapping;
import org.springframework.web.bind.annotation.RequestBody;
import org.springframework.web.bind.annotation.RequestMapping;
import org.springframework.web.bind.annotation.RestController;
import com.ch.ch9_2.entity.Student;
import com.ch.ch9_2.repository.StudentRepository;
@RestController
@RequestMapping("/student")
public class StudentController {
    @Autowired
    private StudentRepository studentRepository;
    /**
     * 保存学生信息
     */
    @PostMapping("/save")
    public String save(@RequestBody Student student) {
        studentRepository.save(student);
```

```
        return "success";
    }
    /**
     * 根据 id 查询学生信息
     */
    @GetMapping("/getOne/{id}")
    public Student getOne(@PathVariable("id") int id){
        return studentRepository.getOne(id);
    }
}
```

7. 创建测试用例

分别使用@WebMvcTest 和@SpringBootTest 两种方式测试控制器类 StudentController 中的请求处理方法。

1）创建基于@WebMvcTest 的测试用例

使用@WebMvcTest 注解测试 Controller 时，带有@Service 以及别的注解组件类不会自动被扫描注册为 Spring 容器管理的 Bean。因此，Controller 所依赖的对象必须使用@MockBean 来模拟实现。

在 src/test/java 目录下的 com.ch.ch9_2 包中，创建基于@WebMvcTest 的测试用例类 WebMvcTestStudentController。具体代码如下：

```
package com.ch.ch9_2;
import static org.springframework.test.web.servlet.request.MockMvcRequestBuilders.get;
import static org.springframework.test.web.servlet.request.MockMvcRequestBuilders.post;
import static org.springframework.test.web.servlet.result.MockMvcResultHandlers.print;
import static org.springframework.test.web.servlet.result.MockMvcResultMatchers.jsonPath;
import static org.springframework.test.web.servlet.result.MockMvcResultMatchers.status;
import org.junit.Test;
import org.junit.runner.RunWith;
import org.mockito.BDDMockito;
import org.springframework.beans.factory.annotation.Autowired;
import org.springframework.boot.test.autoconfigure.web.servlet.WebMvcTest;
import org.springframework.boot.test.mock.mockito.MockBean;
import org.springframework.http.MediaType;
import org.springframework.test.context.junit4.SpringRunner;
import org.springframework.test.web.servlet.MockMvc;
import com.ch.ch9_2.controller.StudentController;
import com.ch.ch9_2.entity.Student;
import com.ch.ch9_2.repository.StudentRepository;
import com.fasterxml.jackson.databind.ObjectMapper;
@RunWith(SpringRunner.class)
/*仅仅扫描这个 StudentController 类，即注入 StudentController 到 Spring 容器*/
@WebMvcTest(StudentController.class)
public class WebMvcTestStudentController {
    //MockMvc 是 Spring 提供的专用于测试 Controller 的类
    @Autowired
    private MockMvc mvc;
    //因为在 StudentController 类依赖 StudentRepository，所以需要 mock(模拟)依赖
```

```java
    @MockBean
    private StudentRepository studentRepository;
    @Test
    public void saveTest() throws Exception {
        Student stu = new Student(1,"5555","陈恒","男");
        ObjectMapper mapper = new ObjectMapper();          //把对象转换成JSON字符串
        mvc.perform(post("/student/save")
                .contentType(MediaType.APPLICATION_JSON_UTF8)   //发送JSON数据格式
                .accept(MediaType.APPLICATION_JSON_UTF8)        //接收JSON数据格式
                .content(mapper.writeValueAsString(stu))        //传递JSON字符串参数
                )
                .andExpect(status().isOk())        //状态响应码为200,如果不是抛出异常,测试不通过
                .andDo(print());                    //输出结果
    }
    @Test
    public void getStudent() throws Exception {
        Student stu = new Student(1,"5555","陈恒","男");
        //模拟StudentRepository,getOne(1)将返回stu对象
        BDDMockito.given(studentRepository.getOne(1)).willReturn(stu);
        mvc.perform(get("/student/getOne/{id}", 1)
                .contentType(MediaType.APPLICATION_JSON_UTF8)
                .accept(MediaType.APPLICATION_JSON_UTF8)
                )
                .andExpect(status().isOk())//状态响应码为200,如果不是抛出异常,测试不通过
                .andExpect(jsonPath("$.sname").value("陈恒"))
                .andDo(print());            //输出结果
    }
}
```

2) 创建基于@SpringBootTest 的测试用例

@SpringBootTest 注解告诉 Spring Boot 去寻找一个主配置类（一个带@SpringBootApplication 的类），并使用它启动 Spring 应用程序的上下文,同时注入所有 Bean。

在 src/test/java 目录下的 com.ch.ch9_2 包中,创建基于@SpringBootTest 的测试用例类 SpringBootTestStudentController。具体代码如下：

```java
package com.ch.ch9_2;
import static org.springframework.test.web.servlet.request.MockMvcRequestBuilders.get;
import static org.springframework.test.web.servlet.request.MockMvcRequestBuilders.post;
import static org.springframework.test.web.servlet.result.MockMvcResultHandlers.print;
import static org.springframework.test.web.servlet.result.MockMvcResultMatchers.jsonPath;
import static org.springframework.test.web.servlet.result.MockMvcResultMatchers.status;
import org.junit.Before;
import org.junit.Test;
import org.junit.runner.RunWith;
import org.springframework.beans.factory.annotation.Autowired;
import org.springframework.boot.test.context.SpringBootTest;
import org.springframework.http.MediaType;
import org.springframework.test.context.junit4.SpringRunner;
import org.springframework.test.web.servlet.MockMvc;
```

```java
import org.springframework.test.web.servlet.setup.MockMvcBuilders;
import org.springframework.transaction.annotation.Transactional;
import org.springframework.web.context.WebApplicationContext;
import com.ch.ch9_2.entity.Student;
import com.fasterxml.jackson.databind.ObjectMapper;
@RunWith(SpringRunner.class)
@SpringBootTest(classes = Ch92Application.class)       //应用的主程序
public class SpringBootTestStudentController {
    //注入 Spring 容器
    @Autowired
    private WebApplicationContext wac;
    //MockMvc 模拟实现对 Controller 的请求
    private MockMvc mvc;
    //在测试前,初始化 MockMvc 对象
    @Before
    public void initMockMvc() {
        mvc = MockMvcBuilders.webAppContextSetup(wac).build();
    }
    @Test
    @Transactional
    public void saveTest() throws Exception {
        Student stu = new Student(1, "5555","陈恒","男");
        ObjectMapper mapper = new ObjectMapper();              //把对象转换成 JSON 字符串
        mvc.perform(post("/student/save")
                .contentType(MediaType.APPLICATION_JSON_UTF8)//发送 JSON 数据格式
                .accept(MediaType.APPLICATION_JSON_UTF8)     //接收 JSON 数据格式
                .content(mapper.writeValueAsString(stu))     //传递 JSON 字符串参数
                )
                .andExpect(status().isOk())       //状态响应码为 200,如果不是抛出异常,测试不通过
                .andDo(print());                  //输出结果
    }
    @Test
    public void getStudent() throws Exception {
        mvc.perform(get("/student/getOne/{id}", 1)
                .contentType(MediaType.APPLICATION_JSON_UTF8)
                .accept(MediaType.APPLICATION_JSON_UTF8)
                )
                .andExpect(status().isOk())  //状态响应码为 200,如果不是抛出异常,测试不通过
                .andExpect(jsonPath("$.sname").value("陈恒"))
                .andDo(print());             //输出结果
    }
}
```

8. 运行

打开 WebMvcTestStudentController 测试类,右击,选择 Run As、JUnit Test 命令,执行结果如图 9.4 所示。

打开 SpringBootTestStudentController 测试类,右击,选择 Run As、JUnit Test 命令,执行结果如图 9.5 所示。

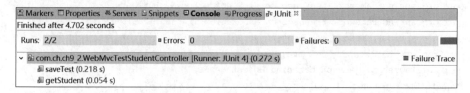

图 9.4　WebMvcTestStudentController 测试类的运行结果

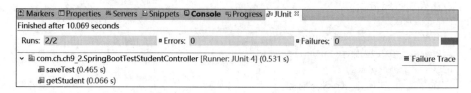

图 9.5　SpringBootTestStudentController 测试类的运行结果

从图 9.4 和图 9.5 可以看出两个测试用例达到测试预期效果。

9.3　本章小结

本章主要介绍了 Spring Boot 应用开发的热部署以及 Spring Boot 的单元测试。通过本章的学习，读者应该了解热部署的目的，掌握测试用例类的编写及原理。

习题 9

1. @SpringBootTest 和 @WebMvcTest 的区别是什么？
2. 什么是热部署？在 Spring Boot 中如何进行热部署？

第10章 监控 Spring Boot 应用

学习目的与要求

本章主要讲解了如何使用 Spring Boot 的 Actuator 功能完成 Spring Boot 的应用监控和管理。通过本章的学习，掌握如何通过 HTTP 进行 Spring Boot 的应用监控和管理。

本章主要内容

- 端点的分类与测试。
- 自定义端点。
- 自定义 HealthIndicator。

Spring Boot 提供了 Actuator 功能，完成运行时的应用监控和管理功能。可以通过 HTTP、JMX（Java Management Extensions，Java 管理扩展）以及 SSH（远程脚本）来进行 Spring Boot 的应用监控和管理功能。本章将学习如何通过 HTTP 进行 Spring Boot 的应用监控和管理功能。

在 Spring Boot 应用中，既然通过 HTTP 使用 Actuator 的监控和管理功能，那么在 pom.xml 文件中，除了引入 spring-boot-starter-web 之外，还需要引入 spring-boot-starter-actuator。具体代码如下：

```xml
<dependency>
    <groupId>org.springframework.boot</groupId>
    <artifactId>spring-boot-starter-actuator</artifactId>
</dependency>
```

10.1 端点的分类与测试

视频讲解

Spring Boot 提供了许多监控和管理功能的端点。根据端点的作用，可以将 Spring Boot 提供的原生端点分为三大类：应用配置端点、度量指标端点和操作控制端点。

10.1.1 端点的开启与暴露

在讲解端点的具体分类以及功能前，先通过实例查看 Spring Boot 默认暴露的端点。

【例 10-1】 查看 Spring Boot 默认暴露的端点。

具体实现步骤如下。

1. 创建基于 Spring Boot Actuator 的 Web 应用 ch10_1

创建基于 Spring Boot Actuator 的 Web 应用 ch10_1，如图 10.1 所示。

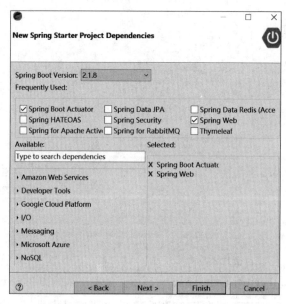

图 10.1　创建基于 Spring Boot Actuator 的 Web 应用 ch10_1

2. 配置 JSON 输出格式

在 Web 应用 ch10_1 的配置文件 application.properties 中，配置 JSON 字符串的输出格式。具体代码如下：

```
#输出的JSON字符串格式更美观
spring.jackson.serialization.indent-output=true
```

3. 启动主程序查看默认暴露的端点

启动 Web 应用 ch10_1 的主程序 Ch101Application 后，通过访问"http://localhost：

8080/actuator"查看默认暴露的端点,运行效果如图 10.2 所示。

图 10.2　查看默认暴露的端点

从图 10.2 可以看出 Spring Boot 默认暴露了 health 和 info 两个端点。如果想暴露 Spring Boot 提供的所有端点,需要在配置文件 application.properties 中配置 "management.endpoints.web.exposure.include=*",配置后重启应用主程序,重新访问 "http://localhost:8080/actuator",就可以查看所有暴露的端点,运行效果如图 10.3 所示。

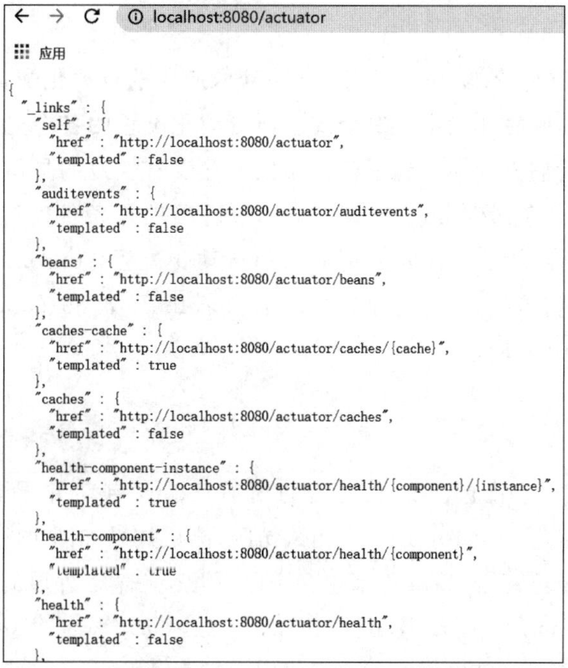

图 10.3　查看暴露的所有端点

默认情况下,除了 shutdown 端点是关闭的,其他端点都是开启的。配置一个端点的开启,使用 management.endpoint..enabled 属性,如启用 shutdown 端点:

management.endpoint.shutdown.enabled = true

在配置文件中可使用"management.endpoints.web.exposure.include"属性列出暴露的端点,示例代码如下:

management.endpoints.web.exposure.include = info,health,env,beans

"*"可用来表示所有的端点,例如,除了 env 和 beans 端点,通过 HTTP 暴露所有端点,示例代码如下:

management.endpoints.web.exposure.include = *
management.endpoints.web.exposure.exclude = env,beans

10.1.2　应用配置端点的测试

Spring Boot 采用了包扫描和自动化配置的机制来加载原本集中于 XML 文件中的各项配置内容,虽然这让代码变得非常简洁,但是整个应用的实例创建和依赖关系等信息都被离散到了各个配置类的注解上,使得我们分析整个应用中资源和实例的各种关系变得非常的困难。而通过应用配置端点就可以帮助我们轻松地获取一系列关于 Spring 应用配置内容的详细报告,例如自动化配置的报告、Bean 创建的报告、环境属性的报告等。

1. conditions

该端点在 1.x 版本中名为 autoconfig,用来获取应用的自动化配置报告,其中包括所有自动化配置的候选项,同时还列出了每个候选项自动化配置的各个先决条件是否满足。所以,该端点可以帮助我们方便地找到一些自动化配置没有生效的具体原因。该报告内容将自动化配置内容分为三部分:positiveMatches 中返回的是条件匹配成功的自动化配置;negativeMatches 中返回的是条件匹配不成功的自动化配置;unconditionalClasses 是无条件配置类。启动并暴露该端点后,可通过"http://localhost:8080/actuator/conditions"测试访问,测试效果如图 10.4 所示。

2. beans

该端点用来获取应用上下文中创建的所有 Bean,启动并暴露该端点后,可通过"http://localhost:8080/actuator/beans"测试访问,测试效果如图 10.5 所示。

从图 10.5 可以看出每个 Bean 中都包含了以下几个信息:外层的 key 是 Bean 的名称;aliases 是 Bean 的别名;scope 是 Bean 的作用域;type 是 Bean 的 Java 类型;resource 是 class 文件的具体路径;dependencies 是依赖的 Bean 名称。

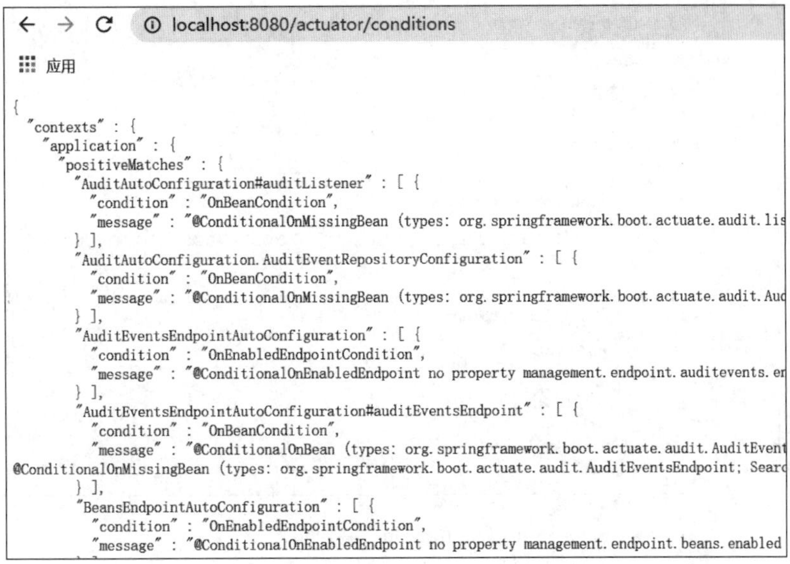

图 10.4　conditions 端点的测试效果

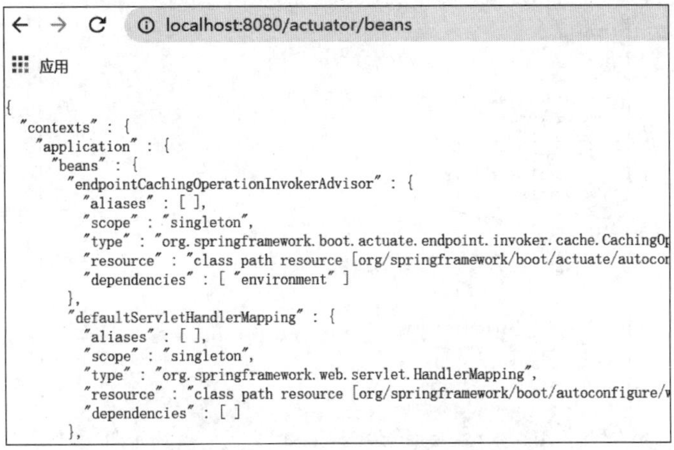

图 10.5　beans 端点的测试效果

3. configprops

该端点用来获取应用中配置的属性信息报告，prefix 属性代表属性的配置前缀，properties 代表各个属性的名称和值，例如可以设置 spring.http.encoding.charset＝"UTF-8"。启动并暴露该端点后，可通过"http://localhost:8080/actuator/configprops"测试访问，测试效果如图 10.6 所示。

4. env

该端点与 configprops 端点不同，它用来获取应用所有可用的环境属性报告，包括环境变量、JVM 属性、应用的配置、命令行中的参数等内容。启动并暴露该端点后，可通过

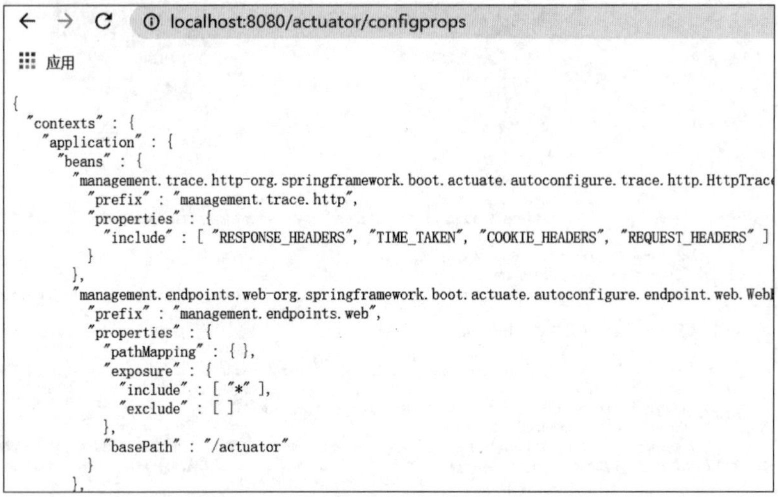

图 10.6 configprops 端点的测试效果

"http://localhost:8080/actuator/env"测试访问,测试效果如图 10.7 所示。

图 10.7 env 端点的测试效果

5．mappings

该端点用来返回所有 Spring MVC 的控制器映射关系报告。启动并暴露该端点后,可通过"http://localhost:8080/actuator/mappings"测试访问,测试效果如图 10.8 所示。

图 10.8　mappings 端点的测试效果

6．info

该端点用来返回一些应用自定义的信息。默认情况下，该端点只会返回一个空的 json 内容。可以在 application.properties 配置文件中通过 info 前缀来设置一些属性。示例代码如下：

```
info.app.name=spring-boot-hello
info.app.version=v1.0.0
```

启动并暴露该端点后，可通过"http://localhost:8080/actuator/info"测试访问，测试效果如图 10.9 所示。

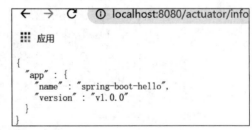

图 10.9　info 端点的测试效果

10.1.3　度量指标端点的测试

通过度量指标端点可获取应用程序运行过程中用于监控的度量指标，例如内存信息、线程信息、HTTP 请求统计等。

1．metrics

该端点用来返回当前应用的各类重要度量指标，例如内存信息、线程信息、垃圾回收信息等。启动并暴露该端点后,可通过"http://localhost:8080/actuator/metrics"测试访问,测试效果如图 10.10 所示。

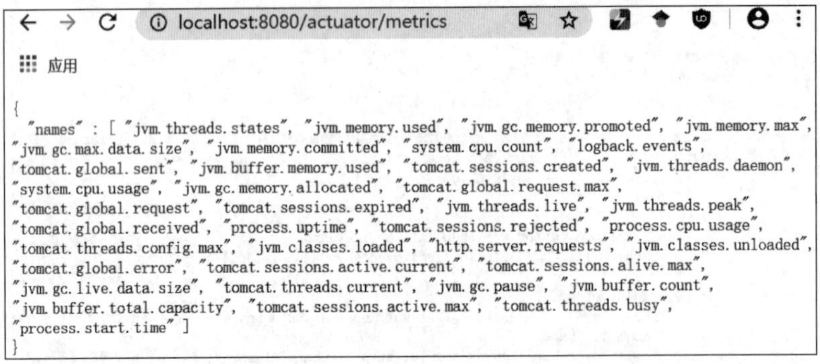

图 10.10 metrics 端点的测试效果

metrics 端点可以提供应用运行状态的完整度量指标报告,这项功能非常实用,但是对于监控系统中的各项监控功能,它们的监控内容、数据收集频率都有所不同,如果每次都通过全量获取报告的方式来收集,略显粗暴。所以,可以通过/metrics/{name}接口来更细粒度地获取度量信息,例如可以通过访问/metrics/jvm.memory.used 来获取当前 JVM 使用的内存数量,如图 10.11 所示。

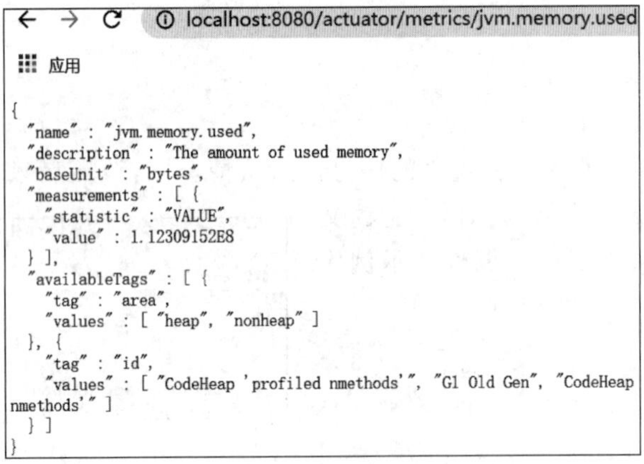

图 10.11 获取当前 JVM 使用的内存数量

2. health

该端点用来获取应用的各类健康指标信息。在 spring-boot-starter-actuator 模块中自带实现了一些常用资源的健康指标检测器。这些检测器都是通过 HealthIndicator 接口实现,并且根据依赖关系的引入实现自动化装配,例如用于检测磁盘的 DiskSpaceHealthIndicator、检测 DataSource 连接是否可用的 DataSourceHealthIndicator 等。

有时还会用到一些 Spring Boot 的 Starter POMs 中还没有封装的产品来进行开发,例如当使用 RocketMQ 作为消息代理时,由于没有自动化配置的检测器,所以需要自己实现一个用来采集健康信息的检测器。

启动并暴露该端点后,可通过"http://localhost:8080/actuator/health"测试访问,测试效果如图 10.12 所示。

图 10.12 中的"UP"表示健康,"DOWN"表示异常。

从图 10.12 可以看出健康指标信息没有显示细节,可以在配置文件中配置属性 management.endpoint.health.show-details=always,将详细健康信息显示给所有用户。再次启动应用后,刷新"http://localhost:8080/actuator/health",显示健康指标详细信息,如图 10.13 所示。

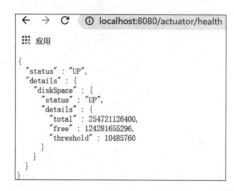

图 10.12 health 端点的测试效果　　图 10.13 健康指标详细信息

3. threaddump

该端点用来暴露程序运行中的线程信息。它使用 java.lang.management.ThreadMXBean 的 dumpAllThreads 方法来返回所有含有同步信息的活动线程详情。启动并暴露该端点后,可通过"http://localhost:8080/actuator/threaddump"测试访问,测试效果如图 10.14 所示。

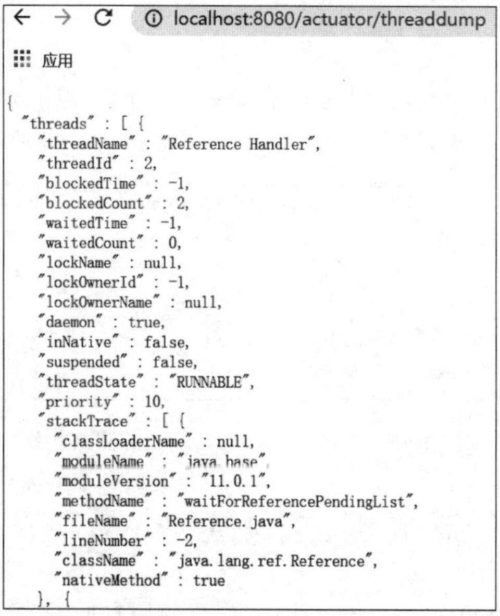

图 10.14 threaddump 端点的测试效果

4. httptrace

该端点用来返回基本的 HTTP 跟踪信息。默认情况下,跟踪信息的存储采用 org.springframework.boot.actuate.trace.InMemoryTraceRepository 实现的内存方式,始终保留最近的 100 条请求记录。启动并暴露该端点后,可通过"http://localhost:8080/actuator/httptrace"测试访问,测试效果如图 10.15 所示。

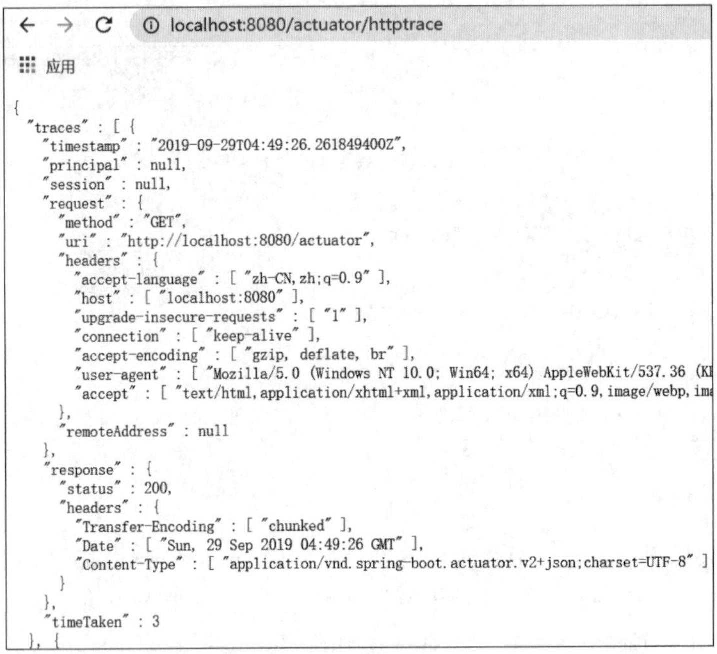

```
← → C  ① localhost:8080/actuator/httptrace
::: 应用
{
  "traces" : [ {
    "timestamp" : "2019-09-29T04:49:26.261849400Z",
    "principal" : null,
    "session" : null,
    "request" : {
      "method" : "GET",
      "uri" : "http://localhost:8080/actuator",
      "headers" : {
        "accept-language" : [ "zh-CN,zh;q=0.9" ],
        "host" : [ "localhost:8080" ],
        "upgrade-insecure-requests" : [ "1" ],
        "connection" : [ "keep-alive" ],
        "accept-encoding" : [ "gzip, deflate, br" ],
        "user-agent" : [ "Mozilla/5.0 (Windows NT 10.0; Win64; x64) AppleWebKit/537.36 (K
        "accept" : [ "text/html,application/xhtml+xml,application/xml;q=0.9,image/webp,ima
      },
      "remoteAddress" : null
    },
    "response" : {
      "status" : 200,
      "headers" : {
        "Transfer-Encoding" : [ "chunked" ],
        "Date" : [ "Sun, 29 Sep 2019 04:49:26 GMT" ],
        "Content-Type" : [ "application/vnd.spring-boot.actuator.v2+json;charset=UTF-8" ]
      }
    },
    "timeTaken" : 3
  }, {
```

图 10.15　httptrace 端点的测试效果

5. scheduledtasks

该端点统计应用程序中调度的任务。启动并暴露该端点后,可通过"http://localhost:8080/actuator/scheduledtasks"测试访问,测试效果如图 10.16 所示。

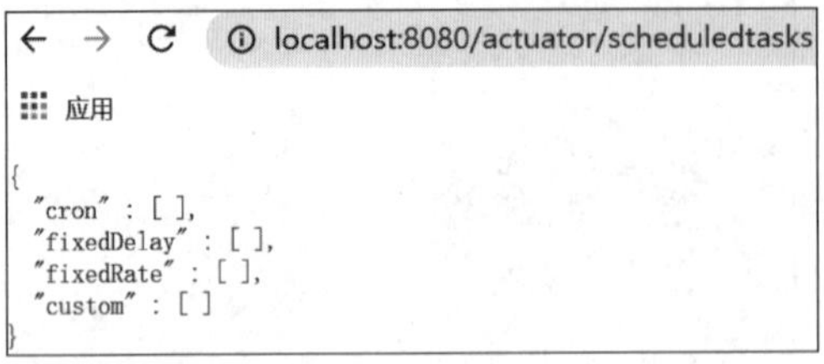

```
← → C  ① localhost:8080/actuator/scheduledtasks
::: 应用
{
  "cron" : [ ],
  "fixedDelay" : [ ],
  "fixedRate" : [ ],
  "custom" : [ ]
}
```

图 10.16　scheduledtasks 端点的测试效果

10.1.4　操作控制端点的测试

操作控制类端点拥有更强大的控制能力,如果使用它们,需要通过属性来配置开启。在原生端点中,只提供了一个用来关闭应用的端点:shutdown,可以通过如下配置开启它:

management.endpoint.shutdown.enabled=true。在配置了上述属性之后，只需要访问该应用的 shutdown 端点就能实现关闭该应用的远程操作。由于开放关闭应用的操作本身是一件非常危险的事，所以真正在线上使用的时候，需要对其加入一定的保护机制，例如定制 Actuator 的端点路径、整合 Spring Security 进行安全校验等。

shutdown 端点不支持 get 提交，不可以直接在浏览器上访问，所以这里可以使用 rest-client-master 来测试。用 post 方式访问"http://localhost：8080/actuator/shutdown"，测试效果如图 10.17 所示。

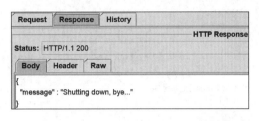

图 10.17　测试 shutdown 端点

10.2　自定义端点

视频讲解

当 Spring Boot 提供的端点不能满足我们的需求时，就需要自定义一个端点对应用进行监控。

Spring Boot 提供了注解 @Endpoint 定义一个端点类，并在端点类的方法上使用 @ReadOperation 注解来显示监控信息（对应 Get 请求），使用 @WriteOperation 来动态更新监控信息（对应 Post 请求，可以是 rest-client-master 访问）。

下面通过一个实例讲解如何自定义一个端点。该实例演示显示数据源的相关信息。

【例 10-2】　自定义端点。

具体实现步骤如下。

1. 创建基于 Spring Data JPA 和 Spring Boot Actuator 的 Web 应用 ch10_2

创建基于 Spring Data JPA 和 Spring Boot Actuator 的 Web 应用 ch10_2，如图 10.18 所示。

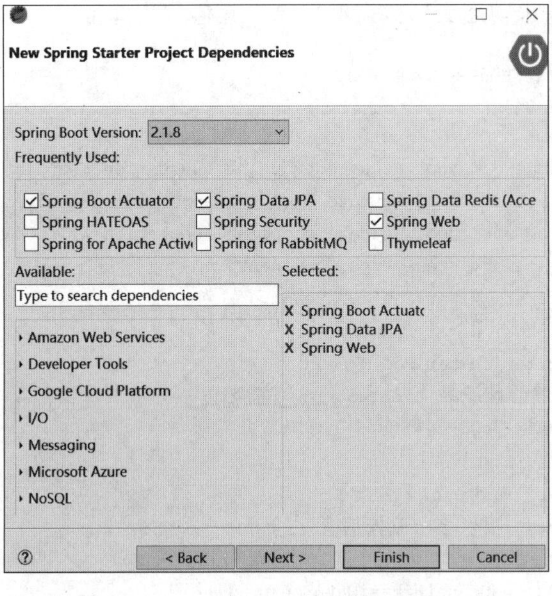

图 10.18　创建 Web 应用 ch10_2

2. 修改 pom.xml 文件添加 MySQL 依赖

修改 pom.xml 文件添加 MySQL 依赖,具体代码如下:

```xml
<dependency>
    <groupId>mysql</groupId>
    <artifactId>mysql-connector-java</artifactId>
    <version>5.1.45</version>
</dependency>
```

3. 配置数据源

因为该实例是监控数据源信息,所以需要在配置文件 application.properties 中配置数据源,具体内容如下:

```
#数据库地址
spring.datasource.url=jdbc:mysql://localhost:3306/springbootjpa?characterEncoding=utf8
#数据库用户名
spring.datasource.username=root
#数据库密码
spring.datasource.password=root
#数据库驱动
spring.datasource.driver-class-name=com.mysql.jdbc.Driver
#输出的 JSON 字符串格式更美观
spring.jackson.serialization.indent-output=true
```

4. 自定义端点

创建名为 com.ch.ch10_2.endPoint 的包,并在该包中使用注解@Endpoint 自定义端点类 DataSourceEndpoint。在该端点类中,使用@ReadOperation 注解来显示数据源信息,使用@WriteOperation 来动态更新数据源信息。具体代码如下:

```java
package com.ch.ch10_2.endPoint;
import java.util.HashMap;
import java.util.Map;
import org.springframework.boot.actuate.endpoint.annotation.Endpoint;
import org.springframework.boot.actuate.endpoint.annotation.ReadOperation;
import org.springframework.boot.actuate.endpoint.annotation.WriteOperation;
import org.springframework.stereotype.Component;
import com.zaxxer.hikari.HikariConfigMXBean;
import com.zaxxer.hikari.HikariDataSource;
import com.zaxxer.hikari.HikariPoolMXBean;
//注册为端点,id不能使用驼峰法(dataSource),需要以-分割
@Endpoint(id = "data-source")
@Component
public class DataSourceEndpoint {
    //HikariDataSource 提供多个监控信息
    HikariDataSource ds;
    public DataSourceEndpoint(HikariDataSource ds) {
```

```
        this.ds = ds;
    }
    @ReadOperation
    public Map<String, Object> info() {
        Map<String, Object> map = new HashMap<String, Object>();
        //连接池配置
        HikariConfigMXBean configBean = ds.getHikariConfigMXBean();
        map.put("max", configBean.getMaximumPoolSize());
        //连接池运行状态
        HikariPoolMXBean mxBean = ds.getHikariPoolMXBean();
        map.put("active", mxBean.getActiveConnections());
        map.put("idle", mxBean.getIdleConnections());
        //连接池无连接时,等待获取连接的线程个数
        map.put("wait", mxBean.getThreadsAwaitingConnection());
        return map;
    }
    @WriteOperation
    public void setMax(int max) {
        ds.getHikariConfigMXBean().setMaximumPoolSize(max);
    }
}
```

5．暴露端点

在配置文件 application.properties 中暴露端点,具体代码如下：

```
#暴露所有端点,当然包括 data-source,也可以只暴露 data-source 端点
management.endpoints.web.exposure.include = *
```

6．测试端点

首先,启动应用程序主类 Ch102Application；然后通过"http://localhost:8080/actuator/data-source"测试端点 data-source。运行效果如图 10.19 所示。

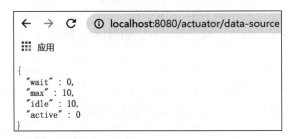

图 10.19　测试端点 data-source

使用 rest-client-master 发送 POST 请求"http://localhost:8080/actuator/data-source?max=20",如图 10.20 所示。

POST 请求"http://localhost:8080/actuator/data-source?max=20"执行后,"http://localhost:8080/actuator/data-source"测试端点 data-source。运行效果如图 10.21 所示。

从图 10.21 可以看出,通过 POST 请求调用端点的 setMax 方法修改了数据源信息。

图 10.20　发送 POST 请求

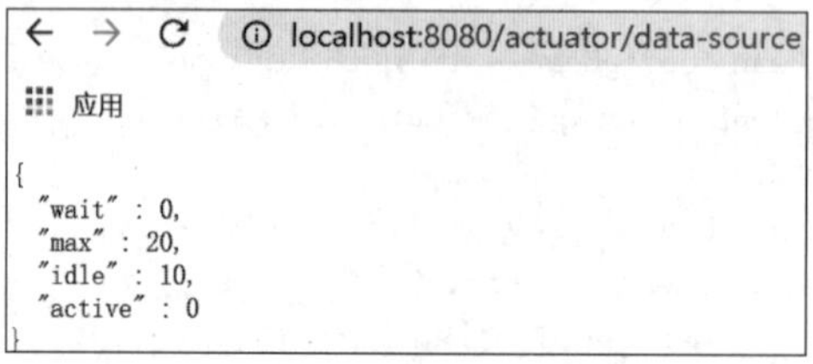

图 10.21　刷新后的效果

10.3　自定义 HealthIndicator

　　我们知道 health 端点用于查看 Spring Boot 应用的健康状态，提供了用于检测磁盘的 DiskSpaceHealthIndicator、检测 DataSource 连接是否可用的 DataSourceHealthIndicator、检测 XXX 内置服务（XXX 代表内置的 Elasticsearch、JMS、Mail、MongoDB、Rabbit、Redis、Solr 等）是否可用的 XXXHealthIndicator 等健康指标检测器。在 Spring Boot 中，这些检测器都是通过 HealthIndicator 接口实现，并且根据依赖关系的引入实现自动化装配。

　　当 Spring Boot 自带的 HealthIndicator 接口实现类不能满足我们的需求时，就需要自定义 HealthIndicator 接口实现类。自定义 HealthIndicator 接口实现类很简单，只需要实现 HealthIndicator 接口，并重写接口方法 health，返回一个 Health 对象。

　　下面通过一个实例讲解如何自定义一个 HealthIndicator 接口实现类。

　　【例 10-3】　自定义 HealthIndicator。

　　具体实现步骤如下。

1. 创建 HealthIndicator 接口实现类 MyHealthIndicator

在 ch10_2 应用中，创建名为 com.ch.ch10_2.health 的包，并在该包中创建一个 HealthIndicator 接口实现类 MyHealthIndicator。在该类中重写接口方法 health，并使用 @Component 注解将该类声明为组件对象。具体代码如下：

```
package com.ch.ch10_2.health;
import org.springframework.boot.actuate.health.Health;
import org.springframework.boot.actuate.health.HealthIndicator;
import org.springframework.stereotype.Component;
@Component
public class MyHealthIndicator implements HealthIndicator{
    @Override
    public Health health() {
        int errorCode = check();
        if(errorCode != 0) {
```

```
        //down 方法表示异常,withDetail 方法添加任意多的异常信息
        return Health.down().withDetail("message", "error:" + errorCode).build();
    }
    //up 方法表示健康
    return Health.up().build();
}
/**
 * 模拟返回一个错误状态
 */
private int check() {
    return 1;
}
```

2. 将详细健康信息显示给所有用户

在配置文件 application.properties 中,配置将详细健康信息显示给所有用户。配置内容如下:

```
#将详细健康信息显示给所有用户
management.endpoint.health.show-details=always
```

3. 测试运行

health 的对象名默认为类名去掉 HealthIndicator 后缀,并且首字母小写,因此该例的 health 对象名为 my。

启动应用程序主类 Ch102Application,并通过"http://localhost:8080/actuator/health/my"测试运行,效果如图 10.22 所示。

图 10.22　测试自定义 HealthIndicator

10.4　本章小结

本章主要介绍了 Spring Boot 的 Actuator 功能,并通过 HTTP 进行 Spring Boot 的应用监控和管理功能的测试。

当 Spring Boot 自带的端点和 HealthIndicator 实现不能满足我们的需要时,我们可以自定义端点和 HealthIndicator 实现。因此,本章还介绍了如何自定义端点和 HealthIndicator 实现。

通过本章的学习,读者应该了解端点的分类与测试,掌握如何自定义端点和 HealthIndicator 实现。

习题 10

1. 默认情况下,Spring Boot 暴露了哪几个端点?又如何暴露所有端点?
2. 如何自定义端点?有哪几个步骤?
3. 如何自定义 HealthIndicator 实现?有哪几个步骤?

第11章

电子商务平台的设计与实现
(Thymeleaf+MyBatis)

视频讲解

📖 学习目的与要求

本章通过一个小型的电子商务平台，讲述如何使用 Spring Boot＋Thymeleaf＋MyBatis 开发一个 Web 应用，其中主要涉及的技术包括 Spring 与 Spring MVC 框架技术、MyBatis 持久层技术和 Thymeleaf 表现层技术。通过本章的学习，掌握基于 Thymeleaf＋MyBatis 的 Spring Boot Web 应用开发的流程、方法以及技术。

📖 本章主要内容

- 系统设计。
- 数据库设计。
- 系统管理。
- 组件设计。
- 系统实现。

本章系统使用 Spring Boot＋Thymeleaf＋MyBatis 实现各个模块，Web 服务器使用内嵌的 Servlet 容器，数据库采用的是 MySQL 5.5，集成开发环境为 Spring Tool Suite(STS)。

11.1 系统设计

电子商务平台分为两个子系统，一是后台管理子系统，一是电子商务子系统。下面分别说明这两个子系统的功能需求与模块划分。

11.1.1 系统功能需求

1. 后台管理子系统

后台管理子系统要求管理员登录成功后,才能对商品进行管理,包括添加商品、查询商品、修改商品以及删除商品。除商品管理外,管理员还需要对商品类型、注册用户以及用户的订单等进行管理。

2. 电子商务子系统

1) 非注册用户

非注册用户或未登录用户具有的功能如下:浏览首页、查看商品详情以及搜索商品。

2) 用户

成功登录的用户除具有未登录用户具有的功能外,还具有购买商品、查看购物车、收藏商品、查看订单、查看收藏以及查看用户个人信息的功能。

11.1.2 系统模块划分

1. 后台管理子系统

管理员登录成功后,进入后台管理主页面(selectGoods.html),可以对商品、商品类型、注册用户以及用户的订单进行管理。后台管理子系统的模块划分如图11.1所示。

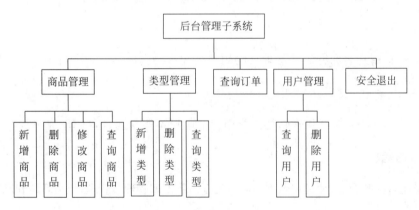

图 11.1 后台管理子系统

2. 电子商务子系统

非注册用户只可以浏览商品、搜索商品,不能购买商品、收藏商品、查看购物车、查看用户中心、查看我的订单和查看我的收藏。成功登录的用户可以拥有电子商务子系统的所有功能,包括购买商品、支付等功能。电子商务子系统的模块划分如图11.2所示。

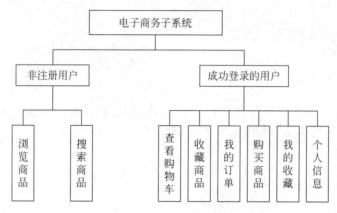

图 11.2　电子商务子系统

11.2　数据库设计

系统采用加载纯 Java 数据库驱动程序的方式连接 MySQL 5.5 数据库。在 MySQL 5.5 中创建数据库 shop，并在 shop 中创建 8 张与系统相关的数据表：ausertable、busertable、carttable、focustable、goodstable、goodstype、orderdetail 和 orderbasetable。

11.2.1　数据库概念结构设计

根据系统设计与分析，可以设计出如下数据结构。

1. 管理员

管理员包括用户名和密码。管理员的用户名和密码由数据库管理员预设，不需要注册。

2. 用户

用户包括用户 ID、邮箱和密码。注册用户的邮箱不能相同，用户 ID 唯一。

3. 商品类型

商品类型包括类型 ID 和类型名称。商品类型由数据库管理员管理，包括新增和删除管理。

4. 商品

商品包括商品编号、名称、原价、现价、库存、图片以及类型。其中，商品编号唯一，类型与"3. 商品类型"关联。

5. 购物车

购物车包括购物车 ID、用户 ID、商品编号以及购买数量。其中，购物车 ID 唯一，用户 ID 与"2. 用户"关联，商品编号与"4. 商品"关联。

6. 收藏商品

收藏商品包括ID、用户ID、商品编号以及收藏时间。其中，ID唯一，用户ID与"2.用户"关联，商品编号与"4.商品"关联。

7. 订单基础信息

订单基础信息包括订单编号、用户ID、订单金额、订单状态以及下单时间。其中，订单编号唯一，用户ID与"2.用户"关联。

8. 订单详情

订单详情包括ID、订单编号、商品编号以及购买数量。其中，订单编号与"7.订单基础信息"关联，商品编号与"4.商品"关联。

根据以上的数据结构，结合数据库设计的特点，可以画出如图11.3所示的数据库概念结构图。

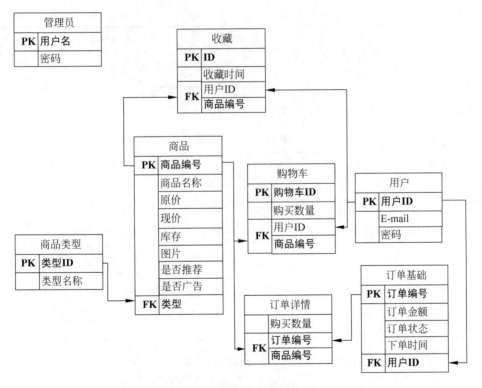

图11.3 数据库概念结构图

11.2.2 数据逻辑结构设计

将数据库概念结构图转换为MySQL数据库所支持的实际数据模型，即数据库的逻辑结构。

管理员信息表(ausertable)的设计如表 11.1 所示。

表 11.1 管理员信息表

字 段	含 义	类 型	长 度	是 否 为 空
aname	用户名(PK)	varchar	50	no
apwd	密码	varchar	50	no

用户信息表(busertable)的设计如表 11.2 所示。

表 11.2 用户信息表

字 段	含 义	类 型	长 度	是 否 为 空
id	用户 ID(PK 自增)	int	11	no
bemail	E-mail	varchar	50	no
bpwd	密码	varchar	50	no

商品类型表(goodstype)的设计如表 11.3 所示。

表 11.3 商品类型表

字 段	含 义	类 型	长 度	是 否 为 空
id	ID(PK 自增)	int	11	no
typename	名称	varchar	50	no

商品信息表(goodstable)的设计如表 11.4 所示。

表 11.4 商品信息表

字 段	含 义	类 型	长 度	是 否 为 空
id	编号(PK 自增)	int	11	no
gname	名称	varchar	50	no
goprice	原价	double		no
grprice	现价	double		no
gstore	库存	int	11	no
gpicture	图片	varchar	50	no
isRecommend	是否推荐	tinyint	2	no
isAdvertisement	是否广告	tinyint	2	no
goodstype_id	类型(FK)	int	11	no

购物车表(carttable)的设计如表 11.5 所示。

表 11.5 购物车表

字 段	含 义	类 型	长 度	是 否 为 空
id	ID(PK 自增)	int	11	no
busertable_id	用户 ID(FK)	int	11	no
goodstable_id	商品编号(FK)	int	11	no
shoppingnum	购买数量	int	11	no

商品收藏表(focustable)的设计如表11.6所示。

表 11.6 商品收藏表

字 段	含 义	类 型	长 度	是否为空
id	ID(PK 自增)	int	11	no
goodstable_id	商品编号(FK)	int	11	no
busertable_id	用户 ID(FK)	int	11	no
focustime	收藏时间	datetime		no

订单基础表(orderbasetable)的设计如表11.7所示。

表 11.7 订单基础表

字 段	含 义	类 型	长 度	是否为空
id	ID(PK 自增)	int	11	no
busertable_id	用户 ID(FK)	int	11	no
amount	订单金额	double		no
status	订单状态	tinyint	4	no
orderdate	下单时间	datetime		no

订单详情表(orderdetail)的设计如表11.8所示。

表 11.8 订单详情表

字 段	含 义	类 型	长 度	是否为空
id	ID(PK 自增)	int	11	no
orderbasetable_id	订单编号(FK)	int	11	no
goodstable_id	商品编号(FK)	int	11	no
shoppingnum	购买数量	int	11	no

11.2.3 创建数据表

根据11.2.2节的逻辑结构,创建数据表。由于篇幅有限,创建数据表的代码请读者参考本书提供的源代码 shop.sql。

11.3 系统管理

11.3.1 添加相关依赖

新建一个基于 Thymeleaf 的 Spring Boot Web 应用 eBusiness,在 eBusiness 应用中开发本系统。除了 STS 快速创建基于 Thymeleaf 的 Spring Boot Web 应用自带的 spring-boot-starter-thymeleaf 和 spring-boot-starter-web 依赖外，还需要向 eBusiness 应用的 pom.xml 文件中添加上传文件依赖 commons-fileupload、MyBatis 与 Spring 整合依赖 mybatis-spring-boot-starter 以及 MySQL 连接器依赖。具体代码如下：

```xml
<!-- 添加 MySQL 连接器依赖 -->
<dependency>
    <groupId>mysql</groupId>
    <artifactId>mysql-connector-java</artifactId>
    <version>5.1.45</version>
    <!-- MySQL 8.x 时,请使用 8.x 的连接器 -->
</dependency>
<!-- MyBatis-Spring,Spring Boot 应用整合 MyBatis 框架的核心依赖配置 -->
<dependency>
    <groupId>org.mybatis.spring.boot</groupId>
    <artifactId>mybatis-spring-boot-starter</artifactId>
    <version>2.1.0</version>
</dependency>
<dependency>
    <groupId>commons-fileupload</groupId>
    <artifactId>commons-fileupload</artifactId>
    <!-- 由于 commons-fileupload 组件不属于 Spring Boot,所以需要加上版本 -->
    <version>1.3.3</version>
</dependency>
```

11.3.2　HTML 页面及静态资源管理

系统由后台管理和电子商务两个子系统组成，为了方便管理，两个子系统的 HTML 页面分开存放。在 src/main/resources/templates/admin 目录下存放与后台管理子系统相关的 HTML 页面；在 src/main/resources/templates/user 目录下存放与电子商务子系统相关的 HTML 页面；在 src/main/resources/static 目录下存放与整个系统相关的 BootStrap 及 jQuery。由于篇幅受限，本章仅附上 HTML 和 Java 文件的核心代码，具体代码请读者参考本书提供的源代码 eBusiness。

1. 后台管理子系统

管理员在浏览器的地址栏中输入"http://localhost:8080/eBusiness/admin/toLogin"访问登录页面，登录成功后，进入后台查询商品页面（selectGoods.html）。selectGoods.html 的运行效果如图 11.4 所示。

2. 电子商务子系统

注册用户或游客在浏览器的地址栏中输入"http://localhost:8080/eBusiness"可以访问电子商务子系统的首页（index.html）。index.html 的运行效果如图 11.5 所示。

第11章 电子商务平台的设计与实现(Thymeleaf+MyBatis)

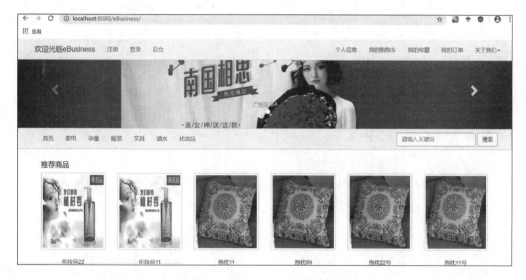

图 11.4　后台查询商品页面

图 11.5　电子商务子系统的首页

11.3.3　应用的目录结构

eBusiness 应用的目录结构如图 11.6 所示。

1. com.ch.ebusiness 包

该包中包括应用的主程序类 EBusinessApplication、统一异常处理类 GlobalException-HandleController 以及自定义异常类 NoLoginException。

2. com.ch.ebusiness.controller 包

系统的控制器类都在该包中,后台管理相关的控制器类在 admin 子包中,电子商务相关的控制器类在 before 子包中。

3. com.ch.ebusiness.entity 包

实体类存放在该包中。

4. com.ch.ebusiness.repository 包

该包中存放的 Java 接口程序是实现数据库的持久化操作。每个接口方法与 SQL 映射文件中的 id 相同。后台管理相关的数据库操作在 admin 子包中，电子商务相关的数据库操作在 before 子包中。

5. com.ch.ebusiness.service 包

service 包中有两个子包：admin 和 before，admin 子包存放后台管理相关业务层的接口与实现类；before 子包存放电子商务相关业务层的接口与实现类。

6. com.ch.ebusiness.util 包

该包中存放的是系统的工具类。

图 11.6　目录结构

11.3.4　配置文件

在配置文件 application.properties 中，配置了数据源等信息，具体内容如下：

```
server.servlet.context-path=/eBusiness
###
## 数据源信息配置
###
# 数据库地址
spring.datasource.url=jdbc:mysql://localhost:3306/shop?characterEncoding=utf8
# 数据库 MySQL 为 8.x 时,url 为 jdbc:mysql://localhost:3306/springbootjpa?useSSL=false&serverTimezone=Asia/Beijing&characterEncoding=utf-8
# 数据库用户名
spring.datasource.username=root
# 数据库密码
spring.datasource.password=root
# 数据库驱动
spring.datasource.driver-class-name=com.mysql.jdbc.Driver
# 设置包别名(在 Mapper 映射文件中直接使用实体类名)
mybatis.type-aliases-package=com.ch.ebusiness.entity
# 告诉系统到哪里去找 mapper.xml 文件(映射文件)
mybatis.mapperLocations=classpath:mappers/*.xml
# 在控制台输出 SQL 语句日志
logging.level.com.ch.ebusiness.repository=debug
# 关闭 Thymeleaf 模板引擎缓存(使页面热部署),默认是开启的
spring.thymeleaf.cache=false
# 上传文件时,默认单个上传文件大小是 1MB,max-file-size 设置单个上传文件大小
```

```
spring.servlet.multipart.max-file-size=50MB
#默认总文件大小是10MB,max-request-size 设置总上传文件大小
spring.servlet.multipart.max-request-size=500MB
```

11.4 组件设计

本系统的组件包括管理员登录权限验证控制器、前台用户登录权限验证控制器、验证码、统一异常处理以及工具类。

11.4.1 管理员登录权限验证

从系统分析得知,管理员成功登录后,才能管理商品、商品类型、用户、订单等功能模块。因此,本系统需要对这些功能模块的操作进行管理员登录权限控制。在com.ch.ebusiness.controller.admin 包中创建 AdminBaseController 控制器类,该类中有一个 @ModelAttribute 注解的方法 isLogin。isLogin 方法的功能是判断管理员是否已成功登录。需要进行管理员登录权限控制的控制器类继承 AdminBaseController 类即可,因为带有 @ModelAttribute 注解的方法首先被控制器执行。AdminBaseController 控制器类的代码如下:

```
package com.ch.ebusiness.controller.admin;
import javax.servlet.http.HttpSession;
import org.springframework.stereotype.Controller;
import org.springframework.web.bind.annotation.ModelAttribute;
import com.ch.ebusiness.NoLoginException;
@Controller
public class AdminBaseController {
    /**
     * 登录权限控制,处理方法执行前执行该方法
     */
    @ModelAttribute
    public void isLogin(HttpSession session) throws NoLoginException {
        if(session.getAttribute("auser") == null){
            throw new NoLoginException("没有登录");
        }
    }
}
```

11.4.2 前台用户登录权限验证

从系统分析得知,用户成功登录后,才能购买商品、收藏商品、查看购物车、查看我的订单以及查看个人信息。与管理员登录权限验证同理,在 com.ch.ebusiness.controller.before 包中创建 BeforeBaseController 控制器类,该类中有一个@ModelAttribute 注解的方法 isLogin。isLogin 方法的功能是判断前台用户是否已成功登录。需要进行前台用户登

录权限控制的控制器类继承 BeforeBaseController 类即可。BeforeBaseController 控制器类的代码如下：

```java
package com.ch.ebusiness.controller.before;
import javax.servlet.http.HttpSession;
import org.springframework.stereotype.Controller;
import org.springframework.web.bind.annotation.ModelAttribute;
import com.ch.ebusiness.NoLoginException;
@Controller
public class BeforeBaseController {
    /**
     * 登录权限控制,处理方法执行前执行该方法
     */
    @ModelAttribute
    public void isLogin(HttpSession session) throws NoLoginException {
        if(session.getAttribute("bUser") == null){
            throw new NoLoginException("没有登录");
        }
    }
}
```

11.4.3 验证码

本系统验证码的使用步骤如下。

1. 创建产生验证码的控制器类

在 com.ch.ebusiness.controller.before 包中，创建产生验证码的控制器类 ValidateCodeController，具体代码参见源程序。

2. 使用验证码

在需要验证码的 HTML 页面中，调用产生验证码的控制器显示验证码。示例代码片段如下：

```html
<div class="form-group has-success">
    <label class="col-sm-2 col-md-2 control-label">验证码</label>
    <div class="col-sm-4 col-md-4">
        <table style="width:100%">
            <tr>
                <td><input type="text" class="form-control"
                    placeholder="请输入验证码" th:field="*{code}"/></td>
                <td>
                <img th:src="@{/validateCode}" id="mycode">
                </td>
                <td>
                    <a href="javascript:refreshCode()">看不清换一张</a>
                </td>
            </tr>
```

```
            </table>
        </div>
</div>
```

11.4.4 统一异常处理

系统对未登录异常、数据库操作异常以及程序未知异常进行了统一异常处理,具体实现步骤如下。

1. 创建未登录自定义异常

创建未登录自定义异常 NoLoginException,具体代码如下:

```
package com.ch.ebusiness;
public class NoLoginException extends Exception{
    private static final long serialVersionUID = 1L;
    public NoLoginException() {
        super();
    }
    public NoLoginException(String message) {
        super(message);
    }
}
```

2. 创建统一异常处理类

使用注解@ControllerAdvice和@ExceptionHandler创建统一异常处理类GlobalExceptionHandleController。使用注解@ControllerAdvice的类是一个增强的Controller类,在增强的控制器类中使用@ExceptionHandler注解的方法对所有控制器类进行统一异常处理。具体代码如下:

```
package com.ch.ebusiness;
import java.sql.SQLException;
import org.springframework.ui.Model;
import org.springframework.web.bind.annotation.ControllerAdvice;
import org.springframework.web.bind.annotation.ExceptionHandler;
/**
 * 统一异常处理
 */
@ControllerAdvice
public class GlobalExceptionHandleController {
    @ExceptionHandler(value = Exception.class)
    public String exceptionHandler(Exception e, Model model) {
        String message = "";
        //数据库异常
        if (e instanceof SQLException) {
            message = "数据库异常";
        } else if (e instanceof NoLoginException) {
            message = "未登录异常";
```

```
        } else {                              //未知异常
            message = "未知异常";
        }
        model.addAttribute("mymessage",message);
        return "myError";
    }
}
```

11.4.5 工具类

本系统使用的工具类有两个：MD5Util 和 MyUtil。
MD5Util 的代码如下：

```
package com.ch.ebusiness.util;
import java.security.MessageDigest;
public class MD5Util {
    /***
     * MD5 加码生成 32 位 md5 码
     */
    public static String string2MD5(String inStr) {
        MessageDigest md5 = null;
        try {
            md5 = MessageDigest.getInstance("MD5");
        } catch (Exception e) {
            System.out.println(e.toString());
            e.printStackTrace();
            return "";
        }
        char[] charArray = inStr.toCharArray();
        byte[] byteArray = new byte[charArray.length];
        for (int i = 0; i < charArray.length; i++)
            byteArray[i] = (byte) charArray[i];
        byte[] md5Bytes = md5.digest(byteArray);
        StringBuffer hexValue = new StringBuffer();
        for (int i = 0; i < md5Bytes.length; i++) {
            int val = ((int) md5Bytes[i]) & 0xff;
            if (val < 16)
                hexValue.append("0");
            hexValue.append(Integer.toHexString(val));
        }
        return hexValue.toString();
    }
    /***
     * 自定义加密规则
     * @param inStr
     * @return
     */
    public static String MD5(String inStr){
        String xy = "abc";
```

```
            String finalStr = "";
            if(inStr!= null){
                String fStr = inStr.substring(0, 1);
                String lStr = inStr.substring(1, inStr.length());
                finalStr = string2MD5( fStr + xy + lStr);
            }else{
                finalStr = string2MD5(xy);
            }
            return finalStr;
        }
    }
```

MyUtil 的代码如下：

```
package com.ch.ebusiness.util;
import java.text.SimpleDateFormat;
import java.util.Date;
import javax.servlet.http.HttpSession;
import com.ch.ebusiness.entity.BUser;
public class MyUtil {
    /**
     * 将实际的文件名重命名
     */
    public static String getNewFileName(String oldFileName) {
        int lastIndex = oldFileName.lastIndexOf(".");
        String fileType = oldFileName.substring(lastIndex);
        Date now = new Date();
        SimpleDateFormat sdf = new SimpleDateFormat("YYYYMMDDHHmmssSSS");
        String time = sdf.format(now);
        String newFileName = time + fileType;
        return newFileName;
    }
    /**
     * 获得用户信息
     */
    public static BUser getUser(HttpSession session) {
        BUser bUser = (BUser)session.getAttribute("bUser");
        return bUser;
    }
}
```

11.5 后台管理子系统的实现

管理员成功登录后，可以对商品及商品类型、注册用户以及用户的订单进行管理。本节将详细讲解管理员的功能实现。

11.5.1 管理员登录

管理员输入用户名和密码后，系统将对管理员的用户名和密码进行验证。如果用户名和密码同时正确,则成功登录,进入后台商品查询页面(selectGoods.html);如果用户名或密码有误,则提示错误。具体实现步骤如下:

1. 编写视图

login.html 页面提供登录信息输入的界面,效果如图 11.7 所示。

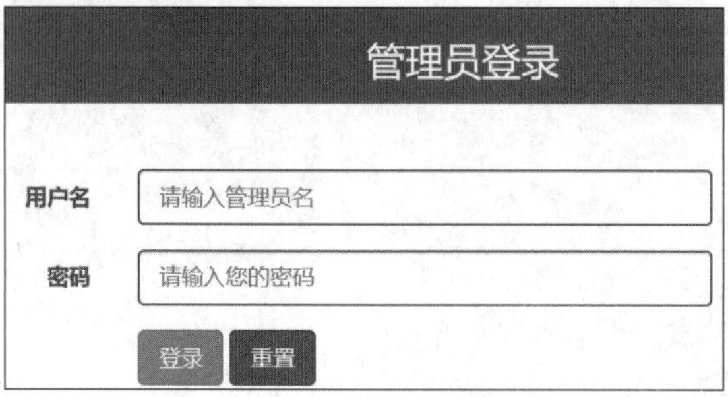

图 11.7 管理员登录界面

在 src/main/resources/templates/admin 目录下,创建 login.html。该页面的代码如下:

```
<!DOCTYPE html>
<html xmlns:th="http://www.thymeleaf.org">
<head>
<meta charset="UTF-8">
<title>管理员登录页面</title>
<link rel="stylesheet" th:href="@{/css/bootstrap.min.css}"/>
<body>
    <div class="container">
        <div class="bg-primary"  style="width:70%;height:60px;padding-top:1px;">
            <h3 align="center">管理员登录</h3>
    </div>
        <br>
        <br>
        <form th:action="@{/admin/login}" name="myform" method="post" th:object="${aUser}"  class="form-horizontal" role="form">
            <div class="form-group has-success">
                <label class="col-sm-2 col-md-2 control-label">用户名</label>
                <div class="col-sm-4 col-md-4">
                    <input type="text" class="form-control"
                     placeholder="请输入管理员名"
                     th:field="*{aname}"/>
                    <span th:errors="*{aname}"></span>
                </div>
            </div>
            <div class="form-group has-success">
                <label class="col-sm-2 col-md-2 control-label">密码</label>
                <div class="col-sm-4 col-md-4">
                    <input type="password" class="form-control"
                     placeholder="请输入您的密码" th:field="*{apwd}"/>
```

```html
                    <span th:errors="*{apwd}"></span>
                </div>
            </div>
            <div class="form-group">
                <div class="col-sm-offset-2 col-sm-10">
                    <button type="submit" class="btn btn-success">登录</button>
                    <button type="reset" class="btn btn-primary">重置</button>
                </div>
            </div>
            <div class="form-group">
                <div class="col-sm-offset-2 col-sm-10">
                    <font size="6" color="red">
                        <span th:text="${errorMessage}"></span>
                    </font>
                </div>
            </div>
        </form>
    </div>
</body>
</html>
```

2. 编写控制器层

视图 Action 的请求路径为"admin/login",系统根据请求路径和@RequestMapping 注解找到对应控制器类 com.ch.ebusiness.controller.admin 的 login 方法处理登录。在控制器类的 login 方法中调用 com.ch.ebusiness.service.admin.AdminService 接口的 login 方法处理登录。登录成功后,首先将登录人信息存入 session,然后转发到查询商品请求;登录失败回到本页面。控制器层的相关代码如下:

```java
@Controller
@RequestMapping("/admin")
public class AdminController {
    @Autowired
    private AdminService adminService;
    @RequestMapping("/toLogin")
    public String toLogin(@ModelAttribute("aUser") AUser aUser) {
        return "admin/login";
    }
    @RequestMapping("/login")
    public String login(@ModelAttribute("aUser") AUser aUser, HttpSession session, Model model) {
        return adminService.login(aUser, session, model);
    }
}
```

3. 编写 Service 层

Service 层由接口 com.ch.ebusiness.service.admin.AdminService 和接口的实现类 com.ch.ebusiness.service.admin.AdminServiceImpl 组成。Service 层是功能模块实现的

核心，Service 层调用数据访问层（Repository）进行数据库操作。管理员登录的业务处理方法 login 的代码如下：

```java
public String login(AUser aUser, HttpSession session, Model model) {
    List<AUser> list = adminRepository.login(aUser);
    if(list.size() > 0) {                           //登录成功
        session.setAttribute("auser", aUser);
        return "forward:/goods/selectAllGoodsByPage?currentPage=1&act=select";
    }else {                                         //登录失败
        model.addAttribute("errorMessage", "用户名或密码错误！");
        return "admin/login";
    }
}
```

4. 编写 SQL 映射文件

数据访问层（Repository）仅由 @Mapper 注解的接口组成，接口方法与 SQL 映射文件中 SQL 语句的 id 相同，不再赘述。管理员登录的 SQL 映射文件为 src/main/resources/mappers 目录下的 AdminMapper.xml，实现的 SQL 语句如下：

```xml
<select id="login" parameterType="AUser" resultType="AUser">
    select * from ausertable where aname = #{aname} and apwd = #{apwd}
</select>
```

11.5.2 类型管理

类型管理分为添加类型和删除类型，如图 11.8 所示。

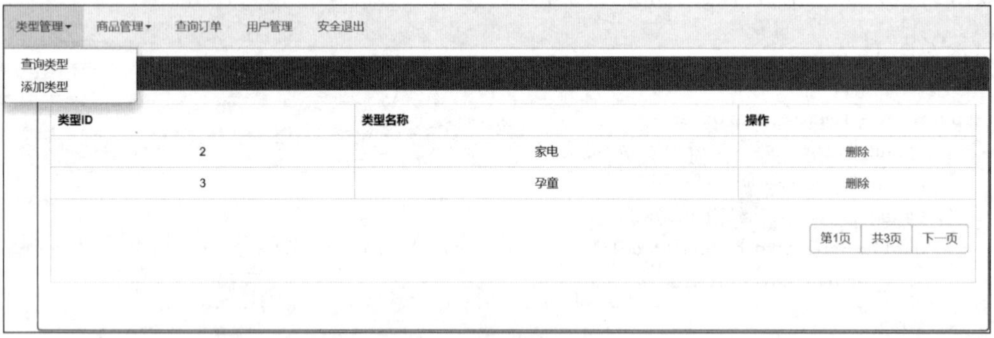

图 11.8 类型管理

1. 添加类型

添加类型的具体实现步骤如下。

1）编写视图

单击图 11.8 中的"添加类型"超链接（type/toAddType），打开如图 11.9 所示的添加页面。

在 src/main/resources/templates/admin 目录下，创建添加类型页面 addType.html。

第11章 电子商务平台的设计与实现(Thymeleaf+MyBatis)

图 11.9 添加类型

该页面的代码如下:

```html
<!DOCTYPE html>
<html xmlns:th="http://www.thymeleaf.org">
<head>
<meta charset="UTF-8">
<title>商品类型添加页面</title>
<link rel="stylesheet" th:href="@{/css/bootstrap.min.css}"/>
<body>
    <div th:include="admin/header"></div>
    <br><br><br>
    <div class="container">
        <div class="bg-primary" style="width:70%;height:60px;padding-top:0.5px;">
            <h3 align="center">添加类型</h3>
        </div><br>
        <form th:action="@{/type/addType}" name="myform" method="post" th:object="${goodsType}" class="form-horizontal" role="form">
            <div class="form-group has-success">
                <label class="col-sm-2 col-md-2 control-label">类型名称</label>
                <div class="col-sm-4 col-md-4">
                    <input type="text" class="form-control"
                     placeholder="请输入类型名"
                     th:field="*{typename}"/>
                </div>
            </div>
            <div class="form-group">
                <div class="col-sm-offset-2 col-sm-10">
                    <button type="submit" class="btn btn-success">添加</button>
                    <button type="reset" class="btn btn-primary">重置</button>
                </div>
            </div>
        </form>
    </div>
</body>
</html>
```

2) 编写控制器层

此功能共有两个处理请求:"添加类型"超链接 type/toAddType 和视图 Action 的请求路径 type/addType。系统根据 @RequestMapping 注解找到对应控制器类 com.ch.ebusiness.controller.admin.TypeController 的 toAddType 和 addType 方法处理请求。在控制器类的处理方法中调用 com.ch.ebusiness.service.admin.TypeService 接口的

addType 方法处理业务。控制器层的相关代码如下：

```
@RequestMapping("/toAddType")
public String toAddType(@ModelAttribute("goodsType") GoodsType goodsType) {
    return "admin/addType";
}
@RequestMapping("/addType")
public String addType(@ModelAttribute("goodsType") GoodsType goodsType) {
    return typeService.addType(goodsType);
}
```

3）编写 Service 层

添加类型 type/addType 的业务处理方法 addType 的代码如下：

```
@Override
public String addType(GoodsType goodsType) {
    typeRepository.addType(goodsType);
    return "redirect:/type/selectAllTypeByPage?currentPage = 1";
}
```

4）编写 SQL 映射文件

实现添加类型 type/addType 的 SQL 语句如下（位于 src/main/resources/mappers/TypeMapper.xml 文件中）：

```
<insert id = "addType" parameterType = "GoodsType">
    insert into goodstype (id, typename) values(null, #{typename})
</insert>
```

2. 删除类型

删除类型的具体实现步骤如下。

1）编写视图

单击图 11.8 中的"查询类型"超链接（type/selectAllTypeByPage?currentPage＝1），打开如图 11.10 所示的查询页面。

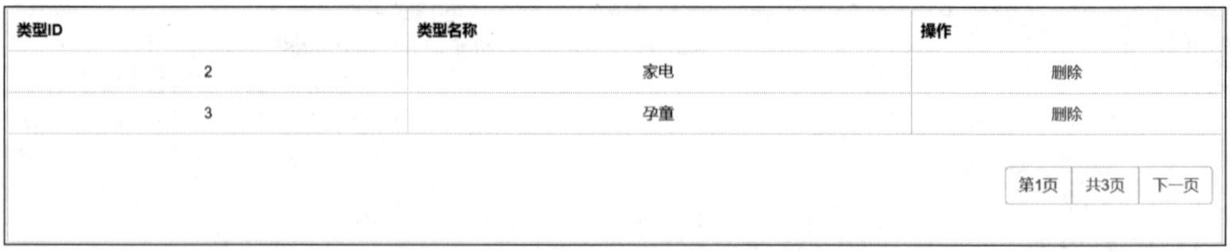

类型ID	类型名称	操作
2	家电	删除
3	孕童	删除

第1页　共3页　下一页

图 11.10　删除类型

在 src/main/resources/templates/admin 目录下，创建查询类型页面 selectGoodsType.html。该页面的代码如下：

```
<!DOCTYPE html>
<html xmlns:th = "http://www.thymeleaf.org">
<head>
<base th:href = "@{/}">
```

```html
<meta charset="UTF-8">
<title>查询类型页面</title>
<link rel="stylesheet" href="css/bootstrap.min.css" />
<script src="js/jquery.min.js"></script>
<script type="text/javascript" th:inline="javascript">
    function deleteType(tid){
        $.ajax(
            {
                //请求路径,要注意的是url和th:inline="javascript"
                url : [[@{/type/deleteType}]],
                //请求类型
                type : "post",
                //data 表示发送的数据
                data : {
                    id : tid
                },
                //成功响应的结果
                success : function(obj){          //obj 响应数据
                    if(obj == "no"){
                        alert("该类型下有商品不允许删除!");
                    }else{
                        if(window.confirm("真的删除该类型吗?")){
                            //获取路径
                            var pathName = window.document.location.pathname;
                            //截取,得到项目名称
                            var projectName = pathName.substring(0,pathName.substr(1).indexOf('/') + 1);
                            window.location.href = projectName + obj;
                        }
                    }
                },
                error : function() {
                    alert("处理异常!");
                }
            }
        );
    }
</script>
</head>
<body>
    <!-- 加载 header.html -->
    <div th:include="admin/header"></div>
    <br><br><br>
    <div class="container">
        <div class="panel panel-primary">
            <div class="panel-heading">
                <h3 class="panel-title">商品类型列表</h3>
            </div>
            <div class="panel-body">
                <div class="table table-responsive">
                    <table class="table table-bordered table-hover">
                        <tbody class="text-center">
                            <tr>
                                <th>类型 ID</th>
                                <th>类型名称</th>
```

```html
                <th>操作</th>
            </tr>
            <tr th:each="qty: ${allTypes}">
                <td th:text="${qty.id}"></td>
                <td th:text="${qty.typename}"></td>
                <td>
                    <a th:href="'javascript:deleteType(' + ${qty.id} + ')'">删除</a>
                </td>
            </tr>
            <tr>
                <td colspan="3" align="right">
                    <ul class="pagination">
                        <li><a>第<span th:text="${currentPage}"></span>页</a></li>
                        <li><a>共<span th:text="${totalPage}"></span>页</a></li>
                        <li>
                            <span th:if="${currentPage} != 1">
                                <a th:href="'type/selectAllTypeByPage?currentPage=' + ${currentPage - 1}">上一页</a>
                            </span>
                            <span th:if="${currentPage} != ${totalPage}">
                                <a th:href="'type/selectAllTypeByPage?currentPage=' + ${currentPage + 1}">下一页</a>
                            </span>
                        </li>
                    </ul>
                </td>
            </tr>
        </tbody>
    </table>
            </div>
        </div>
    </div>
</div>
</body>
</html>
```

2）编写控制器层

此功能模块共有两个处理请求：“查询类型”超链接"type/selectAllTypeByPage?currentPage=1"与视图"删除"的请求路径"type/deleteType"。系统根据@RequestMapping注解找到对应控制器类com.ch.ebusiness.controller.admin.TypeController的selectAllTypeByPage和delete方法处理请求。在控制器类的处理方法中调用com.ch.ebusiness.service.admin.TypeService接口的selectAllTypeByPage和delete方法处理业务。控制器层的相关代码如下：

```java
@RequestMapping("/selectAllTypeByPage")
public String selectAllTypeByPage(Model model, int currentPage) {
    return typeService.selectAllTypeByPage(model, currentPage);
```

```
}
@RequestMapping("/deleteType")
@ResponseBody                              //返回字符串数据而不是视图
public String delete(int id) {
    return typeService.delete(id);
}
```

3）编写 Service 层

超链接"type/selectAllTypeByPage?currentPage=1"的业务处理方法 selectAllTypeByPage 的代码如下：

```
public String selectAllTypeByPage(Model model, int currentPage) {
    //共多少个类型
    int totalCount = typeRepository.selectAll();
    //计算共多少页
    int pageSize = 2;
    int totalPage = (int)Math.ceil(totalCount * 1.0/pageSize);
    List<GoodsType> typeByPage =
     typeRepository.selectAllTypeByPage((currentPage - 1) * pageSize, pageSize);
    model.addAttribute("allTypes", typeByPage);
    model.addAttribute("totalPage", totalPage);
    model.addAttribute("currentPage", currentPage);
    return "admin/selectGoodsType";
}
```

删除"type/deleteType"的业务处理方法 delete 的代码如下：

```
public String delete(int id) {
    List<Goods> list = typeRepository.selectGoods(id);
    if(list.size() > 0) {
        //该类型下有商品不允许删除
        return "no";
    }else {
        typeRepository.deleteType(id);
        //删除后回到查询页面
        return "/type/selectAllTypeByPage?currentPage = 1";
    }
}
```

4）编写 SQL 映射文件

实现超链接"type/selectAllTypeByPage? currentPage=1"的 SQL 语句如下：

```
<select id = "selectAll"  resultType = "integer">
    select count(*) from goodstype
</select>
<!-- 分页查询 -->
<select id = "selectAllTypeByPage"  resultType = "GoodsType">
    select * from goodstype  limit #{startIndex}, #{perPageSize}
</select>
```

实现删除"type/deleteType"的 SQL 语句如下：

```
<!-- 删除类型 -->
```

```xml
<delete id="deleteType" parameterType="integer">
    delete from goodstype where id=#{id}
</delete>
<!-- 查询该类型下是否有商品 -->
<select id="selectGoods" parameterType="integer" resultType="Goods">
    select * from goodstable where goodstype_id = #{goodstype_id}
</select>
```

11.5.3 添加商品

单击图 11.11 中的"添加商品"超链接，打开如图 11.12 所示的"添加商品"页面。添加商品的具体实现步骤如下。

图 11.11　商品管理

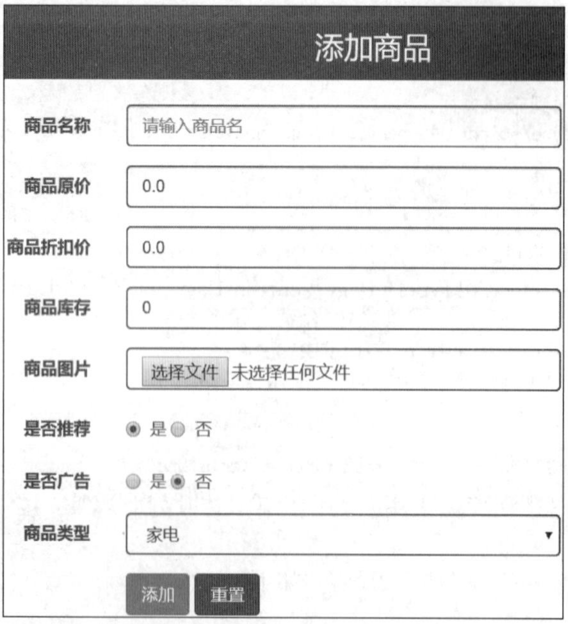

图 11.12　添加商品

第11章 电子商务平台的设计与实现(Thymeleaf+MyBatis)

1. 编写视图

在 src/main/resources/templates/admin 目录下,创建添加商品页面 addGoods.html。该页面的代码如下:

```html
<!DOCTYPE html>
<html xmlns:th="http://www.thymeleaf.org">
<head>
<meta charset="UTF-8">
<title>商品类型添加页面</title>
<link rel="stylesheet" th:href="@{/css/bootstrap.min.css}" />
<body>
    <div th:include="admin/header"></div>
    <br><br><br>
    <div class="container">
        <div class="bg-primary" style="width:70%; height: 60px;padding-top: 0.5px;">
            <h3 align="center">添加商品</h3>
        </div><br>
        <form th:action="@{/goods/addGoods?act=add}"
          name="myform" method="post"
          th:object="${goods}"
          class="form-horizontal"
          enctype="multipart/form-data">
            <div class="form-group has-success">
                <label class="col-sm-2 col-md-2 control-label">商品名称</label>
                <div class="col-sm-4 col-md-4">
                    <input type="text" class="form-control"
                     placeholder="请输入商品名"
                     th:field="*{gname}"/>
                </div>
            </div>
            <div class="form-group has-success">
                <label class="col-sm-2 col-md-2 control-label">商品原价</label>
                <div class="col-sm-4 col-md-4">
                    <input type="number" class="form-control"
                     placeholder="请输入商品原价"
                     th:field="*{goprice}"/>
                </div>
            </div>
            <div class="form-group has-success">
                <label class="col-sm-2 col-md-2 control-label">商品折扣价</label>
                <div class="col-sm-4 col-md-4">
                    <input type="number" class="form-control"
                     placeholder="请输入商品折扣价"
                     th:field="*{grprice}"/>
                </div>
            </div>
            <div class="form-group has-success">
                <label class="col-sm-2 col-md-2 control-label">商品库存</label>
                <div class="col-sm-4 col-md-4">
```

```html
                    <input type="number" class="form-control"
                     placeholder="请输入商品库存"
                     th:field="*{gstore}"/>
                </div>
            </div>
            <div class="form-group has-success">
                <label class="col-sm-2 col-md-2 control-label">商品图片</label>
                <div class="col-sm-4 col-md-4">
                    <input type="file" placeholder="请选择商品图片" class="form-control" name="fileName"/>2
                </div>
            </div>
            <div class="form-group has-success">
                <label class="col-sm-2 col-md-2 control-label">是否推荐</label>
                <div class="col-sm-4 col-md-4 radio">
                    <label>
                        <input type="radio" th:field="*{isRecommend}" value="1">是
                    </label>
                    <label>
                        <input type="radio" th:field="*{isRecommend}" value="0">否
                    </label>
                </div>
            </div>
            <div class="form-group has-success">
                <label class="col-sm-2 col-md-2 control-label">是否广告</label>
                <div class="col-sm-4 col-md-4 radio">
                    <label>
                        <input type="radio" th:field="*{isAdvertisement}" value="1">是
                    </label>
                    <label>
                        <input type="radio" th:field="*{isAdvertisement}"s value="0">否
                    </label>
                </div>
            </div>
            <div class="form-group has-success">
                <label class="col-sm-2 col-md-2 control-label">商品类型</label>
                <div class="col-sm-4 col-md-4">
                    <select class="form-control" th:field="*{goodstype_id}">
                        <option th:each="gty:${goodsType}" th:value="${gty.id}" th:text="${gty.typename}">
                    </select>
                </div>
            </div>
            <div class="form-group">
                <div class="col-sm-offset-2 col-sm-10">
                    <button type="submit" class="btn btn-success">添加</button>
                    <button type="reset" class="btn btn-primary">重置</button>
                </div>
            </div>
        </form>
```

```
        </div>
    </body>
</html>
```

2. 编写控制器层

此功能模块共有两个处理请求:"添加商品"超链接"goods/toAddGoods"与视图"添加"的请求路径"goods/addGoods?act=add"。系统根据@RequestMapping 注解找到对应控制器类 com.ch.ebusiness.controller.admin.GoodsController 的 toAddGoods 和 addGoods 方法处理请求。在控制器类的处理方法中调用 com.ch.ebusiness.service.admin.GoodsService 接口的 toAddGoods 和 addGoods 方法处理业务。控制器层的相关代码如下:

```java
@RequestMapping("/toAddGoods")
public String toAddGoods(@ModelAttribute("goods") Goods goods, Model model) {
    goods.setIsAdvertisement(0);
    goods.setIsRecommend(1);
    return goodsService.toAddGoods(goods, model);
}
@RequestMapping("/addGoods")
public String addGoods(@ModelAttribute("goods") Goods goods, HttpServletRequest request,
String act) throws IllegalStateException, IOException {
    return goodsService.addGoods(goods, request, act);
}
```

3. 编写 Service 层

添加商品的 Service 层相关代码如下:

```java
@Override
public String addGoods (Goods goods, HttpServletRequest request, String act) throws
IllegalStateException, IOException {
    MultipartFile myfile = goods.getFileName();
    //如果选择了上传文件,将文件上传到指定的目录 images
    if(!myfile.isEmpty()) {
        //上传文件路径(生产环境)
        //String path = request.getServletContext().getRealPath("/images/");
        //获得上传文件原名
        //上传文件路径(开发环境)
String path = "C:\\workspace-spring-tool-suite-4-4.1.1.RELEASE\\eBusiness\\src\\main\\resources\\static\\images";
        //获得上传文件原名
        String fileName = myfile.getOriginalFilename();
        //对文件重命名
        String fileNewName = MyUtil.getNewFileName(fileName);
        File filePath = new File(path + File.separator + fileNewName);
        //如果文件目录不存在,创建目录
        if(!filePath.getParentFile().exists()) {
            filePath.getParentFile().mkdirs();
        }
```

```
            //将上传文件保存到一个目标文件中
            myfile.transferTo(filePath);
            //将重命名后的图片名存到 goods 对象中,添加时使用
            goods.setGpicture(fileNewName);
        }
        if("add".equals(act)) {
            int n = goodsRepository.addGoods(goods);
            if(n > 0)                              //成功
                return "redirect:/goods/selectAllGoodsByPage?currentPage = 1&act = select";
            //失败
            return "admin/addGoods";
        }else {                                    //修改
            int n = goodsRepository.updateGoods(goods);
            if(n > 0)                              //成功
                return "redirect:/goods/selectAllGoodsByPage?currentPage = 1&act = updateSelect";
            //失败
            return "admin/UpdateAGoods";
        }
    }
    @Override
    public String toAddGoods(Goods goods, Model model) {
        model.addAttribute("goodsType", goodsRepository.selectAllGoodsType());
        return "admin/addGoods";
    }
```

4. 编写 SQL 映射文件

添加商品的 SQL 语句如下:

```
<!-- 添加商品 -->
< insert id = "addGoods" parameterType = "Goods">
    insert into goodstable ( id, gname, goprice, grprice, gstore, gpicture, isRecommend,
isAdvertisement,goodstype_id)
    values ( null, #{gname}, #{goprice}, #{grprice}, #{gstore}, #{gpicture},
#{isRecommend}, #{isAdvertisement}, #{goodstype_id})
</insert>
<!-- 查询商品类型 -->
< select id = "selectAllGoodsType"   resultType = "GoodsType">
    select * from goodstype
</select>
```

11.5.4 查询商品

管理员登录成功后,进入后台管理子系统的主页面,在主页面中初始显示查询商品页面。也可以通过单击图 11.11 中的"查询商品"超链接显示查询商品页面。查询页面运行效果如图 11.13 所示。单击图 11.13 中的"详情"链接,显示如图 11.14 所示的商品详情页面。查询商品的具体实现步骤如下。

第11章 电子商务平台的设计与实现 (Thymeleaf+MyBatis)

商品列表			
商品ID	商品名称	商品类型	详情
35	化妆品22	化妆品	详情
34	化妆品11	化妆品	详情
33	抱枕11	服装	详情
32	抱枕99	服装	详情
29	抱枕22号	文具	详情

第1页　共2页　下一页

图 11.13　查询商品

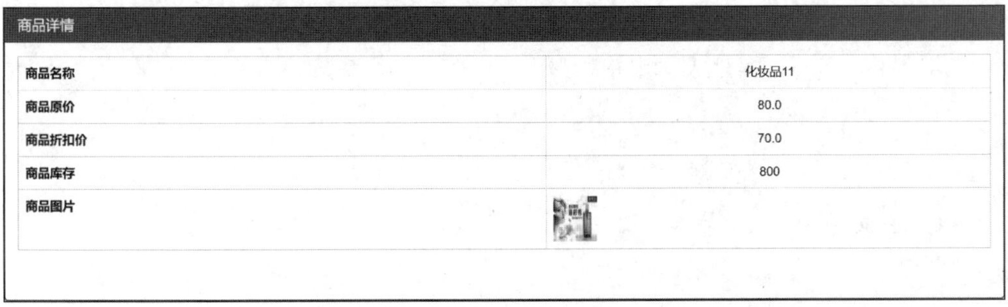

商品详情	
商品名称	化妆品11
商品原价	80.0
商品折扣价	70.0
商品库存	800
商品图片	

图 11.14　商品详情

1. 编写视图

在 src/main/resources/templates/admin 目录下，创建查询商品页面 selectGoods.html，该页面显示查询商品、修改商品查询以及删除商品查询的结果。具体代码如下：

```html
<!DOCTYPE html>
<html xmlns:th="http://www.thymeleaf.org">
<head>
<base th:href="@{/}">
<meta charset="UTF-8">
<title>主页</title>
<link rel="stylesheet" href="css/bootstrap.min.css" />
<script src="js/jquery.min.js"></script>
<script type="text/javascript" th:inline="javascript">
    function deleteGoods(tid){
        $.ajax(
            {
                //请求路径,要注意的是url和th:inline="javascript"
                url:[[@{/goods/delete}]],
                //请求类型
                type:"post",
                //data表示发送的数据
                data:{
```

```
                        id : tid
                },
                //成功响应的结果
                success : function(obj){      //obj 响应数据
                    if(obj == "no"){
                        alert("该商品有关联不允许删除!");
                    }else{
                        if(window.confirm("真的删除该商品吗?")){
                            //获取路径
                            var pathName = window.document.location.pathname;
                            //截取,得到项目名称
                            var projectName = pathName.substring(0,pathName.substr(1).indexOf('/') + 1);
                            window.location.href = projectName + obj;
                        }
                    }
                },
                error : function() {
                    alert("处理异常!");
                }
            }
        );
    }
</script>
</head>
<body>
    <!-- 加载 header.html -->
    <div th:include = "admin/header"></div>
    <br><br><br>
    <div class = "container">
        <div class = "panel panel-primary">
            <div class = "panel-heading">
                <h3 class = "panel-title">商品列表</h3>
            </div>
            <div class = "panel-body">
                <div class = "table table-responsive">
                    <table class = "table table-bordered table-hover">
                        <tbody class = "text-center">
                            <tr>
                                <th>商品 ID</th>
                                <th>商品名称</th>
                                <th>商品类型</th>
                                <th>详情</th>
                                <th th:if = "${act} == 'updateSelect'">操作</th>
                                <th th:if = "${act} == 'deleteSelect'">操作</th>
                            </tr>
                            <tr th:each = "gds:${allGoods}">
                                <td th:text = "${gds.id}"></td>
                                <td th:text = "${gds.gname}"></td>
                                <td th:text = "${gds.typename}"></td>
                                <td>
                                    <a th:href = "@{goods/detail(id = ${gds.id},act = detail)}" target = "_
```

```html
blank">详情</a>
                                    </td>
                                    <td th:if="${act} == 'updateSelect'">
                        <a th:href="@{goods/detail(id=${gds.id},act=update)}" target="_blank">修改</a>
                                    </td>
                                    <td th:if="${act} == 'deleteSelect'">
                        <a th:href="'javascript:deleteGoods(' + ${gds.id} + ')'">删除</a>
                                    </td>
                                </tr>
                                <tr th:if="${act} == 'select'">
                                    <td colspan="4" align="right">
                                        <ul class="pagination">
                            <li><a>第<span th:text="${currentPage}"></span>页</a></li>
                            <li><a>共<span th:text="${totalPage}"></span>页</a></li>
                                            <li>
                                <span th:if="${currentPage} != 1">
<a th:href="@{goods/selectAllGoodsByPage(act=select,currentPage=${currentPage}-1)}">上一页</a>
                                                </span>
                                <span th:if="${currentPage} != ${totalPage}">
<a th:href="@{goods/selectAllGoodsByPage(act=select,currentPage=${currentPage}+1)}">下一页</a>
                                                </span>
                                            </li>
                                        </ul>
                                    </td>
                                </tr>
                                <tr th:if="${act} == 'updateSelect'">
                                    <td colspan="5" align="right">
                                        <ul class="pagination">
                            <li><a>第<span th:text="${currentPage}"></span>页</a></li>
                            <li><a>共<span th:text="${totalPage}"></span>页</a></li>
                                            <li>
                                                <span th:if="${currentPage} != 1">
<a th:href="@{goods/selectAllGoodsByPage(act=updateSelect,currentPage=${currentPage}-1)}">上一页</a>
                                                </span>
                                                <span th:if="${currentPage} != ${totalPage}">
<a th:href="@{goods/selectAllGoodsByPage(act=updateSelect,currentPage=${currentPage}+1)}">下一页</a>
                                                </span>
                                            </li>
                                        </ul>
                                    </td>
                                </tr>
                                <tr th:if="${act} == 'deleteSelect'">
                                    <td colspan="5" align="right">
                                        <ul class="pagination">
                            <li><a>第<span th:text="${currentPage}"></span>页</a></li>
                            <li><a>共<span th:text="${totalPage}"></span>页</a></li>
```

```html
                                        <li>
                                            <span th:if="${currentPage} != 1">
<a th:href="@{goods/selectAllGoodsByPage(act=deleteSelect,currentPage=${currentPage}-1)}">上一页</a>
                                            </span>
                                            <span th:if="${currentPage} != ${totalPage}">
<a th:href="@{goods/selectAllGoodsByPage(act=deleteSelect,currentPage=${currentPage}+1)}">下一页</a>
                                            </span>
                                        </li>
                                    </ul>
                                </td>
                            </tr>
                        </tbody>
                    </table>
                </div>
            </div>
        </div>
    </div>
</body>
</html>
```

在 src/main/resources/templates/admin 目录下，创建商品详情页面 detail.html。具体代码如下：

```html
<!DOCTYPE html>
<html xmlns:th="http://www.thymeleaf.org">
<head>
<base th:href="@{/}">
<meta charset="UTF-8">
<title>商品详情页</title>
<link rel="stylesheet" href="css/bootstrap.min.css" />
</head>
<body>
    <!-- 加载 header.html -->
    <div th:include="admin/header"></div>
    <br><br><br>
    <div class="container">
        <div class="panel panel-primary">
            <div class="panel-heading">
                <h3 class="panel-title">商品详情</h3>
            </div>
            <div class="panel-body">
                <div class="table table-responsive">
                    <table class="table table-bordered table-hover">
                        <tbody class="text-center">
                            <tr>
                                <th>商品名称</th>
                                <td th:text="${goods.gname}"></td>
                            </tr>
```

```html
                    <tr>
                        <th>商品原价</th>
                        <td th:text=""${goods.goprice}"></td>
                    </tr>
                    <tr>
                        <th>商品折扣价</th>
                        <td th:text=""${goods.grprice}"></td>
                    </tr>
                    <tr>
                        <th>商品库存</th>
                        <td th:text=""${goods.gstore}"></td>
                    </tr>
                    <tr>
                        <th>商品图片</th>
                        <td>
                            <img th:src=""'images/'+${goods.gpicture}"
                                style="height: 50px; width: 50px; display: block;">
                        </td>
                    </tr>
                </tbody>
            </table>
        </div>
      </div>
    </div>
  </div>
</body>
</html>
```

2. 编写控制器层

此功能模块共有两个处理请求：goods/selectAllGoodsByPage?currentPage＝1&act＝select 和@{goods/detail(id＝${gds.id},act＝detail)}。系统根据@RequestMapping 注解找到对应控制器类 com.ch.ebusiness.controller.admin.GoodsController 的 selectAllGoodsByPage 和 detail 方法处理请求。在控制器类的处理方法中调用 com.ch.ebusiness.service.admin.GoodsService 接口的 selectAllGoodsByPage 和 detail 方法处理业务。控制器层的相关代码如下：

```java
@RequestMapping("/selectAllGoodsByPage")
public String selectAllGoodsByPage(Model model, int currentPage, String act) {
    return goodsService.selectAllGoodsByPage(model, currentPage, act);
}
@RequestMapping("/detail")
public String detail(Model model, Integer id, String act) {
    return goodsService.detail(model, id, act);
}
```

3. 编写 Service 层

查询商品和查看详情的 Service 层相关代码如下：

```java
@Override
public String selectAllGoodsByPage(Model model, int currentPage, String act) {
    //共多少个商品
        int totalCount = goodsRepository.selectAllGoods();
        //计算共多少页
        int pageSize = 5;
        int totalPage = (int)Math.ceil(totalCount * 1.0/pageSize);
 List< Goods > typeByPage = goodsRepository.selectAllGoodsByPage((currentPage - 1) * pageSize, pageSize);
    model.addAttribute("allGoods", typeByPage);
    model.addAttribute("totalPage", totalPage);
    model.addAttribute("currentPage", currentPage);
    model.addAttribute("act", act);
    return "admin/selectGoods";
}
@Override
public String detail(Model model, Integer id, String act) {
    model.addAttribute("goods", goodsRepository.selectAGoods(id));
    if("detail".equals(act))
        return "admin/detail";
    else {
        model.addAttribute("goodsType", goodsRepository.selectAllGoodsType());
        return "admin/updateAGoods";
    }
}
```

4. 编写 SQL 映射文件

查询商品和查看详情的 SQL 语句如下：

```xml
< select id = "selectAllGoods"    resultType = "integer">
    select count( * ) from goodstable
</select >
<!-- 分页查询 -->
< select id = "selectAllGoodsByPage"    resultType = "Goods">
    select gt. * ,gy.typename
     from goodstable gt,goodstype gy
    where gt.goodstype_id = gy.id
    order by id desc limit #{startIndex}, #{perPageSize}
</select >
<!-- 查询商品详情 -->
< select id = "selectAGoods"    resultType = "Goods">
    select gt. * , gy.typename
    from  goodstable gt,goodstype gy
    where gt.goodstype_id = gy.id  and gt.id = #{id}
</select >
```

11.5.5 修改商品

单击图 11.11 中的"修改商品"超链接(goods/selectAllGoodsByPage?currentPage=

1&act=updateSelect),打开修改查询页面selectGoods.html,如图11.15所示。

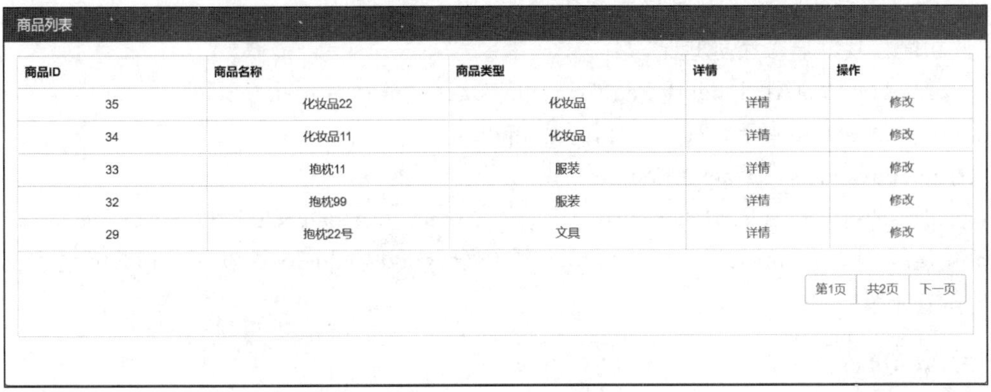

图11.15 修改查询页面

单击图11.15中的"修改"超链接(goods/detail(id=${gds.id},act=update)),打开修改商品信息页面updateAGoods.html,如图11.16所示。在图11.16中输入要修改的信息后,单击"修改"按钮,将商品信息提交给goods/addGoods?act=update处理。"修改商品"的具体实现步骤如下。

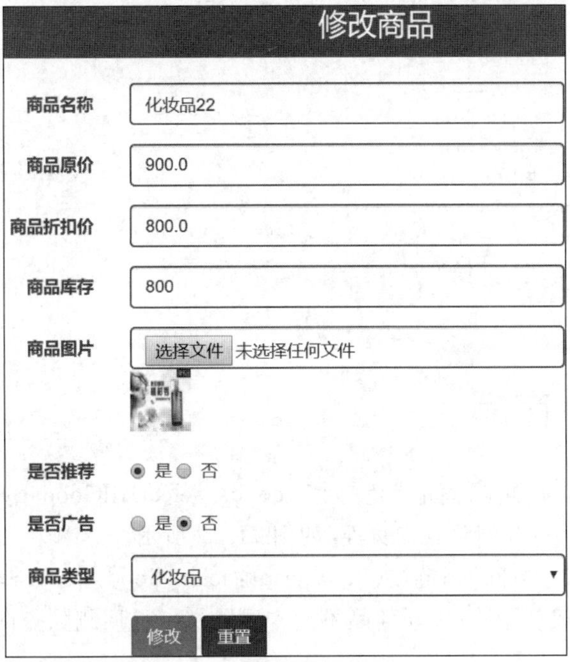

图11.16 修改商品页面

1. 编写视图

在src/main/resources/templates/admin目录下,创建修改商品信息页面updateAGoods.html。updateAGoods.html与添加商品页面内容基本一样,不再赘述。

2. 编写控制器层

此功能模块共有 3 个处理请求：

- goods/selectAllGoodsByPage?currentPage=1&act=updateSelect；
- goods/detail(id=${gds.id},act=update)；
- goods/addGoods?act=update。

goods/selectAllGoodsByPage?currentPage=1&act=updateSelect 和 goods/detail（id=${gds.id},act=update)请求已在 11.5.4 节介绍，goods/addGoods?act=update 请求已在 11.5.3 节介绍。

3. 编写 Service 层

同理，Service 层请参考 11.5.3 节与 11.5.4 节。

4. 编写 SQL 映射文件

"修改商品"的 SQL 语句如下：

```
<!-- 修改一个商品 -->
<update id="updateGoods" parameterType="Goods">
update goodstable set
    gname = #{gname},
    goprice = #{goprice},
    grprice = #{grprice},
    gstore = #{gstore},
    gpicture = #{gpicture},
    isRecommend = #{isRecommend},
    isAdvertisement = #{isAdvertisement},
    goodstype_id = #{goodstype_id}
    where id = #{id}
</update>
```

11.5.6 删除商品

单击图 11.11 中的"删除商品"超链接（goods/selectAllGoodsByPage?currentPage=1&act=deleteSelect），打开删除查询页面，如图 11.17 所示。

单击图 11.17 中的"删除"超链接（'javascript:deleteGoods('+${gds.id}+')'）可实现单个商品的删除。成功删除（关联商品不允许删除）后，返回删除查询页面。

1. 编写视图

删除查询页面就是查询商品页面，代码见 11.5.4 节。

2. 编写控制器层

此功能模块共有两个处理请求：goods/selectAllGoodsByPage?currentPage=1&act=deleteSelect 和 goods/delete。

商品ID	商品名称	商品类型	详情	操作
35	化妆品22	化妆品	详情	删除
34	化妆品11	化妆品	详情	删除
33	抱枕11	服装	详情	删除
32	抱枕99	服装	详情	删除
29	抱枕22号	文具	详情	删除

第1页　共2页　下一页

图 11.17　删除查询页面

goods/selectAllGoodsByPage?currentPage＝1&act＝deleteSelect 请求已在 11.5.4 节介绍，不再赘述。goods/delete 请求的相关控制器层代码如下：

```
@RequestMapping("/delete")
@ResponseBody
public String delete(Integer id) {
    return goodsService.delete(id);
}
```

3. 编写 Service 层

删除商品的相关业务处理代码如下：

```
@Override
public String delete(Integer id) {
    if(goodsRepository.selectCartGoods(id).size() > 0
            || goodsRepository.selectFocusGoods(id).size() > 0
            || goodsRepository.selectOrderGoods(id).size() > 0)
        return "no";
    else {
        goodsRepository.deleteAGoods(id);
        return "/goods/selectAllGoodsByPage?currentPage = 1&act = deleteSelect";
    }
}
```

4. 编写 SQL 映射文件

"删除商品"功能模块的相关 SQL 语句如下：

```
<select id = "selectFocusGoods" parameterType = "integer" resultType = "map">
    select * from focustable where goodstable_id = #{id}
</select>
<select id = "selectCartGoods" parameterType = "integer" resultType = "map">
    select * from carttable where goodstable_id = #{id}
</select>
<select id = "selectOrderGoods" parameterType = "integer" resultType = "map">
```

```
        select * from orderdetail where goodstable_id = #{id}
</select>
<delete id="deleteAGoods" parameterType="Integer">
        delete from goodstable where id = #{id}
</delete>
```

11.5.7 查询订单

单击后台管理主页面中的"查询订单"超链接(selectOrder?currentPage=1),打开查询订单页面 allOrder.html,如图 11.18 所示。

图 11.18 查询订单页面

1. 编写视图

在 src/main/resources/templates/admin 目录下,创建查询订单页面 allOrder.html。该页面的代码如下:

```
<!DOCTYPE html>
<html xmlns:th="http://www.thymeleaf.org">
<head>
<base th:href="@{/}">
<meta charset="UTF-8">
<title>主页</title>
<link rel="stylesheet" href="css/bootstrap.min.css"/>
</head>
<body>
    <!-- 加载 header.html -->
    <div th:include="admin/header"></div>
    <br><br><br>
    <div class="container">
        <div class="panel panel-primary">
            <div class="panel-heading">
                <h3 class="panel-title">订单列表</h3>
            </div>
            <div class="panel-body">
                <div class="table table-responsive">
                    <table class="table table-bordered table-hover">
                        <tbody class="text-center">
                            <tr>
                                <th>订单 ID</th>
```

```html
                    <th>用户邮箱</th>
                    <th>订单金额</th>
                    <th>订单状态</th>
                    <th>下单日期</th>
                </tr>
                <tr th:each="ao:${allOrders}">
                    <td th:text="${ao.id}"></td>
                    <td th:text="${ao.bemail}"></td>
                    <td th:text="${ao.amount}"></td>
                    <td th:text="(${ao.status} == 1)?'已支付':'未支付'"></td>
                    <td th:text="${ao.orderdate}"></td>
                </tr>
                <tr>
                    <td colspan="5" align="right">
                        <ul class="pagination">
                <li><a>第<span th:text="${currentPage}"></span>页</a></li>
                <li><a>共<span th:text="${totalPage}"></span>页</a></li>
                            <li>
                                <span th:if="${currentPage} != 1">
                <a th:href="@{selectOrder?currentPage=${currentPage - 1}}">上一页</a>
                                </span>
                <span th:if="${currentPage} != ${totalPage}">
<a th:href="@{selectOrder?currentPage=${currentPage + 1}}">下一页</a>
                                </span>
                            </li>
                        </ul>
                    </td>
                </tr>
            </tbody>
        </table>
    </div>
  </div>
 </div>
</div>
</body>
</html>
```

2. 编写控制器层

此功能模块有一个处理请求：selectOrder?currentPage=1。系统根据@RequestMapping 注解找到对应控制器类 com.ch.ebusiness.controller.admin.UserAndOrderAndOutController 的 selectOrder 方法处理请求。在控制器类的处理方法中调用 com.ch.ebusiness.service.admin.UserAndOrderAndOutService 接口的 selectOrder 方法处理业务。相关控制器层代码如下：

```
@RequestMapping("/selectOrder")
public String selectOrder(Model model, int currentPage) {
    return userAndOrderAndOutService.selectOrder(model, currentPage);
}
```

3. 编写 Service 层

"查询订单"功能模块的相关 Service 层代码如下：

```
@Override
public String selectOrder(Model model, int currentPage) {
    //共多少个订单
        int totalCount = userAndOrderAndOutRepository.selectAllOrder();
    //计算共多少页
        int pageSize = 5;
        int totalPage = (int)Math.ceil(totalCount * 1.0/pageSize);
    List<Map<String, Object>> orderByPage = userAndOrderAndOutRepository.selectOrderByPage
((currentPage-1) * pageSize, pageSize);
        model.addAttribute("allOrders", orderByPage);
        model.addAttribute("totalPage", totalPage);
        model.addAttribute("currentPage", currentPage);
        return "admin/allOrder";
}
```

4. 编写 SQL 映射文件

"查询订单"功能模块的相关 SQL 语句如下：

```
<select id="selectAllOrder" resultType="integer">
    select count(*) from orderbasetable
</select>
<!-- 分页查询 -->
<select id="selectOrderByPage" resultType="map">
    select obt.*, bt.bemail from orderbasetable obt, busertable bt where obt.busertable_id =
bt.id limit #{startIndex}, #{perPageSize}
</select>
```

11.5.8 用户管理

单击后台管理主页面中的"用户管理"超链接（selectUser?currentPage=1），打开用户管理页面 allUser.html，如图 11.19 所示。

图 11.19 用户管理页面

单击图 11.19 中的"删除"超链接('javascript:deleteUsers('＋＄{u.id}＋')')可删除未关联的用户。

"用户管理"与 11.5.7 节"查询订单"的实现方式基本一样,不再赘述。

11.5.9 安全退出

单击后台管理主页面中的"安全退出"超链接(loginOut),将返回后台登录页面。系统根据@RequestMapping 注解找到对应控制器类 com.ch.ebusiness.controller.admin.UserAndOrderAndOutController 的 loginOut 方法处理请求。在 loginOut 方法中执行 session.invalidate()将使 session 失效,并返回后台登录页面。具体代码如下:

```
@RequestMapping("/loginOut")
public String loginOut(@ModelAttribute("aUser") AUser aUser, HttpSession session) {
    session.invalidate();
    return "admin/login";
}
```

11.6 前台电子商务子系统的实现

游客具有浏览首页、查看商品详情和搜索商品等权限。成功登录的用户除具有游客所具有的权限外,还具有购买商品、查看购物车、收藏商品、查看我的订单以及用户信息的权限。本节将详细讲解前台电子商务子系统的实现。

11.6.1 导航栏及首页搜索

在前台每个 HTML 页面中,都引入了一个名为 header.html 的页面,引入代码如下:

```
<div th:include="user/header"></div>
```

header.html 中的商品类型以及广告区域的商品信息都是从数据库中获取。header.html 页面的运行效果如图 11.20 所示。

图 11.20 导航栏

在导航栏的搜索框中输入信息,单击"搜索"按钮,将搜索信息提交给 search 请求处理,系统根据@RequestMapping 注解找到 com.ch.ebusiness.controller.before.IndexController 控制器类的 search 方法处理请求,并将搜索到的商品信息转发给 searchResult.html。

searchResult.html 页面的运行效果如图 11.21 所示。

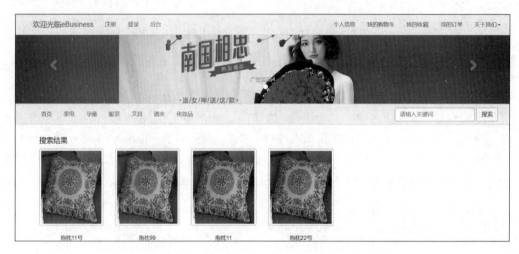

图 11.21 搜索结果

1. 编写视图

该模块的视图涉及 src/main/resources/templates/user 目录下的两个 HTML 页面：header.html 和 searchResult.html。

head.html 的代码如下：

```
<!DOCTYPE html>
<html xmlns:th="http://www.thymeleaf.org">
<head>
<meta charset="UTF-8">
<title>导航页</title>
<base th:href="@{/}"><!-- 不用base就使用th:src="@{/js/jquery.min.js}" -->
<!-- Bootstrap -->
<link href="css/bootstrap.min.css" rel="stylesheet">
<script src="js/jquery.min.js"></script>
<script src="js/bootstrap.min.js"></script>
<style type="text/css">
    .carousel{
        height: 200px;
        background-color: #000;
    }
    .carousel .item{
        height: 200px;
        background-color: #000;
    }
    .carousel img{
        width: 100%;
    }
</style>
</head>
<body>
    <div class="container-fruid">
```

第11章 电子商务平台的设计与实现(Thymeleaf+MyBatis)

```html
<div class="navbar navbar-default navbar-fixed-top" role="navigation"
    style="padding-left:30px;">
    <div class="navbar-header">
        <span class="navbar-brand">欢迎光临 eBusiness</span>
    </div>
    <ul class="nav navbar-nav">
        <li><a th:href="@{user/toRegister}">注册</a></li>
        <li>
            <a th:href="(${session.bUser} == null)?'user/toLogin':'#'">
                <span th:if="${session.bUser} == null">
                    登录
                </span>
                <span th:if="${session.bUser} != null">
                    欢迎<span th:text="${session.bUser.bemail}"></span>
                </span>
            </a>
        </li>
        <li><a th:href="@{admin/toLogin}">后台</a></li>
    </ul>
    <ul class="nav navbar-nav navbar-right" style="padding-right:30px;">
        <li><a href="cart/userInfo">个人信息</a></li>
        <li><a href="cart/selectCart">我的购物车</a></li>
        <li><a href="cart/myFocus">我的收藏</a></li>
        <li><a href="cart/myOder">我的订单</a></li>
        <li class="dropdown"><a href="##" data-toggle="dropdown"
            class="dropdown-toggle">关于我们<span class="caret"></span></a>
            <ul class="dropdown-menu">
                <li><a href="##">联系我们</a></li>
                <li><a href="##">投诉建议</a></li>
            </ul>
        </li>
    </ul>
</div>
<!-- *************************************************** -->
<div id="carousel-example-generic" class="carousel slide"
    data-ride="carousel" style="margin-top:20px;">
    <!-- Indicators 小圆圈 -->
    <ol class="carousel-indicators">
        <li data-target="#carousel-example-generic"
            th:each="advertise,adstat:${advertisementGoods}"
            th:data-slide-to="${adstat.index}"
            th:class="(${adstat.index} == 0)? 'active' : ''"></li>
    </ol>
    <!-- 滚动广告图片 -->
    <div class="carousel-inner" role="listbox">
        <div th:each="advertise,adstat:${advertisementGoods}" th:class=
        "(${adstat.index} == 0)? 'item active' : 'item'">
            <img th:src="'images/' + ${advertise.gpicture}" th:alt="${adstat.index + 1}">
            <div class="carousel-caption"><span th:text="${advertise.gname}">
            </span></div>
```

```html
            </div>
        </div>
        <!-- Controls -->
        <a class="left carousel-control" href="#carousel-example-generic"
            role="button" data-slide="prev"><span
            class="glyphicon glyphicon-chevron-left" aria-hidden="true"></span>
            <span class="sr-only">Previous</span>
        </a><a class="right carousel-control" href="#carousel-example-generic"
            role="button" data-slide="next"><span
            class="glyphicon glyphicon-chevron-right" aria-hidden="true"></span>
            <span class="sr-only">Next</span>
        </a>
    </div>
    <!-- ****************************** -->
    <div class="navbar navbar-default" role="navigation">
        <ul class="nav navbar-nav" style="padding-left: 50px;">
            <li><a th:href="@{/}">首页</a></li>
            <li th:each="gty: ${goodsType}"><a th:href="'?tid=' + ${gty.id}">
            <span th:text="${gty.typename}"></span></a></li>
        </ul>
        <form action="search" class="navbar-form navbar-right"
            style="padding-right: 50px;">
            <div class="form-group">
                <input type="text" class="form-control" name="mykey" placeholder="请输入关键词" />
            </div>
            <button type="submit" class="btn btn-default">搜索</button>
        </form>
    </div>
    </div>
</body>
</html>
```

searchResult.html 的代码如下：

```html
<!DOCTYPE html>
<html xmlns:th="http://www.thymeleaf.org">
<head>
<base th:href="@{/}"><!-- 不用 base 就使用 th:src="@{/js/jquery.min.js}" -->
<meta charset="UTF-8">
<title>主页</title>
<link rel="stylesheet" href="css/bootstrap.min.css" />
<script src="js/jquery.min.js"></script>
<script type="text/javascript" th:inline="javascript">
    function focus(gid){
        $.ajax(
            {
                //请求路径,要注意的是 url 和 th:inline="javascript"
                url : [[@{/cart/focus}]],
                //请求类型
                type : "post",
```

```
                    contentType : "application/json",
                    //data 表示发送的数据
                    data : JSON.stringify({
                        id : gid
                    }),
                    //成功响应的结果
                    success : function(obj){ //obj 响应数据
                        if(obj == "no"){
                            alert("您已收藏该商品!");
                        }else if(obj == "ok"){
                            alert("成功收藏该商品");
                        }else{
                            alert("您没有登录,请登录!");
                        }
                    },
                    error : function() {
                        alert("处理异常!");
                    }
                }
            );
        }
</script>
</head>
<body>
    <!-- 加载 header.html -->
    <div th:include="user/header"></div>
    <div class="container">
        <div>
            <h4>搜索结果</h4>
        </div>
        <div class="row">
            <div class="col-xs-6 col-md-2" th:each="rGoods: ${searchgoods}">
                <a th:href="'goodsDetail?id=' + ${rGoods.id}" class="thumbnail"><img
                    alt="100%x180" th:src="'images/' + ${rGoods.gpicture}"
                    style="height: 180px; width: 100%; display: block;">
                </a>
                <div class="caption" style="text-align: center;">
                    <div>
                        <span th:text="${rGoods.gname}"></span>
                    </div>
                    <div>
                        <span style="color: red;">&yen;
                            <span th:text="${rGoods.grprice}"></span>
                        </span>
                        <span class="text-dark" style="text-decoration: line-through;">&yen;
                            <span th:text="${rGoods.goprice}"></span>
                        </span>
                    </div>
                    <a th:href="'javascript:focus('+ ${rGoods.id} +')'" class="btn btn-primary"
```

```
                       style="font-size:10px;">加入收藏</a>
                </div>
            </div>
        </div>
    </div>
</body>
</html>
```

2. 编写控制器层

该功能模块的控制器层涉及 com.ch.ebusiness.controller.before.IndexController 控制器类的处理方法 search,具体代码如下:

```
@RequestMapping("/search")
public String search(Model model, String mykey) {
    return indexService.search(model, mykey);
}
```

3. 编写 Service 层

该功能模块的 Service 层代码如下:

```
@Override
public String search(Model model, String mykey) {
    //广告区商品
    model.addAttribute("advertisementGoods", indexRepository.selectAdvertisementGoods());
    //导航栏商品类型
    model.addAttribute("goodsType", indexRepository.selectGoodsType());
    //商品搜索
    model.addAttribute("searchgoods", indexRepository.search(mykey));
    return "user/searchResult";
}
```

4. 编写 SQL 映射文件

该功能模块涉及的 SQL 语句如下:

```
<!-- 查询广告商品 -->
<select id="selectAdvertisementGoods" resultType="Goods">
    select
        gt.*, gy.typename
    from
        goodstable gt, goodstype gy
    where
        gt.goodstype_id = gy.id
        and gt.isAdvertisement = 1
    order by gt.id desc limit 5
</select>
<!-- 查询商品类型 -->
<select id="selectGoodsType" resultType="GoodsType">
    select * from goodstype
```

```xml
</select>
<!-- 首页搜索 -->
<select id="search" resultType="Goods" parameterType="String">
    select gt.*, gy.typename from GOODSTABLE gt,GOODSTYPE gy where gt.goodstype_id = gy.id
    and gt.gname like concat('%',#{mykey},'%')
</select>
```

11.6.2 推荐商品及最新商品

推荐商品是根据商品表中的字段 isRecommend 值判断的。最新商品是以商品 ID 排序的,因为商品 ID 是用 MySQL 自动递增产生的。具体实现步骤如下。

1. 编写视图

该模块的视图涉及 src/main/resources/templates/user 目录下的 index.html 页面,其核心代码如下:

```html
<div class="container">
    <div>
        <h4>推荐商品</h4>
    </div>
    <div class="row">
        <div class="col-xs-6 col-md-2" th:each="rGoods: ${recommendGoods}">
            <a th:href="'goodsDetail?id=' + ${rGoods.id}" class="thumbnail"><img
                alt="100%x180" th:src="'images/' + ${rGoods.gpicture}"
                style="height: 180px; width: 100%; display: block;">
            </a>
            <div class="caption" style="text-align: center;">
                <div>
                    <span th:text="${rGoods.gname}"></span>
                </div>
                <div>
                    <span style="color: red;">&yen;
                        <span th:text="${rGoods.grprice}"></span>
                    </span>
                    <span class="text-dark" style="text-decoration: line-through;">&yen;
                        <span th:text="${rGoods.goprice}"></span>
                    </span>
                </div>
                <a th:href="''javascript:focus(' + ${rGoods.id} + ')'" class="btn btn-primary"
                    style="font-size: 10px;">加入收藏</a>
            </div>
        </div>
    </div>
    <!-- ********************************************************** -->
    <div>
        <h4>最新商品</h4>
    </div>
    <div class="row">
```

```html
<div class="col-xs-6 col-md-2" th:each="lGoods:${lastedGoods}">
    <a th:href="'goodsDetail?id='+${lGoods.id}" class="thumbnail"><img alt="100%x180"
        th:src="'images/'+${lGoods.gpicture}"
        style="height: 180px; width: 100%; display: block;">
    </a>
    <div class="caption" style="text-align: center;">
        <div>
            <span th:text="${lGoods.gname}"></span>
        </div>
        <div>
            <span style="color: red;">&yen;
                <span th:text="${lGoods.grprice}"></span>
            </span>
            <span class="text-dark" style="text-decoration: line-through;">&yen;
                <span th:text="${lGoods.goprice}"></span>
            </span>
        </div>
        <a th:href="'javascript:focus('+${lGoods.id}+')'" class="btn btn-primary" style="font-size: 10px;">加入收藏</a>
    </div>
</div>
</div>
</div>
```

2. 编写控制器层

该功能模块的控制器层涉及 com.ch.ebusiness.controller.before.IndexController 控制器类的处理方法 index，具体代码如下：

```java
@RequestMapping("/")
public String index(Model model, Integer tid) {
    return indexService.index(model, tid);
}
```

3. 编写 Service 层

该功能模块的 Service 层代码如下：

```java
@Override
public String index(Model model, Integer tid) {
    if(tid == null)
        tid = 0;
    //广告区商品
    model.addAttribute("advertisementGoods", indexRepository.selectAdvertisementGoods());
    //导航栏商品类型
    model.addAttribute("goodsType", indexRepository.selectGoodsType());
    //推荐商品
    model.addAttribute("recommendGoods", indexRepository.selectRecommendGoods(tid));
    //最新商品
    model.addAttribute("lastedGoods", indexRepository.selectLastedGoods(tid));
    return "user/index";
}
```

第11章 电子商务平台的设计与实现(Thymeleaf+MyBatis)

4. 编写 SQL 映射文件

该功能模块涉及的 SQL 语句如下:

```xml
<!-- 查询推荐商品 -->
<select id="selectRecommendGoods" resultType="Goods" parameterType="integer">
    select
        gt.*, gy.typename
    from
        goodstable gt,goodstype gy
    where
        gt.goodstype_id = gy.id
        and gt.isRecommend = 1
        <if test="tid != 0">
            and gy.id = #{tid}
        </if>
    order by gt.id desc limit 6
</select>
<!-- 查询最新商品 -->
<select id="selectLastedGoods" resultType="Goods" parameterType="integer">
    select
        gt.*, gy.typename
    from
        goodstable gt,goodstype gy
    where
        gt.goodstype_id = gy.id
        <if test="tid != 0">
            and gy.id = #{tid}
        </if>
    order by gt.id desc limit 6
</select>
```

11.6.3 用户注册

单击导航栏的"注册"超链接(user/toRegister),将打开注册页面 register.html,如图 11.22 所示。

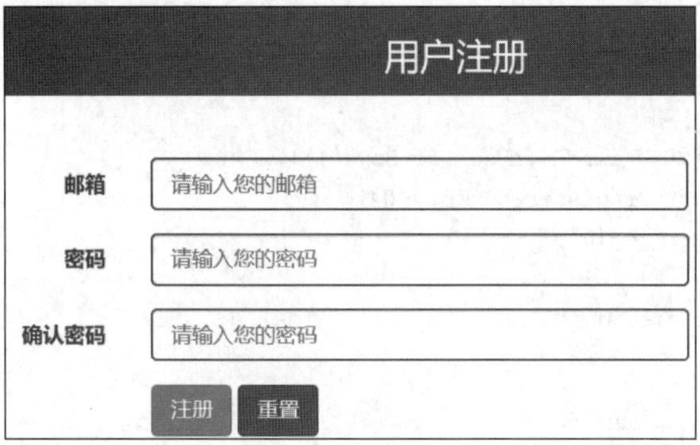

图 11.22 注册页面

输入用户信息,单击"注册"按钮,将用户信息提交给 user/register 处理请求,系统根据 @RequestMapping 注解找到 com.ch.ebusiness.controller.before.UserController 控制器类的 toRegister 和 register 方法处理请求。注册模块的具体实现步骤如下。

1. 编写视图

该模块的视图涉及 src/main/resources/templates/user 目录下的 register.html,其代码与后台登录页面代码类似,不再赘述。

2. 编写控制器层

该功能模块涉及 com.ch.ebusiness.controller.before.UserController 控制器类的 toRegister 和 register 方法。具体代码如下:

```
@RequestMapping("/toRegister")
public String toRegister(@ModelAttribute("bUser") BUser bUser) {
    return "user/register";
}
@RequestMapping("/register")
public String register(@ModelAttribute("bUser") @Validated BUser bUser,BindingResult rs) {
    if(rs.hasErrors()){                        //验证失败
        return "user/register";
    }
    return userService.register(bUser);
}
```

3. 编写 Service 层

该功能模块的 Service 层代码如下:

```
@Override
public String isUse(BUser bUser) {
    if(userRepository.isUse(bUser).size() > 0) {
        return "no";
    }
    return "ok";
}
@Override
public String register(BUser bUser) {
    //对密码 MD5 加密
    bUser.setBpwd(MD5Util.MD5(bUser.getBpwd()));
    if(userRepository.register(bUser) > 0) {
        return "user/login";
    }
    return "user/register";
}
```

4. 编写 SQL 映射文件

该功能模块涉及的 SQL 语句如下:

```xml
<select id = "isUse" parameterType = "BUser" resultType = "BUser">
    select * from busertable where bemail = #{bemail}
</select>
<insert id = "register" parameterType = "BUser">
    insert into busertable (id, bemail, bpwd) values(null, #{bemail}, #{bpwd})
</insert>
```

11.6.4 用户登录

用户注册成功后,跳转到登录页面 login.html,如图 11.23 所示。

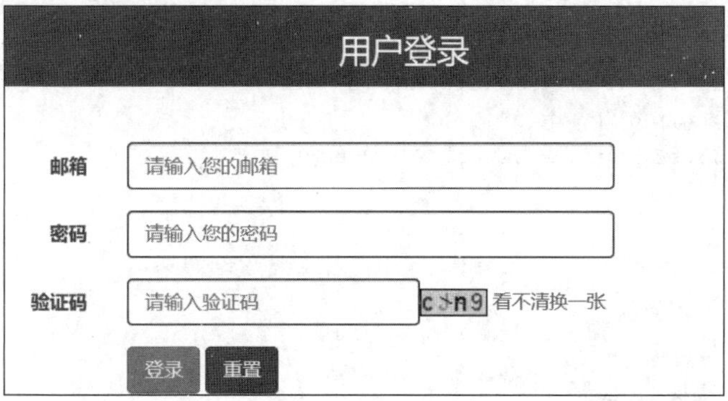

图 11.23 登录页面

在图 11.23 中,输入信息后单击"登录"按钮,将用户输入的 E-mail、密码以及验证码提交给 user/login 请求处理。系统根据 @RequestMapping 注解找到 com.ch.ebusiness.controller.before.UserController 控制器类的 login 方法处理请求。登录成功后,将用户的登录信息保存在 session 对象中,然后回到网站首页。具体实现步骤如下。

1. 编写视图

该模块的视图涉及 src/main/resources/templates/user 目录下的 login.html。其代码与后台登录页面代码类似,不再赘述。

2. 编写控制器层

该功能模块涉及 com.ch.ebusiness.controller.before.UserController 控制器类的 login 方法。具体代码如下:

```java
@RequestMapping("/login")
public String login(@ModelAttribute("bUser") @Validated BUser bUser,
        BindingResult rs, HttpSession session, Model model) {
    if(rs.hasErrors()){                    //验证失败
        return "user/login";
    }
    return userService.login(bUser, session, model);
}
```

3. 编写 Service 层

该功能模块的 Service 层代码如下：

```java
@Override
public String login(BUser bUser, HttpSession session, Model model) {
    //对密码 MD5 加密
    bUser.setBpwd(MD5Util.MD5(bUser.getBpwd()));
    String rand = (String)session.getAttribute("rand");
    if(!rand.equalsIgnoreCase(bUser.getCode())) {
        model.addAttribute("errorMessage", "验证码错误!");
        return "user/login";
    }
    List<BUser> list = userRepository.login(bUser);
    if(list.size() > 0) {
        session.setAttribute("bUser", list.get(0));
        return "redirect:/";                //到首页
    }
    model.addAttribute("errorMessage", "用户名或密码错误!");
    return "user/login";
}
```

4. 编写 SQL 映射文件

该功能模块的 SQL 语句如下：

```xml
<select id="login" parameterType="BUser" resultType="BUser">
    select * from busertable where bemail = #{bemail} and bpwd = #{bpwd}
</select>
```

11.6.5 商品详情

可以从推荐商品、最新商品、广告商品以及搜索商品结果等位置处，单击商品图片进入商品详情页面 goodsDetail.html，如图 11.24 所示。

图 11.24　商品详情页面

第11章 电子商务平台的设计与实现(Thymeleaf+MyBatis)

商品详情的具体实现步骤如下。

1. 编写视图

该模块的视图涉及 src/main/resources/templates/user 目录下的 goodsDetail.html,其核心代码如下:

```html
<body>
    <!-- 加载 header.html -->
    <div th:include="user/header"></div>
    <div class="container">
        <div class="row">
            <div class="col-xs-6 col-md-3">
                <img
                    th:src="'images/' + ${goods.gpicture}"
                    style="height: 220px; width: 280px; display: block;">
            </div>
            <div class="col-xs-6 col-md-3">
                <p>商品名:<span th:text="${goods.gname}"></span></p>
                <p>
                    商品折扣价:<span style="color: red;">&yen;
                        <span th:text="${goods.grprice}"></span>
                    </span>
                </p>
                <p>
                    商品原价:
                    <span class="text-dark" style="text-decoration: line-through;">&yen;
                        <span th:text="${goods.goprice}"></span>
                    </span>
                </p>
                <p>
                    商品类型:<span th:text="${goods.typename}"></span>
                </p>
                <p>
                    库存:<span id="gstore" th:text="${goods.gstore}"></span>
                </p>
                <p>
                    <input type="text" size="12" class="form-control" placeholder="请输入购买量" id="buyNumber" name="buyNumber"/>
                    <input type="hidden" name="gid" id="gid" th:value="${goods.id}"/>
                </p>
                <p>
                    <a href="javascript:focus()" class="btn btn-primary"
                        style="font-size: 10px;">加入收藏</a>
                    <a href="javascript:putCart()" class="btn btn-success"
                        style="font-size: 10px;">加入购物车</a>
                </p>
            </div>
        </div>
```

```
        </div>
    </body>
```

2. 编写控制器层

该功能模块涉及 com.ch.ebusiness.controller.before.IndexController 控制器类的 goodsDetail 方法。具体代码如下：

```
@RequestMapping("/goodsDetail")
public String goodsDetail(Model model, Integer id) {
    return indexService.goodsDetail(model, id);
}
```

3. 编写 Service 层

该功能模块的 Service 层代码如下：

```
@Override
public String goodsDetail(Model model, Integer id) {
    //广告区商品
    model.addAttribute("advertisementGoods", indexRepository.selectAdvertisementGoods());
    //导航栏商品类型
    model.addAttribute("goodsType", indexRepository.selectGoodsType());
    //商品详情
    model.addAttribute("goods", indexRepository.selectAGoods(id));
    return "user/goodsDetail";
}
```

4. 编写 SQL 映射文件

该功能模块的 SQL 语句如下：

```
<!-- 查询商品详情 -->
<select id="selectAGoods" resultType="Goods">
    select
        gt.*, gy.typename
    from
        goodstable gt,goodstype gy
    where
        gt.goodstype_id = gy.id
        and gt.id = #{id}
</select>
```

11.6.6 收藏商品

成功登录的用户可以在商品详情页面、首页以及搜索商品结果页面，单击"加入收藏"按钮收藏该商品。此时，请求路径为 cart/focus(Ajax 实现)。系统根据 @RequestMapping 注解找到 com.ch.ebusiness.controller.before.CartController 控制器类的 focus 方法处理请求。具体实现步骤如下：

1. 编写控制器层

该功能模块涉及 com.ch.ebusiness.controller.before.CartController 控制器类的 focus 方法。具体代码如下：

```
@RequestMapping("/focus")
@ResponseBody
public String focus(@RequestBody Goods goods, Model model, HttpSession session) {
    return cartService.focus(model, session, goods.getId());
}
```

2. 编写 Service 层

该功能模块的 Service 层代码如下：

```
@Override
public String focus(Model model, HttpSession session, Integer gid) {
    Integer uid = MyUtil.getUser(session).getId();
    List<Map<String,Object>> list = cartRepository.isFocus(uid, gid);
    //判断是否已收藏
    if(list.size() > 0) {
        return "no";
    }else {
        cartRepository.focus(uid, gid);
        return "ok";
    }
}
```

3. 编写 SQL 映射文件

该功能模块的 SQL 语句如下：

```
<!-- 处理加入收藏 -->
<select id="isFocus" resultType="map">
    select * from focustable where goodstable_id = #{gid} and busertable_id = #{uid}
</select>
<insert id="focus">
    insert into focustable (id, goodstable_id, busertable_id, focustime) values(null, #{gid}, #{uid}, now())
</insert>
```

11.6.7 购物车

单击商品详情页面中的"加入购物车"按钮或导航栏中的"我的购物车"超链接，打开购物车页面 cart.html，如图 11.25 所示。

与购物车有关的处理请求有 cart/putCart（加入购物车）、cart/clearCart（清空购物车）、cart/selectCart（查询购物车）和 cart/deleteCart（删除购物车）。系统根据 @RequestMapping 注解分别找到 com.ch.ebusiness.controller.before.CartController 控

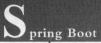

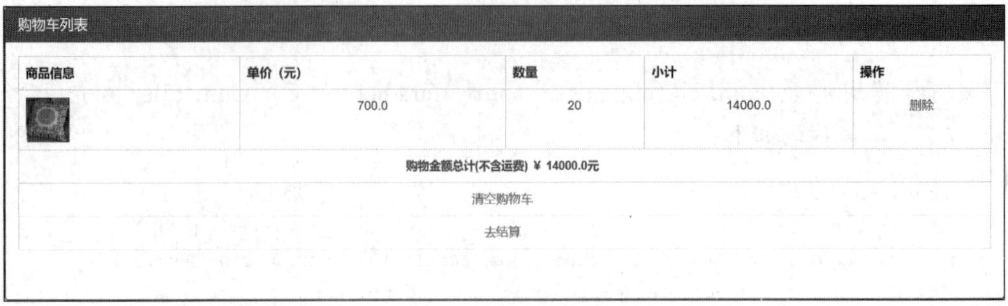

图 11.25 购物车

制器类的 putCart、clearCart、selectCart、deleteCart 等方法处理请求。具体实现步骤如下。

1. 编写视图

该模块的视图涉及 src/main/resources/templates/user 目录下的 cart.html,具体代码如下:

```html
<!DOCTYPE html>
<html xmlns:th="http://www.thymeleaf.org">
<head>
<base th:href="@{/}"><!-- 不用base就使用 th:src="@{/js/jquery.min.js} -->
<meta charset="UTF-8">
<title>购物车页面</title>
<link rel="stylesheet" href="css/bootstrap.min.css" />
<script src="js/jquery.min.js"></script>
<script type="text/javascript">
    function deleteCart(obj){
        if(window.confirm("确认删除吗?")){
            //获取路径
            var pathName = window.document.location.pathname;
            //截取,得到项目名称
            var projectName = pathName.substring(0,pathName.substr(1).indexOf('/')+1);
            window.location.href = projectName + "/cart/deleteCart?gid=" + obj;
        }
    }
    function clearCart(){
        if(window.confirm("确认清空吗?")){
            //获取路径
            var pathName = window.document.location.pathname;
            //截取,得到项目名称
            var projectName = pathName.substring(0,pathName.substr(1).indexOf('/')+1);
            window.location.href = projectName + "/cart/clearCart";
        }
    }
</script>
</head>
<body>
<div th:include="user/header"></div>
```

```html
<div class = "container">
    <div class = "panel panel-primary">
        <div class = "panel-heading">
            <h3 class = "panel-title">购物车列表</h3>
        </div>
        <div class = "panel-body">
            <div class = "table table-responsive">
                <table class = "table table-bordered table-hover">
                    <tbody class = "text-center">
                        <tr>
                            <th>商品信息</th>
                            <th>单价(元)</th>
                            <th>数量</th>
                            <th>小计</th>
                            <th>操作</th>
                        </tr>
                        <tr th:each = "cart:${cartlist}">
                            <td>
                                <a th:href = "'goodsDetail?id=' + ${cart.id}">
                                    <img th:src = "'images/' + ${cart.gpicture}"
                                        style = "height: 50px; width: 50px; display: block;">
                                </a>
                            </td>
                            <td th:text = "${cart.grprice}"></td>
                            <td th:text = "${cart.shoppingnum}"></td>
                            <td th:text = "${cart.smallsum}"></td>
                            <td>
                                <a th:href = "'javascript:deleteCart(' + ${cart.id} + ')'">删除</a>
                            </td>
                        </tr>
                        <tr>
                            <td colspan = "5">
                                <font style = "color: #a60401; font-size: 13px; font-weight: bold; letter-spacing: 0px;">
                                    购物金额总计(不含运费) ¥ <span th:text = "${total}"></span>元
                                </font>
                            </td>
                        </tr>
                        <tr>
                            <td colspan = "5">
                                <a href = "javascript:clearCart()">清空购物车</a>
                            </td>
                        </tr>
                        <tr>
                            <td colspan = "5">
                                <a href = "cart/selectCart?act = toCount">去结算</a>
                            </td>
                        </tr>
                    </tbody>
                </table>
```

```html
            </div>
        </div>
    </div>
</div>
</body>
</html>
```

2. 编写控制器层

该功能模块涉及 com.controller.before.CartController 控制器类的 putCart、clearCart、selectCart、deleteCart 等方法。具体代码如下:

```java
@RequestMapping("/putCart")
public String putCart(Goods goods, Model model, HttpSession session) {
    return cartService.putCart(goods, model, session);
}
@RequestMapping("/selectCart")
public String selectCart(Model model, HttpSession session, String act) {
    return cartService.selectCart(model, session, act);
}
@RequestMapping("/deleteCart")
public String deleteCart(HttpSession session, Integer gid) {
    return cartService.deleteCart(session, gid);
}
@RequestMapping("/clearCart")
public String clearCart(HttpSession session) {
    return cartService.clearCart(session);
}
```

3. 编写 Service 层

该功能模块的 Service 层代码如下:

```java
@Override
public String putCart(Goods goods, Model model, HttpSession session) {
    Integer uid = MyUtil.getUser(session).getId();
    //如果商品已在购物车,只更新购买数量
    if(cartRepository.isPutCart(uid, goods.getId()).size() > 0) {
        cartRepository.updateCart(uid, goods.getId(), goods.getBuyNumber());
    }else {                                    //新增到购物车
        cartRepository.putCart(uid, goods.getId(), goods.getBuyNumber());
    }
    //跳转到查询购物车
    return "forward:/cart/selectCart";
}
@Override
public String selectCart(Model model, HttpSession session, String act) {
    List<Map<String, Object>> list = cartRepository.selectCart(MyUtil.getUser(session).getId());
    double sum = 0;
    for (Map<String, Object> map : list) {
```

```java
            sum = sum + (Double)map.get("smallsum");
        }
        model.addAttribute("total", sum);
        model.addAttribute("cartlist", list);
        //广告区商品
        model.addAttribute("advertisementGoods", indexRepository.selectAdvertisementGoods());
        //导航栏商品类型
        model.addAttribute("goodsType", indexRepository.selectGoodsType());
        if("toCount".equals(act)) {          //去结算页面
            return "user/count";
        }
        return "user/cart";
    }
    @Override
    public String deleteCart(HttpSession session, Integer gid) {
        Integer uid = MyUtil.getUser(session).getId();
        cartRepository.deleteAgoods(uid, gid);
        return "forward:/cart/selectCart";
    }
    @Override
    public String clearCart(HttpSession session) {
        cartRepository.clear(MyUtil.getUser(session).getId());
        return "forward:/cart/selectCart";
    }
```

4. 编写 SQL 映射文件

该功能模块的 SQL 语句如下：

```xml
<!-- 是否已添加购物车 -->
<select id="isPutCart" resultType="map">
    select * from carttable where goodstable_id = #{gid} and busertable_id = #{uid}
</select>
<!-- 添加购物车 -->
<insert id="putCart">
    insert into carttable (id, busertable_id, goodstable_id, shoppingnum) values(null, #{uid}, #{gid}, #{bnum})
</insert>
<!-- 更新购物车 -->
<update id="updateCart">
    update carttable set shoppingnum = shoppingnum + #{bnum} where busertable_id = #{uid} and goodstable_id = #{gid}
</update>
<!-- 查询购物车 -->
<select id="selectCart" parameterType="Integer" resultType="map">
    select gt.id, gt.gname, gt.gpicture, gt.grprice, ct.shoppingnum, ct.shoppingnum * gt.grprice smallsum from GOODSTABLE gt, CARTTABLE ct where gt.id = ct.goodstable_id and ct.busertable_id = #{uid}
</select>
<!-- 删除购物车 -->
```

```
    <delete id = "deleteAgoods">
        delete from carttable where busertable_id = #{uid} and goodstable_id = #{gid}
    </delete>
    <!-- 清空购物车 -->
    <delete id = "clear" parameterType = "Integer">
        delete from carttable where busertable_id = #{uid}
    </delete>
```

11.6.8　下单

在购物车页面单击"去结算"按钮,进入订单确认页面 count.html,如图 11.26 所示。

图 11.26　订单确认

在订单确认页面单击"提交订单"按钮,完成订单提交。订单完成时,提示页面效果如图 11.27 所示。

图 11.27　订单提交完成页面

单击图 11.27 中的"去支付"完成订单支付。

具体实现步骤如下。

1. 编写视图

该模块的视图涉及 src/main/resources/templates/user 目录下的 count.html 和 pay.html。count.html 的代码与购物车页面代码基本一样,不再赘述。pay.html 的代码如下:

```
<!DOCTYPE html>
<html xmlns:th = "http://www.thymeleaf.org">
<head>
<base th:href = "@{/}"><!-- 不用 base 就使用 th:src = "@{/js/jquery.min.js}" -->
<meta charset = "UTF-8">
<title>支付页面</title>
<link rel = "stylesheet" href = "css/bootstrap.min.css" />
<script src = "js/jquery.min.js"></script>
<script type = "text/javascript" th:inline = "javascript">
```

第11章 电子商务平台的设计与实现(Thymeleaf+MyBatis)

```
            function pay(){
                $.ajax(
                    {
                        //请求路径,要注意的是 url 和 th:inline = "javascript"
                        url:[[@{/cart/pay}]],
                        //请求类型
                        type:"post",
                        contentType:"application/json",
                        //data 表示发送的数据
                        data:JSON.stringify({
                            id:$("#oid").text()
                        }),
                        //成功响应的结果
                        success:function(obj){    //obj 响应数据
                            alert("支付成功");
                            //获取路径
                            var pathName = window.document.location.pathname;
                            //截取,得到项目名称
                            var projectName = pathName.substring(0,pathName.substr(1).indexOf('/') + 1);
                            window.location.href = projectName;
                        },
                        error:function() {
                            alert("处理异常!");
                        }
                    }
                );
            }
        </script>
    </head>
    <body>
        <div class = "container">
            <div class = "panel panel-primary">
                <div class = "panel-heading">
                    <h3 class = "panel-title">订单提交成功</h3>
                </div>
                <div class = "panel-body">
                    <div>
        您的订单编号为<font color = "red" size = "5"><span id = "oid" th:text = "${order.id}">
</span></font>  。<br><br>
                        <a href = "javascript:pay()">去支付</a>
                    </div>
                </div>
            </div>
        </div>
    </body>
</html>
```

2. 编写控制器层

该功能模块涉及 com.ch.ebusiness.controller.before.CartController 控制器类的

submitOrder 和 pay 方法。具体代码如下：

```java
@RequestMapping("/submitOrder")
public String submitOrder(Order order, Model model, HttpSession session) {
    return cartService.submitOrder(order, model, session);
}
@RequestMapping("/pay")
@ResponseBody
public String pay(@RequestBody Order order) {
    return cartService.pay(order);
}
```

3. 编写 Service 层

该功能模块的 Service 层代码如下：

```java
@Override
@Transactional
public String submitOrder(Order order, Model model, HttpSession session) {
    order.setBusertable_id(MyUtil.getUser(session).getId());
    //生成订单
    cartRepository.addOrder(order);
    //生成订单详情
    cartRepository.addOrderDetail(order.getId(), MyUtil.getUser(session).getId());
    //减少商品库存
    List<Map<String, Object>> listGoods = cartRepository.selectGoodsShop(MyUtil.getUser(session).getId());
    for (Map<String, Object> map : listGoods) {
        cartRepository.updateStore(map);
    }
    //清空购物车
    cartRepository.clear(MyUtil.getUser(session).getId());
    model.addAttribute("order", order);
    return "user/pay";
}
@Override
public String pay(Order order) {
    cartRepository.pay(order.getId());
    return "ok";
}
```

4. 编写 SQL 映射文件

该功能模块涉及的 SQL 语句如下：

```xml
<!-- 添加一个订单,成功后将主键值回填给 id(实体类的属性) -->
<insert id="addOrder" parameterType="Order" keyProperty="id" useGeneratedKeys="true">
    insert into orderbasetable (busertable_id, amount, status, orderdate) values (#{busertable_id}, #{amount}, 0, now())
</insert>
```

```xml
<!-- 生成订单详情 -->
<insert id="addOrderDetail">
    insert into ORDERDETAIL (orderbasetable_id, goodstable_id, SHOPPINGNUM) select #{ordersn}, goodstable_id, SHOPPINGNUM from CARTTABLE where busertable_id = #{uid}
</insert>
<!-- 查询商品购买量,以便更新库存 -->
<select id="selectGoodsShop" parameterType="Integer" resultType="map">
    select shoppingnum gshoppingnum, goodstable_id gid from carttable where busertable_id = #{uid}
</select>
<!-- 更新商品库存 -->
<update id="updateStore" parameterType="map">
    update GOODSTABLE set GSTORE = GSTORE - #{gshoppingnum} where id = #{gid}
</update>
<!-- 支付订单 -->
<update id="pay" parameterType="Integer">
    update orderbasetable set status = 1 where id = #{ordersn}
</update>
```

11.6.9 个人信息

成功登录的用户,在导航栏的上方,单击"个人信息"超链接(cart/userInfo),进入用户修改密码页面 userInfo.html,如图 11.28 所示。

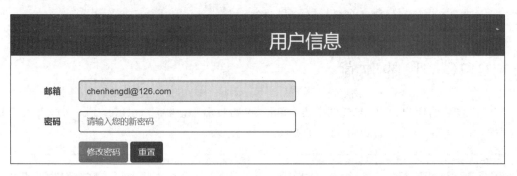

图 11.28 用户修改密码页面

具体实现步骤如下。

1. 编写视图

该模块的视图涉及 src/main/resources/templates/user 目录下的 userInfo.html,其代码与登录页面类似,不再赘述。

2. 编写控制器层

该功能模块涉及 com.ch.ebusiness.controller.before.CartController 控制器类的 userInfo 和 updateUpwd 方法。具体代码如下:

```
@RequestMapping("/userInfo")
public String userInfo() {
    return "user/userInfo";
}
@RequestMapping("/updateUpwd")
public String updateUpwd(HttpSession session, String bpwd) {
    return cartService.updateUpwd(session, bpwd);
}
```

3. 编写 Service 层

该功能模块的 Service 层代码如下：

```
@Override
public String updateUpwd(HttpSession session, String bpwd) {
    Integer uid = MyUtil.getUser(session).getId();
    cartRepository.updateUpwd(uid, MD5Util.MD5(bpwd));
    return "forward:/user/toLogin";
}
```

4. 编写 SQL 映射文件

该功能模块的 SQL 语句如下：

```
<!-- 修改密码 -->
<update id = "updateUpwd">
    update busertable set bpwd = #{bpwd} where id = #{uid}
</update>
```

11.6.10 我的收藏

成功登录的用户，在导航栏的上方，单击"我的收藏"超链接（cart/myFocus），进入用户收藏页面 myFocus.html，如图 11.29 所示。

图 11.29 用户收藏页面

第11章 电子商务平台的设计与实现(Thymeleaf+MyBatis)

具体实现步骤如下。

1. 编写视图

该模块的视图涉及 src/main/resources/templates/user 目录下的 myFocus.html。具体代码如下：

```html
<!DOCTYPE html>
<html xmlns:th="http://www.thymeleaf.org">
<head>
<base th:href="@{/}"><!-- 不用base就使用 th:src="@{/js/jquery.min.js}" -->
<meta charset="UTF-8">
<title>收藏页面</title>
<link rel="stylesheet" href="css/bootstrap.min.css" />
</head>
<body>
<div th:include="user/header"></div>
<div class="container">
    <div class="panel panel-primary">
        <div class="panel-heading">
            <h3 class="panel-title">收藏列表</h3>
        </div>
        <div class="panel-body">
            <div class="table table-responsive">
                <table class="table table-bordered table-hover">
                    <tbody class="text-center">
                        <tr>
                            <th>商品图片</th>
                            <th>商品名称</th>
                            <th>原价</th>
                            <th>现价</th>
                        </tr>
                        <tr th:each="focus:${myFocus}">
                            <td>
                                <a th:href="'goodsDetail?id='+${focus.id}">
                                    <img th:src="'images/'+${focus.gpicture}"
                                        style="height:50px;width:50px;display:block;">
                                </a>
                            </td>
                            <td th:text=""${focus.gname}"></td>
                            <td th:text=""${focus.goprice}"></td>
                            <td th:text=""${focus.grprice}"></td>
                        </tr>
                    </tbody>
                </table>
            </div>
        </div>
    </div>
</div>
</body>
</html>
```

2. 编写控制器层

该功能模块涉及 com.ch.ebusiness.controller.before.CartController 控制器类的 myFocus 方法。具体代码如下：

```java
@RequestMapping("/myFocus")
public String myFocus(Model model, HttpSession session) {
    return cartService.myFocus(model, session);
}
```

3. 编写 Service 层

该功能模块的 Service 层代码如下：

```java
@Override
public String myFocus(Model model, HttpSession session) {
    //广告区商品
    model.addAttribute("advertisementGoods", indexRepository.selectAdvertisementGoods());
    //导航栏商品类型
    model.addAttribute("goodsType", indexRepository.selectGoodsType());
    model.addAttribute("myFocus", cartRepository.myFocus(MyUtil.getUser(session).getId()));
    return "user/myFocus";
}
```

4. 编写 SQL 映射文件

该功能模块的 SQL 语句如下：

```xml
<!-- 我的收藏 -->
<select id="myFocus" resultType="map" parameterType="Integer">
    select gt.id, gt.gname, gt.goprice, gt.grprice, gt.gpicture from FOCUSTABLE ft, GOODSTABLE gt
    where ft.goodstable_id = gt.id and ft.busertable_id = #{uid}
</select>
```

11.6.11 我的订单

成功登录的用户，在导航栏的上方，单击"我的订单"超链接（cart/myOder），进入用户订单页面 myOrder.html，如图 11.30 所示。

单击图 11.30 中的"查看详情"超链接（'cart/orderDetail? id=' + ${order.id}），进入订单详情页面 orderDetail.html，如图 11.31 所示。

具体实现步骤如下。

1. 编写视图

该模块的视图涉及 src/main/resources/templates/user 目录下的 myOrder.html 和 orderDetail.html。

第11章 电子商务平台的设计与实现(Thymeleaf+MyBatis)

图11.30 用户订单页面

图11.31 订单详情页面

myOrder.html的代码如下：

```html
<!DOCTYPE html>
<html xmlns:th="http://www.thymeleaf.org">
<head>
<base th:href="@{/}"><!-- 不用base就使用 th:src="@{/js/jquery.min.js} -->
<meta charset="UTF-8">
<title>订单页面</title>
<link rel="stylesheet" href="css/bootstrap.min.css" />
<script src="js/jquery.min.js"></script>
<script type="text/javascript" th:inline="javascript">
    function pay(){
        $.ajax(
            {
                //请求路径,要注意的是url和th:inline="javascript"
                url : [[@{/cart/pay}]],
                //请求类型
                type : "post",
                contentType : "application/json",
                //data 表示发送的数据
                data : JSON.stringify({
                    id : $("#oid").text()
                }),
                //成功响应的结果
```

```html
                    success : function(obj){ //obj 响应数据
                        alert("支付成功");
                        //获取路径
                        var pathName = window.document.location.pathname;
                        //截取,得到项目名称
                    var projectName = pathName.substring(0,pathName.substr(1).indexOf('/') + 1);
                        window.location.href = projectName;
                    },
                    error : function() {
                        alert("处理异常!");
                    }
                }
            );
        }
    </script>
</head>
<body>
<div th:include = "user/header"></div>
<div class = "container">
    <div class = "panel panel-primary">
        <div class = "panel-heading">
            <h3 class = "panel-title">订单列表</h3>
        </div>
        <div class = "panel-body">
            <div class = "table table-responsive">
                <table class = "table table-bordered table-hover">
                    <tbody class = "text-center">
                        <tr>
                            <th>订单编号</th>
                            <th>订单金额</th>
                            <th>订单状态</th>
                            <th>下单时间</th>
                            <th>查看详情</th>
                        </tr>
                        <tr th:each = "order: ${myOrder}">
                            <td th:text = "${order.id}"></td>
                            <td th:text = "${order.amount}"></td>
                            <td>
                                <span th:if = "${order.status} == 0">
                                    未支付  <a href = "javascript:pay()">去支付</a>
                                </span>
                                <span th:if = "${order.status} == 1">
                                    已支付
                                </span>
                            </td>
                            <td th:text = "${order.orderdate}"></td>
                            <td>
            <a th:href = "'cart/orderDetail?id=' + ${order.id}" target = "_blank">查看详情</a>
                            </td>
                        </tr>
```

```html
                </tbody>
            </table>
        </div>
    </div>
</div>
</body>
</html>
```

orderDetail.html 的代码如下：

```html
<!DOCTYPE html>
<html xmlns:th="http://www.thymeleaf.org">
<head>
<base th:href="@{/}"><!-- 不用 base 就使用 th:src="@{/js/jquery.min.js}" -->
<meta charset="UTF-8">
<title>订单详情页面</title>
<link rel="stylesheet" href="css/bootstrap.min.css" />
</head>
<body>
<div class="container">
    <div class="panel panel-primary">
        <div class="panel-heading">
            <h3 class="panel-title">订单详情</h3>
        </div>
        <div class="panel-body">
            <div class="table table-responsive">
                <table class="table table-bordered table-hover">
                    <tbody class="text-center">
                        <tr>
                            <th>商品编号</th>
                            <th>商品图片</th>
                            <th>商品名称</th>
                            <th>商品购买价</th>
                            <th>购买数量</th>
                        </tr>
                        <tr th:each="od:${orderDetail}">
                            <td th:text="${od.id}"></td>
                            <td>
                                <a th:href="'goodsDetail?id='+${od.id}">
                                    <img th:src="'images/'+${od.gpicture}"
                                        style="height:50px;width:50px;display:block;">
                                </a>
                            </td>
                            <td th:text="${od.gname}"></td>
                            <td th:text="${od.grprice}"></td>
                            <td th:text="${od.shoppingnum}"></td>
                        </tr>
                    </tbody>
                </table>
            </div>
```

```
            </div>
        </div>
</div>
</body>
</html>
```

2. 编写控制器层

该功能模块涉及 com.ch.ebusiness.controller.before.CartController 控制器类的 myOrder 和 orderDetail 方法。具体代码如下：

```
@RequestMapping("/myOrder")
public String myOrder(Model model, HttpSession session) {
    return cartService.myOrder(model, session);
}
@RequestMapping("/orderDetail")
public String orderDetail(Model model, Integer id) {
    return cartService.orderDetail(model, id);
}
```

3. 编写 Service 层

该功能模块的 Service 层代码如下：

```
@Override
public String myOrder(Model model, HttpSession session) {
    //广告区商品
    model.addAttribute("advertisementGoods", indexRepository.selectAdvertisementGoods());
    //导航栏商品类型
    model.addAttribute("goodsType", indexRepository.selectGoodsType());
    model.addAttribute("myOrder", cartRepository.myOrder(MyUtil.getUser(session).getId()));
    return "user/myOrder";
}
@Override
public String orderDetail(Model model, Integer id) {
    model.addAttribute("orderDetail", cartRepository.orderDetail(id));
    return "user/orderDetail";
}
```

4. 编写 SQL 映射文件

该功能模块的 SQL 语句如下：

```
<!-- 我的订单 -->
<select id="myOrder" resultType="map" parameterType="Integer">
select id, amount, busertable_id, status, orderdate from ORDERBASETABLE where busertable_id = #{uid}
</select>
<!-- 订单详情 -->
<select id="orderDetail" resultType="map" parameterType="Integer">
select gt.id, gt.gname, gt.goprice, gt.grprice, gt.gpicture, odt.shoppingnum from GOODSTABLE
```

```
gt, ORDERDETAIL odt
    where  odt.orderbasetable_id = #{id} and gt.id = odt.goodstable_id
</select>
```

11.7　本章小结

　　本章讲述了电子商务平台通用功能的设计与实现。通过本章的学习，读者不仅应该掌握 Spring Boot 应用开发的流程、方法和技术，还应该熟悉电子商务平台的业务需求、设计以及实现。

第12章

名片系统的设计与实现 (Vue.js+JPA)

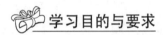

视频讲解

本章以名片系统的设计与实现为综合案例,讲述如何使用 Spring Boot＋Vue.js＋Spring Data JPA 开发一个前后端分离的应用程序。通过本章的学习,掌握基于 Vue.js＋Spring Data JPA 的前后端分离的应用程序的开发流程、方法以及技术。

本章主要内容

- 名片系统介绍。
- 使用 IntelliJ IDEA 构建名片后端系统。
- 使用 IntelliJ IDEA 构建名片前端系统。

前后端分离的核心思想是前端页面通过 Ajax 调用后端的 RESTful API 进行数据交互。本章将使用 Spring Boot＋Spring Data JPA 实现后端系统,使用 Vue.js 实现前端系统,数据库采用的是 MySQL 5.x,集成开发环境为 IntelliJ IDEA。

12.1 名片系统功能介绍

名片系统是针对注册用户使用的系统。系统提供的功能如下:
(1) 非注册用户可以注册为注册用户。
(2) 成功注册的用户,可以登录系统。

(3) 成功登录的用户,可以添加、修改、删除以及浏览自己客户的名片信息。

12.2 使用 IntelliJ IDEA 构建名片后端系统 cardmis

本节使用 IntelliJ IDEA 构建名片后端系统,当然也可以使用 STS 或 Eclipse 构建该系统。读者可以根据自己的使用习惯,选择集成开发环境。

12.2.1 构建基于 JPA 的 Spring Boot Web 应用

打开 IDEA,新建项目,选择 Spring Initializr,如图 12.1 所示。

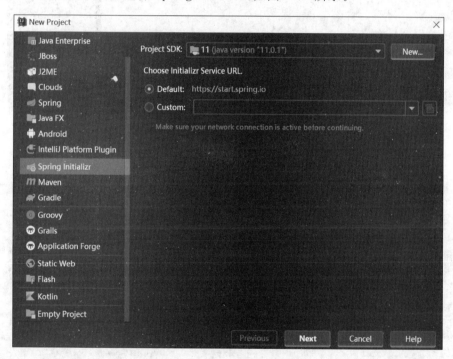

图 12.1 选择 Spring Initializr

单击图 12.1 中的 Next 按钮,打开如图 12.2 所示的项目元数据输入对话框。

单击图 12.2 中的 Next 按钮,打开项目依赖选择对话框,选择 Web|Spring Web,SQL|Spring Data JPA,如图 12.3 所示。

单击图 12.3 中的 Next 按钮,再单击 Finish 按钮后,等待项目自动初始化即可。

12.2.2 修改 pom.xml

修改名片后端系统 cardmis 的 pom.xml 文件,添加 MySQL 连接依赖和热部署依赖,具

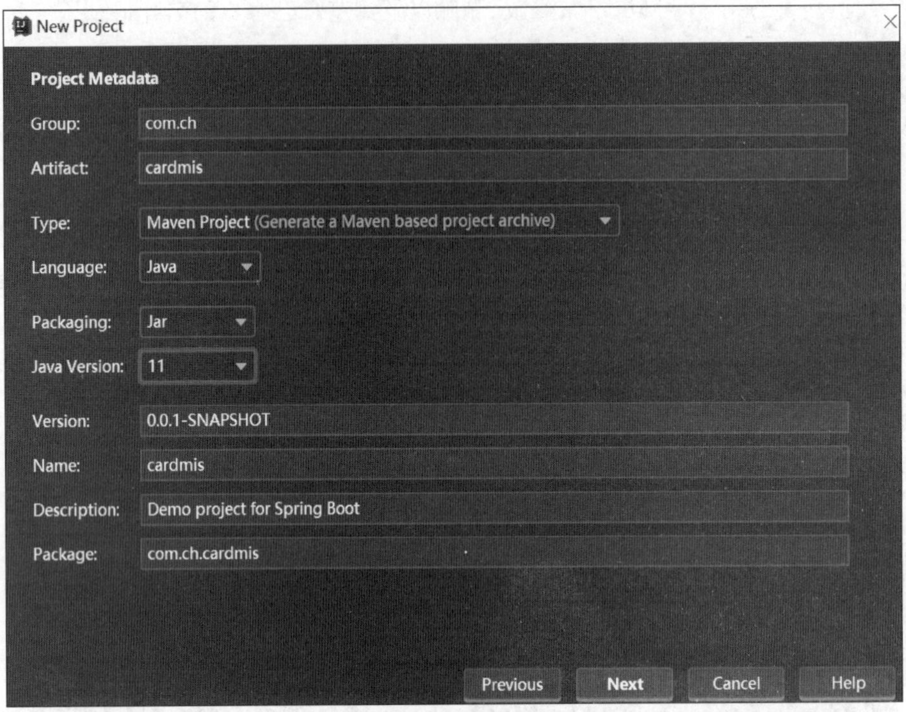

图 12.2 输入项目元数据

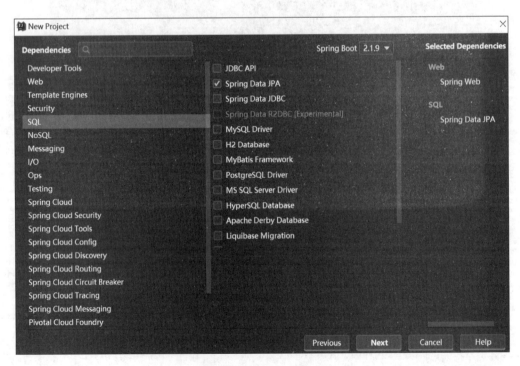

图 12.3 项目依赖选择界面

体代码如下：

```xml
<!-- MySQL 连接依赖 -->
<dependency>
    <groupId>mysql</groupId>
    <artifactId>mysql-connector-java</artifactId>
    <version>5.1.45</version>
</dependency>
<!-- 热部署依赖 -->
<dependency>
    <groupId>org.springframework.boot</groupId>
    <artifactId>spring-boot-devtools</artifactId>
    <optional>true</optional>
</dependency>
```

12.2.3 配置数据源等信息

在后端系统 cardmis 的配置文件 application.properties 中，配置端口号、数据源以及文件上传等信息，具体内容如下：

```
server.port=8443
###
## 数据源信息配置
###
# 数据库地址
spring.datasource.url=jdbc:mysql://localhost:3306/card?characterEncoding=utf8
# 数据库用户名
spring.datasource.username=root
# 数据库密码
spring.datasource.password=root
# 数据库驱动
spring.datasource.driver-class-name=com.mysql.jdbc.Driver
# 数据库 MySQL 为 8.x 时,驱动类为 com.mysql.cj.jdbc.Driver
####
# JPA 持久化配置
####
# 指定数据库类型
spring.jpa.database=MYSQL
# 指定是否在日志中显示 SQL 语句
spring.jpa.show-sql=true
# 指定自动创建、更新数据库表等配置
# update 表示如果数据库中存在持久化类对应的表就不创建,不存在就创建
spring.jpa.hibernate.ddl-auto=update
# 上传文件时,默认单个上传文件大小是 1MB,max-file-size 设置单个上传文件大小
spring.servlet.multipart.max-file-size=50MB
# 默认总文件大小是 10MB,max-request-size 设置总上传文件大小
spring.servlet.multipart.max-request-size=500MB
```

12.2.4 创建持久化实体类

根据名片系统功能可知,名片系统共有两个实体:用户和卡片,并且是一对多的关系。因此,后端系统 cardmis 的持久化实体类共有两个,分别为 UserEntity 和 CardEntity。

UserEntity 实体类的代码如下:

```java
package com.ch.cardmis.entity;
import java.util.List;
import java.io.Serializable;
import javax.persistence.CascadeType;
import javax.persistence.Entity;
import javax.persistence.FetchType;
import javax.persistence.GeneratedValue;
import javax.persistence.GenerationType;
import javax.persistence.Id;
import javax.persistence.OneToMany;
import javax.persistence.Table;
import com.fasterxml.jackson.annotation.JsonIgnoreProperties;
@Entity
@Table(name = "user_table")
@JsonIgnoreProperties(value = { "hibernateLazyInitializer"})
public class UserEntity implements Serializable{
    private static final long serialVersionUID = 1L;
    @Id
    @GeneratedValue(strategy = GenerationType.IDENTITY)
    private int id;                    //主键
    private String uname;
    private String upwd;
    //名片列表,用户与名片是一对多的关系
    @OneToMany(
        mappedBy = "user",
        cascade = CascadeType.ALL,
        targetEntity = CardEntity.class,
        fetch = FetchType.LAZY
    )
    private List<CardEntity> cardEntityList;
    //省略了 set 和 get 方法
}
```

CardEntity 实体类的代码如下:

```java
package com.ch.cardmis.entity;
import java.io.Serializable;
import javax.persistence.*;
import com.fasterxml.jackson.annotation.JsonIgnore;
import com.fasterxml.jackson.annotation.JsonIgnoreProperties;
@Entity
@Table(name = "card_table")
@JsonIgnoreProperties(value = { "hibernateLazyInitializer"})
```

```java
public class CardEntity implements Serializable {
    private static final long serialVersionUID = 1L;
    @Id
    @GeneratedValue(strategy = GenerationType.IDENTITY)
    private int id;                       //主键
    private String name;
    private String telephone;
    private String email;
    private String company;
    private String post;
    private String address;
    private String logo;
    //所属用户,名片与用户是多对一的关系
    @ManyToOne(cascade = {CascadeType.MERGE,CascadeType.REFRESH},optional = false)
    //可选属性 optional = false,表示用户不能为空。删除名片,不影响用户
    @JoinColumn(name = "id_user_id") //设置在 card_table 表中的关联字段(外键)
    @JsonIgnore
    private UserEntity user;
    //省略 set 和 get 方法
}
```

12.2.5 创建 Repository 持久层

后端系统 cardmis 的数据访问是基于 Spring Data JPA 的,数据访问接口需要继承 Repository 接口。与持久化实体类一样,本层涉及两个数据访问接口:UserRepository 和 CardRepository。

UserRepository 接口的代码如下:

```java
package com.ch.cardmis.repository;
import com.ch.cardmis.entity.UserEntity;
import org.springframework.data.jpa.repository.JpaRepository;
public interface UserRepository extends JpaRepository<UserEntity, Integer> {
    /**
     * 查询用户名是否已存在
     */
    public UserEntity findByUname(String uname);

    /**
     * 登录
     */
    public UserEntityfindByUnameAndUpwd(String uname, String upwd);
}
```

CardRepository 接口的代码如下:

```java
package com.ch.cardmis.repository;
import com.ch.cardmis.entity.CardEntity;
import org.springframework.data.domain.Sort;
import org.springframework.data.jpa.repository.JpaRepository;
```

```
import java.util.List;
public interface CardRepository extends JpaRepository<CardEntity, Integer> {
    //根据用户id查询该用户的名片
    public List<CardEntity> findByUser_id(Integer id, Sort sort);
}
```

12.2.6 创建业务层

在 Spring 框架中，提倡使用接口。因此，在后端系统 cardmis 的业务层中，涉及 Service 接口和 Service 实现类。Service 接口代码极其简单，请读者参考源程序，这里只列出 Service 实现类的代码。

UserServiceImpl 实现类的代码如下：

```java
package com.ch.cardmis.service;
import com.ch.cardmis.entity.UserEntity;
import com.ch.cardmis.repository.UserRepository;
import org.springframework.beans.factory.annotation.Autowired;
import org.springframework.stereotype.Service;
import javax.servlet.http.HttpSession;
@Service
public class UserServiceImpl implements UserService {
    @Autowired
    private UserRepository userRepository;
    /**
     * 注册
     */
    @Override
    public String register(UserEntity requestUser) {
        UserEntity ue = userRepository.findByUname(requestUser.getUname());
        if(ue != null)
            return "no";
        else{
            userRepository.save(requestUser);
            return "yes";
        }
    }
    /**
     * 登录
     */
    @Override
    public String login(UserEntity requestUser, HttpSession session) {
        UserEntity ue =
            userRepository.findByUnameAndUpwd(requestUser.getUname(), requestUser.getUpwd());
        if(ue != null){            //登录成功
            session.setAttribute("user", ue);
            return "ok";
        }
        return "no";
    }
}
```

第12章 名片系统的设计与实现(Vue.js+JPA)

CardServiceImpl 实现类的代码如下：

```java
package com.ch.cardmis.service;
import com.ch.cardmis.entity.CardEntity;
import com.ch.cardmis.entity.UserEntity;
import com.ch.cardmis.repository.CardRepository;
import com.ch.cardmis.util.MyUtil;
import org.springframework.beans.factory.annotation.Autowired;
import org.springframework.data.domain.Sort;
import org.springframework.stereotype.Service;
import org.springframework.web.multipart.MultipartFile;
import javax.servlet.http.HttpSession;
import java.io.File;
@Service
public class CardServiceImpl implements CardService {
    @Autowired
    private CardRepository cardRepository;
    /**
     * 添加或修改名片
     */
    @Override
    public String add(CardEntity cardEntity, HttpSession session, MultipartFile file) {
        if("noLogin".equals(isLogin(session))){
            return "noLogin";
        }else{
            UserEntity user = (UserEntity)session.getAttribute("user");
            cardEntity.setUser(user);
            //防止文件名重名
            String newFileName = "";
            if(file != null){
                String fileName = file.getOriginalFilename();
                newFileName = MyUtil.getNewFileName(fileName);
                String realpath
                    = "C:\\Users\\ChenHeng\\IdeaProjects\\cardmis-vue\\cardmis-vue\\static";
                File targetFile = new File(realpath, newFileName);
                if(!targetFile.exists()){
                    targetFile.mkdirs();
                }
                //设置文件名
                cardEntity.setLogo("static/" + newFileName);
                //上传
                try {
                    file.transferTo(targetFile);
                } catch (Exception e) {
                    e.printStackTrace();
                }
            }
            cardRepository.save(cardEntity);
            return "ok";
        }
    }
```

```java
/**
 * 查询登录用户的名片
 */
@Override
public Object cards(HttpSession session) {
    if("noLogin".equals(isLogin(session))){
        return "noLogin";
    }else {
        Sort sort = new Sort(Sort.Direction.DESC, "id");
        UserEntity user = (UserEntity)session.getAttribute("user");
        return cardRepository.findByUser_id(user.getId(), sort);
    }
}
/**
 * 根据id删除名片
 */
@Override
public String delete(HttpSession session, Integer cid) {
    if("noLogin".equals(isLogin(session))){
        return "noLogin";
    }else {
        cardRepository.deleteById(cid);
        return "ok";
    }
}
/**
 * 根据id获得一个名片信息
 */
@Override
public Object aCard(HttpSession session, Integer cid) {
    if("noLogin".equals(isLogin(session))){
        return "noLogin";
    }else {
        return cardRepository.getOne(cid);
    }
}
/**
 * 操作名片时,检查用户是否登录
 */
@Override
public String isLogin(HttpSession session) {
    Object user = session.getAttribute("user");
    if(user == null)
        return "noLogin";
    return "yes";
}
```

12.2.7 创建控制器层

本章将名片系统的前后端进行分离,因此,需要在控制器的请求处理方法上加上"@

CrossOrigin"注解启用跨域访问。

UserController 的代码如下：

```java
package com.ch.cardmis.controller;
import com.ch.cardmis.entity.UserEntity;
import com.ch.cardmis.service.UserService;
import org.springframework.beans.factory.annotation.Autowired;
import org.springframework.web.bind.annotation.*;
import javax.servlet.http.HttpSession;
@RestController
public class UserController {
    @Autowired
    private UserService userService;
    @CrossOrigin                                        //跨域访问
    @PostMapping(value = "cardmis/register")            //注册
    public String register(@RequestBody UserEntity requestUser) {
        return userService.register(requestUser);
    }
    @CrossOrigin                                        //跨域访问
    @PostMapping(value = "cardmis/login")               //登录
    public String login(@RequestBody UserEntity requestUser, HttpSession session) {
        return userService.login(requestUser, session);
    }
}
```

CardController 的代码如下：

```java
package com.ch.cardmis.controller;
import com.ch.cardmis.entity.CardEntity;
import com.ch.cardmis.service.CardService;
import org.springframework.beans.factory.annotation.Autowired;
import org.springframework.web.bind.annotation.*;
import org.springframework.web.multipart.MultipartFile;
import javax.servlet.http.HttpSession;
@RestController
public class CardController {
    @Autowired
    private CardService cardService;
    @CrossOrigin                                        //跨域访问
    @PostMapping(value = "cardmis/add")                 //添加或修改名片
    public String add(CardEntity cardEntity, HttpSession session, MultipartFile file) {
        return cardService.add(cardEntity, session, file);
    }
    @CrossOrigin                                        //跨域访问
    @GetMapping(value = "cardmis/cards")                //查询登录用户的名片
    public Object cards(HttpSession session) {
        return cardService.cards(session);
    }
    @CrossOrigin                                        //跨域访问
    @PostMapping(value = "cardmis/delete")              //根据 id 删除名片
    public String delete(HttpSession session,Integer cid) {
```

```java
        return cardService.delete(session,cid);
    }
    @CrossOrigin                                    //跨域访问
    @GetMapping(value = "cardmis/aCard")            //根据 id 获得一个名片信息
    public Object aCard(HttpSession session,Integer cid) {
        return cardService.aCard(session,cid);
    }
}
```

12.2.8　创建跨域响应头设置过滤器

跨域访问涉及请求域名、请求方式、发送的内容类型以及携带证书式访问等问题。在后端系统中，将这些设置放在过滤器中完成。过滤器的具体代码如下：

```java
package com.ch.cardmis.filter;
import javax.servlet.*;
import javax.servlet.annotation.WebFilter;
import javax.servlet.http.HttpServletRequest;
import javax.servlet.http.HttpServletResponse;
import java.io.IOException;
@WebFilter(urlPatterns = "/*",filterName = "headerFilter")
public class HeaderFilter implements Filter {
    @Override
    public void doFilter(ServletRequest servletRequest, ServletResponse servletResponse,
FilterChain filterChain) throws IOException, ServletException {
        HttpServletResponse response = (HttpServletResponse) servletResponse;
        HttpServletRequest request = (HttpServletRequest) servletRequest;
        /**
         * 解决跨域访问 start
         */
        response.setContentType("text/html;charset=UTF-8");
        //允许跨域请求的域名
        response.setHeader("Access-Control-Allow-Origin", request.getHeader("Origin"));
        //response.setHeader("Access-Control-Allow-Origin", "*");
        //Access-Control-Allow-Methods:跨域请求允许的请求方式
        response.setHeader("Access-Control-Allow-Methods","POST, GET, OPTIONS, DELETE");
        //response.setHeader("Access-Control-Allow-Methods","*");
        /*浏览器的同源策略就是出于安全考虑,浏览器会限制从脚本发起的跨域 HTTP 请求(例如
异步请求 GET、POST、PUT、DELETE、OPTIONS 等),所以浏览器会向所请求的服务器发起两次请求,第一次
是浏览器使用 OPTIONS 方法发起一个预检请求,第二次才是真正的异步请求,第一次的预检请求获知
服务器是否允许该跨域请求:如果允许,才发起第二次真实的请求;如果不允许,则拦截第二次请求。
Access-Control-Max-Age 用来指定本次预检请求的有效期,单位为秒,在此期间不用发出另一条
预检请求。例如 response.addHeader("Access-Control-Max-Age", "0"),表示每次异步请求都发
起预检请求,也就是说,发送两次请求; response.addHeader("Access-Control-Max-Age",
"1800"),表示隔 30 分钟才发起预检请求,也就是说,发送两次请求 */
        response.setHeader("Access-Control-Max-Age","0");
        //Access-Control-Allow-Headers:允许发送的内容类型
        response.setHeader("Access-Control-Allow-Headers","Origin, No-Cache, X-
Requested-With, If-Modified-Since, Pragma, Last-Modified, Cache-Control, Expires,
```

```
                Content-Type, X-E4M-With,token");
        //response.setHeader("Access-Control-Allow-Headers","*");
        //服务器端通过在响应 header 中设置 Access-Control-Allow-Credentials = true 来
        //运行客户端携带证书式访问
        //通过对 Credentials 参数的设置,就可以保持跨域 Ajax 时的 Cookie
        //这里需要注意的是:服务器端 Access-Control-Allow-Credentials = true 时
        //参数 Access-Control-Allow-Origin 的值不能为 '*'
        response.setHeader("Access-Control-Allow-Credentials","true");
        filterChain.doFilter(request, response);
        /**
         * 解决跨域访问 end
         */
    }
}
```

12.2.9 创建工具类

在后端系统 cardmis 中,我们使用了工具类 MyUtil 的 getNewFileName 方法对文件进行重命名。具体代码如下:

```
package com.ch.cardmis.util;
import java.text.SimpleDateFormat;
import java.util.Date;
public class MyUtil {
    /**
     * 将实际的文件重命名
     */
    public static String getNewFileName(String oldFileName) {
        int lastIndex = oldFileName.lastIndexOf(".");
        String fileType = oldFileName.substring(lastIndex);
        Date now = new Date();
        SimpleDateFormat sdf = new SimpleDateFormat("YYYYMMDDHHmmssSSS");
        String time = sdf.format(now);
        String newFileName = time + fileType;
        return newFileName;
    }
}
```

12.3 使用 IntelliJ IDEA 构建名片前端系统 cardmis-vue

本节只是实现简单的名片前端系统,注重功能的实现,旨在让读者了解前后端分离的应用程序的实现原理及开发流程。

12.3.1 安装 Node.js

我们利用 Vue CLI(Vue 脚手架)搭建名片前端系统 cardmis-vue。因为需要使用 npm

安装 Vue CLI,而 npm 是集成在 Node.js 中的,所以需要首先安装 Node.js。通过访问官网"https://nodejs.org/en/"即可下载对应版本的 Node.js,本书下载的是"10.16.3 LTS",如图 12.4 所示。

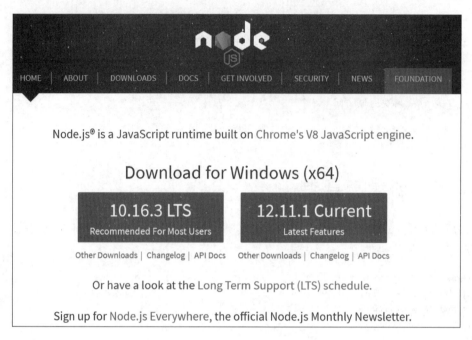

图 12.4　选择对应版本的 Node.js

下载完成后运行安装包 node-v10.16.3-x64.msi,一直单击"下一步"按钮即可完成安装。然后在命令行窗口中输入命令"node -v",检查是否安装成功,如图 12.5 所示。

图 12.5 中出现了版本号,说明 Node.js 已安装成功。同时,npm 包也已经安装成功,可以输入"npm -v"查看版本号,如图 12.6 所示。

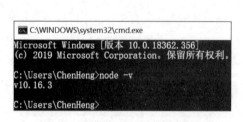

图 12.5　查看 Node.js 的版本

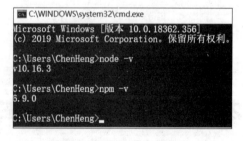

图 12.6　查看 npm 的版本

输入"npm -g install npm",将 npm 更新至最新版本,如图 12.7 所示。

12.3.2　安装 Vue CLI

在 IDEA 中,使用"npm i -g vue-cli"命令安装 Vue 脚手架。具体过程如下:
(1) 打开 IDEA 新建项目(New|Project|Static Web),填写项目名(cardmis-vue)和选择

第12章 名片系统的设计与实现(Vue.js+JPA)

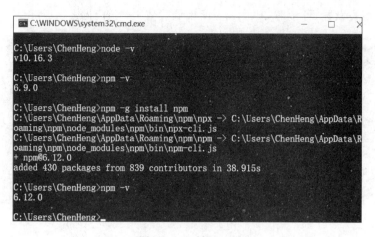

图 12.7　更新 npm

保存的工作空间后，单击 Finish 按钮完成项目创建。

（2）打开 IDEA 的 Terminal，输入命令"npm i -g vue-cli"安装 Vue 脚手架，输入命令"vue -V"查看版本（测试是否安装成功），如图 12.8 所示。

图 12.8　安装 Vue 脚手架

12.3.3　构建前端 Vue 项目 cardmis-vue

Vue 脚手架安装完成后，即可初始化 webpack 包结构。在 Terminal 中继续输入命令"vue init webpack cardmis-vue"初始化包结构，cardmis-vue 为前面新建的项目名。初始化过程中将进行相关设置，可参考如图 12.9 所示的设置。

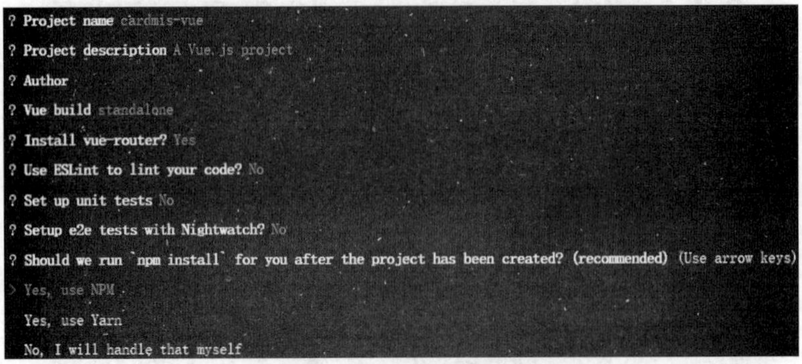

? Project name cardmis-vue
? Project description A Vue.js project
? Author
? Vue build standalone
? Install vue-router? Yes
? Use ESLint to lint your code? No
? Set up unit tests No
? Setup e2e tests with Nightwatch? No
? Should we run `npm install` for you after the project has been created? (recommended) (Use arrow keys)
> Yes, use NPM
Yes, use Yarn
No, I will handle that myself

图 12.9　初始化设置

初始化完成后，显示如图 12.10 所示的信息。

cardmis-vue [C:\Users\ChenHeng\IdeaProjects\cardmis-vue] - IntelliJ IDEA
File Edit View Navigate Code Analyze Refactor Build Run Tools VCS
cardmis-vue
Project
cardmis-vue C:\Users\ChenHeng\IdeaProject
.idea
cardmis-vue
cardmis-vue.iml
1: Project
Terminal: Local
run `npm audit fix` to fix them, or `npm audit` for details
Project initialization finished!
========================
To get started:
cd cardmis-vue
npm run dev
Documentation can be found at https://vuejs-templates.github.io/webpack
2: Favorites
7: Structure
C:\Users\ChenHeng\IdeaProjects\cardmis-vue>

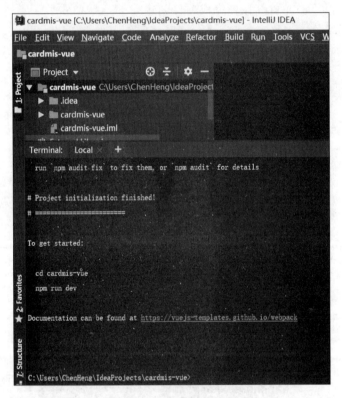

图 12.10　初始化完成

在 Terminal 中，依次输入命令“cd cardmis-vue”和“npm run dev”启动 Vue 项目。完成后，将提示通过哪个端口进行访问，如图 12.11 所示。

打开浏览器，在地址栏输入“localhost:8081”，出现如图 12.12 所示的页面效果。至此，简单的前端项目 cardmis-vue 已搭建完成。

当然，还可以通过 IDEA 的用户界面启动 Vue 项目，具体过程为：在 package.json 文件上右击，选择 Show npm Scripts 命令，如图 12.13 所示。

第12章 名片系统的设计与实现（Vue.js+JPA）

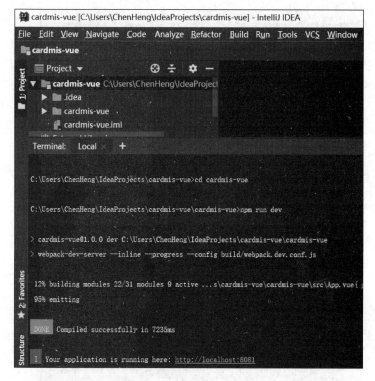

图 12.11　启动 Vue 项目

图 12.12　运行 Vue

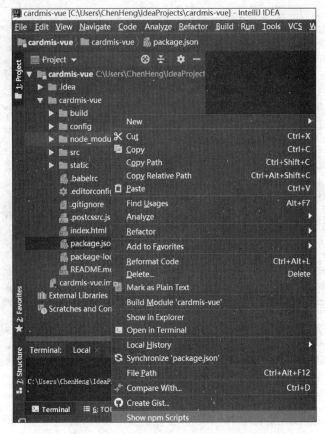

图 12.13　选择 Show npm Scripts 命令

然后弹出 npm 命令窗口,需要执行哪个命令直接双击运行即可,如图 12.14 所示。这些命令都预定义在 package.json 文件中。dev 和 start 是一样的,即执行 npm run dev 命令。

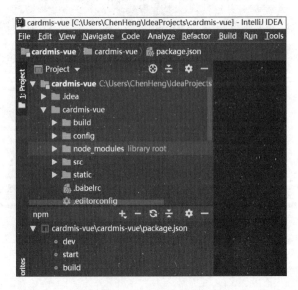

图 12.14　npm 命令窗口

12.3.4 分析 Vue 项目结构

使用 Vue CLI 构建的 Vue 项目结构如图 12.15 所示。

图 12.15 Vue 项目结构

需要我们重点关注的目录或文件，参考 Evan-Nightly 的博客（https://learner.blog.csdn.net/article/details/88926242）使用"小红旗"标注，如图 12.16 所示。

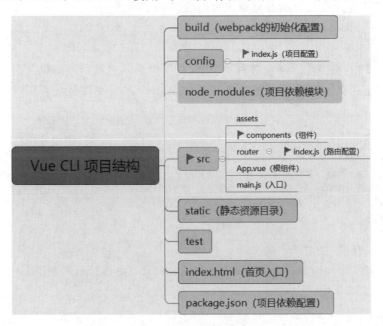

图 12.16 标注重点关注的目录或文件

其中，components（组件）文件夹是我们修改最频繁的部分，几乎所有手动编写的代码都位于此处。下面分析几个文件，目的是理解各个部分是如何联系的。

1. index.html

index.html 文件（首页）的初始代码如下：

```html
<!DOCTYPE html>
<html>
  <head>
    <meta charset="utf-8">
    <meta name="viewport" content="width=device-width,initial-scale=1.0">
    <title>cardmis-vue</title>
  </head>
  <body>
    <div id="app"></div>
    <!-- built files will be auto injected -->
  </body>
</html>
```

从上述代码可以看出，index.html 是一个普通的 html 文件，让它与众不同的是"<div id="app"></div>"这句程序，下面有一行注释，意思是："构建的文件将会被自动注入"，也就是说我们编写的其他的内容都将在这个 div 中展示。另外，整个项目只有这一个 html 文件，所以这是一个单页面应用，当打开这个应用时，表面上可以看到很多页面，实际上它们都在这一个 div 中显示。单页面应用(Single Page Web Application，SPA)，即只有一张页面，并在用户与应用程序交互时动态更新该页面的 Web 应用程序。

2. App.vue

在图 12.16 中，将 App.vue 文件称为"根组件"，因为其他的组件都包含在此组件中。.vue 文件是一种自定义文件类型，在结构上类似于 html，一个 .vue 文件是一个 Vue 组件。App.vue 的初始代码如下：

```vue
<template>
  <div id="app">
    <img src="./assets/logo.png">
    <router-view/>
  </div>
</template>
<script>
export default {
  name: 'App'
}
</script>
<style>
#app {
  font-family: 'Avenir', Helvetica, Arial, sans-serif;
  -webkit-font-smoothing: antialiased;
  -moz-osx-font-smoothing: grayscale;
  text-align: center;
  color: #2c3e50;
  margin-top: 60px;
```

```
}
</style>
```

在上述代码中,需要注意的是,也有一句< div id="app">,但与 index.html 中的< div id="app">是没有关系的。此处的 id=app 与下面的 CSS 对应。

< script >标签里的内容即该组件的脚本,export default 是 ES6(ECMAScript 6.0,是 JavaScript 的下一个版本标准)的语法,意思是将该组件整体导出,然后可以使用 import 导入该组件。大括号里的内容是该组件的相关属性。

App.vue 中最关键的一点是第四行< router-view/>,它是一个容器,即"路由视图",意思是当前路由(URL)指向的内容将显示在该容器中。也就是说,其他的组件即使拥有自己的路由(URL,需要在 router 文件夹的 index.js 文件中定义,参见 12.3.7 节),也只不过表面上是一个单独的页面,实际上只是在根组件 App.vue 中。

3. main.js

前面我们说 App.vue 的< div id="app">和 index.html 的< div id="app">没有关系,那么这两个文件是如何建立联系的呢?现在,让我们看看程序的入口文件 main.js 的代码:

```
import Vue from 'vue'
import App from './App'
import router from './router'
Vue.config.productionTip = false
new Vue({
  el: '#app',
  router,
  components: { App },
  template: '<App/>'
})
```

上述代码的 import 语句是导入的模块,其中 vue 模块在 node_modules 中,App 即 App.vue 定义的组件,router 即 router 文件夹中定义的路由。"Vue.config.productionTip = false"的作用是阻止 vue 在启动时生成生产提示。

在 main.js 文件中,我们创建了一个 Vue 对象(实例),el 属性提供一个在页面上已存在的 DOM 元素作为 Vue 对象的挂载目标(值可以是 CSS 选择符,或实际 HTML 元素,或返回 HTML 元素的函数。这里就是通过 index.html 中的< div id="app">< div >的 id="app"和"#app"进行挂载);router 代表该对象包含 Vue Router,并使用项目中定义的路由(在 router 文件夹的 index.js 文件中定义);components 表示该对象包含的 Vue 组件;template 是用一个字符串模板作为 Vue 实例的标识使用,类似于定义一个 html 标签,模板将会替换挂载的元素,template:'< App/>'表示用< app ></app >替换 index.html 里面的 < div id="app"></div >。

综上所述,main.js 与 index.html 是项目启动的首加载页面资源与 js 资源,App.vue 则是 vue 页面资源的首加载项。具体操作过程如下:首先启动项目,找到 index.html 与 main.js,执行 main.js,根据 import 加载 App.vue 文件;然后用 vue 渲染 index.html 中的 id="app" DOM 元素(el:'#app')新建 Vue,渲染规则是 template:'< App/>',直接将其渲

染为上一步 components 调用的局部组件"App"。

12.3.5 设置 IntelliJ IDEA 支持创建 *.vue 文件及打开 *.vue 文件

设置 IntelliJ IDEA 支持创建 *.vue 文件及打开 *.vue 文件，具体操作实现步骤如下。

1. 安装 vue 插件

选择 File | Settings | Plugins 命令，然后输入"vue.js"进行搜索，单击搜索结果中的 vue.js 的 Install 按钮，如图 12.17 所示。安装成功后重启 IDEA，即可识别 .vue 文件。

图 12.17 安装 vue 插件

2. 设置 vue 新建文件模板

选择 File | Setting | Editor | File and Code Templates 命令，选择 Vue Single File Component 选项，然后在右边框中编辑默认模板内容，编辑完单击 OK 按钮，即可通过 new 选项创建 .vue 文件，如图 12.18 所示。

3. 添加 *.vue 文件类型

经过前两步的操作，还不能打开 *.vue 文件。需要添加 File Types，选择 HTML 选项，添加 *.vue 文件类型，如图 12.19 所示。

12.3.6 开发前端页面

在开发的时候，前端用前端的服务器（如 Ngix）开发，后端用后端的服务器（如 Tomcat）开发。当开发前端内容时，可以把前端的请求通过前端服务器转发给后端，即可实时观察结果，并且不需要知道后端怎么实现，而只需要知道接口提供的功能，前后端的开发人员各司其职。

后端的开发已在 12.2 节中完成，本节开发前端页面，具体操作步骤如下。

第12章 名片系统的设计与实现（Vue.js+JPA）

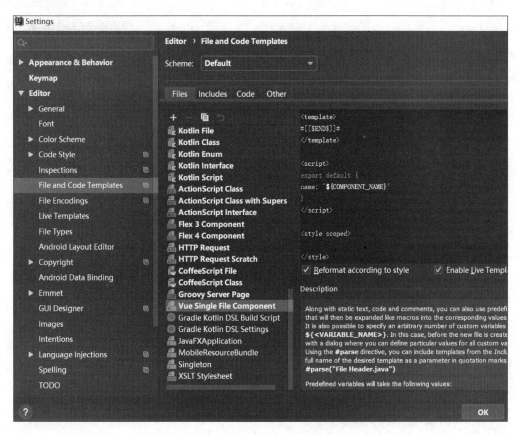

图 12.18　设置 vue 文件模板

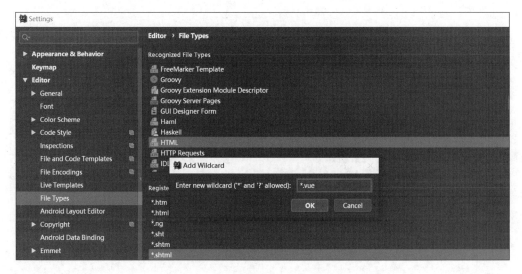

图 12.19　添加 *.vue 文件类型

1. 首页组件

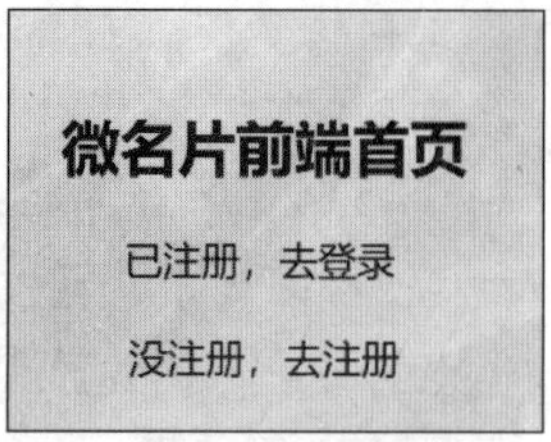

右击 src\components 文件夹,新建一个 Appindex.vue,即首页组件。在该组件中,路由跳转到登录和注册组件。通过"http://localhost:xxxx/"运行首页组件,效果如图 12.20 所示。

Appindex.vue 的代码如下:

```
<!-- 每个.vue 文件包含三种类型的顶级语言块 <template>、<script> 和 <style>。这三个部分分别代表了 html、js、css。-->
<template>
  <div class = "active">
    <br>
    <h2>微名片前端首页</h2>
    <!-- exact 精确匹配 -->
    已注册,<router-link to = "/login" :exact = "true">去登录</router-link><br><br>
    没注册,<router-link to = "/register" :exact = "true">去注册</router-link><br><br>
  </div>
</template>
<script>
  export default {
    name: 'AppIndex'
  }
</script>
<style scoped>
  a {text-decoration: none}
  .active {
    background-image: url("../assets/bb.jpg");
  }
</style>
```

图 12.20 首页组件

单击图 12.20 中的"去登录"和"去注册"超链接,根据 12.3.7 节的路由配置,分别跳转到"登录"和"注册"组件。

2. 注册组件

右击 src\components 文件夹,新建一个 Register.vue,即注册组件。注册用户时,检查用户名是否已注册。注册组件运行效果如图 12.21 所示。

图 12.21 注册组件

Register.vue 的代码如下:

```
<template>
  <div class = "active">
    <br>
    <h2>注册页面</h2>
    <form>
      用户名:<input type = "text" v-model = "registerForm.uname" placeholder = "请输入用户名"/><br><br>
```

```
      密码:<input type="password" v-model="registerForm.upwd" placeholder="请输入密码
"/><br><br>
      确认密码:<input type="password" v-model="registerForm.reupwd" placeholder="请再次
输入密码"/><br><br>
      <button type="button" @click="register" :disabled="isDisable">注册</button>
      <button type="reset">重置</button>
    </form>
    <br>
  </div>
</template>
<script>
  export default {
    name: 'Register',
    data() {
      return {
        isDisable:false,
        registerForm: {
          uname: '',
          upwd: '',
          reupwd: ''
        }
      }
    },
    methods: {
      register() {
        this.isDisable = true;
        if(this.registerForm.upwd != this.registerForm.reupwd) {
          alert("两次密码不一致,重新输入!");
          return false;
        }
        this.$axios
          .post('/register', {
            uname: this.registerForm.uname,
            upwd: this.registerForm.upwd
          })
          .then(successResponse => {
            if (successResponse.data === "no") {
              alert("用户名已存在!")
              this.isDisable = false
            }else{
              alert("注册成功!")
              this.$router.replace({path: '/login'})
            }
          })
          .catch(failResponse => {
            alert("响应异常")
          })
      }
    }
  }
</script>
```

```
<style>
  .active {
    background-image: url("../assets/bb.jpg");
  }
</style>
```

在 Register.vue 的<template>标签中编写了一个注册界面，methods 中定义了"注册"按钮的处理方法 register，即向后端/register 接口发送数据，获得成功的响应后，页面跳转到/login。在前端组件中，都是通过 axios 向后端提交 Ajax 请求。axios 模块的安装请参见12.3.8节。

3. 登录组件

成功注册的用户，可通过登录组件登录名片系统。右击 src\components 文件夹，新建一个 Login.vue，即登录组件。登录组件运行效果如图12.22所示。

图12.22 登录组件

Login.vue 的代码如下：

```
<template>
  <div class="active">
    <br>
    <h2>登录页面</h2>
    <form>
      用户名:<input type="text" v-model="loginForm.uname" placeholder="请输入用户名"/>
      <br><br>
      密码:<input type="password" v-model="loginForm.upwd" placeholder="请输入密码"/>
      <br><br>
      <button type="button" @click="login"  :disabled="isDisable">登录</button>
      <button type="reset">重置</button>
    </form>
    <br>
  </div>
</template>
<script>
  export default {
    name: 'Login',
    data() {
      return {
        isDisable:false,
        loginForm: {
          uname: '',
          upwd: ''
        }
      }
    },
    methods: {
      login() {
        this.isDisable = true;
        var formData = JSON.stringify(this.loginForm);
```

```
            let config = {
                headers:{'Content-Type':'application/json; charset=UTF-8'}
            };
            this.$axios
                .post('/login', formData, config)          //直接提交表单
                .then(successResponse => {
                    if (successResponse.data === "ok") {
                        alert("登录成功")
                        this.$store.commit('changeLogin',this.loginForm.uname)
                        let path = this.$route.query.redirect
                        this.$router.replace({path: path === '/' || path === undefined ? '/main': path})
                    }else {
                        alert("用户名或密码错误!")
                        this.isDisable = false
                    }
                })
                .catch(failResponse => {
                    alert("响应异常")
                })
        }
    }
</script>
<style>
    .active {
        background-image: url("../assets/bb.jpg");
    }
</style>
```

在 Login.vue 的 methods 中定义了"登录"按钮的处理方法 login，即向后端/login 接口发送数据，登录成功后，页面跳转到/main。

4. 名片查询组件

右击 src\components 文件夹，新建一个 Main.vue，即名片查询组件。名片查询组件运行效果如图 12.23 所示。

图 12.23　名片查询组件

Main.vue 的代码如下:

```vue
<template>
    <div class="active">
      <NavMain></NavMain>
      <br>
      <table border="1" align="center">
        <tr>
          <th>ID</th>
          <th>姓名</th>
          <th>公司</th>
          <th>地址</th>
          <th>邮箱</th>
          <th>电话</th>
          <th>邮编</th>
          <th>照片</th>
        </tr>
        <tr v-for="card in cards">
          <td>{{ card.id }}</td>
          <td>{{ card.name }}</td>
          <td>{{ card.company }}</td>
          <td>{{ card.address }}</td>
          <td>{{ card.email }}</td>
          <td>{{ card.telephone }}</td>
          <td>{{ card.post }}</td>
          <!-- assets:在项目编译的过程中会被webpack处理解析为模块依赖,只支持相对路径的形
          式,如<img src="./logo.png">和background:url(./logo.png),"./logo.png"是相对资源路径,将
          由webpack解析为模块依赖。static:在这个目录下文件不会被webpack处理,就是说存放第三方文
          件的地方,不会被webpack解析,它会直接被复制到最终的打包目录(默认是dist/static)下。必须
          使用绝对路径引用这些文件,这是通过config.js文件中的build.assetsPublic和build
          .assertsSubDirectory链接来确定的。任何放在static/中的文件需要以绝对路径的形式引用
          :/static[filename] -->
          <td><img :src="card.logo"></td>
        </tr>
      </table>
      <br>
    </div>
</template>
<script>
  import NavMain from '@/components/NavMain'
  export default {
    name: 'Main',
    components: {NavMain},
    data() {
      return {
        cards: []                              //保存数据
      }
    },
    //created:在模板渲染成html前调用,即通常初始化某些属性值,然后再渲染成视图
    //mounted:在模板渲染成html后调用
    //通常是初始化页面完成后,再对html的dom节点进行一些需要的操作
```

```
      created: function() {
        this.loadCards()
      },
      methods: {
        loadCards() {
          let _this = this
          this.$axios
            .get('/cards')
            .then(successResponse => {
              if(successResponse.data === "noLogin"){
                alert("没有登录,请登录!")
                _this.$router.replace({path: '/login'})
              }else{
                _this.cards = successResponse.data
              }
            })
            .catch(failResponse => {
              alert("响应异常")
            })
        }
      }
    }
</script>
<style>
  .active {
    background-image: url("../assets/bb.jpg");
  }
  table{
    font-size:12px;
    border-collapse:collapse
  }
  img{
    height: 20px;
    width: 20px;
  }
</style>
```

在Main.vue的methods中定义了页面查询结果初始化方法loadCards,即从后端/cards接口获得登录用户的名片信息。另外,在Main.vue中使用了导航组件NavMain.vue。

NavMain.vue的代码如下:

```
<template>
  <div>
    <br>
    <h2>微名片主页面</h2>
    <router-link to="/toAdd" :exact="true">添加名片</router-link>  
    <router-link to="/toDeleteSelect" :exact="true">删除名片</router-link>  
    <router-link to="/toUpdateSelect" :exact="true">修改名片</router-link>  
    <router-link to="/main" :exact="true">查询名片</router-link><br>
```

```
    </div>
</template>
<script>
  export default {
    name: 'NavMain'
  }
</script>
<style>
  a {text-decoration: none}
</style>
```

5. 添加名片组件

单击图12.23中的"添加名片"超链接，根据路由配置打开添加名片组件Add.vue。右击src\components文件夹，新建一个Add.vue，即添加名片组件。添加名片组件运行效果如图12.24所示。

图12.24 添加名片组件

Add.vue的代码如下：

```
<template>
  <div class="active">
    <NavMain></NavMain>
    <h3>添加名片</h3>
    <form>
      姓名:<input type="text" v-model="name" placeholder="请输入用户名"/><br><br>
      电话:<input type="text" v-model="telephone" placeholder="请输入电话"/><br><br>
      邮箱:<input type="text" v-model="email" placeholder="请输入邮箱"/><br><br>
      单位:<input type="text" v-model="company" placeholder="请输入单位"/><br><br>
      邮编:<input type="text" v-model="post1" placeholder="请输入邮编"/><br><br>
      地址:<input type="text" v-model="address" placeholder="请输入地址"/><br><br>
      头像:
      <input type="file"  @change="getFile($event)"/><br><br>
      <button type="button" @click="add($event)" :disabled="isDisable">添加</button>
      <button type="reset">重置</button>
    </form>
    <br>
  </div>
</template>
<script>
  import NavMain from '@/components/NavMain'
  export default {
    name: 'Add',
    components: {NavMain},
    data() {
      return {
        isDisable:false
      }
    },
    methods: {
      //获得文件对象
      getFile(event) {
        this.file = event.target.files[0];
```

```js
      },
      add (event) {
        this.isDisable = true;
        event.preventDefault();
        let formData = new FormData();
        formData.append('name', this.name === undefined ? '': this.name);
        formData.append('telephone', this.telephone === undefined ? '': this.telephone);
        formData.append('email', this.email === undefined ? '': this.email);
        formData.append('company', this.company === undefined ? '': this.company);
        formData.append('post', this.post1 === undefined ? '': this.post1);
        formData.append('address', this.address === undefined ? '': this.address);
        formData.append('file', this.file === undefined ? null: this.file);
        let config = {
          headers:{'Content-Type':'multipart/form-data'}
        };
        this.$axios
          .post('/add', formData, config)         //直接提交表单
          .then(successResponse => {
            if (successResponse.data === "ok") {
              alert("添加成功")
              this.$router.replace({path: '/main'})
            }else if(successResponse.data === "noLogin"){
              alert("没有登录,请登录!")
              this.$router.replace({path: '/login'})
            }else {
              alert("添加失败")
              this.isDisable = false
            }
          })
          .catch(failResponse => {
            alert("响应异常")
          })
      }
    }
  }
</script>
<style>
  .active {
    background-image: url("../assets/bb.jpg");
  }
</style>
```

在 Add.vue 的 methods 中定义了添加按钮的处理方法 add,即向后端/add 接口发送数据,添加成功后,页面跳转到/main。

6. 删除名片组件

单击图 12.23 中的"删除名片"超链接,根据路由配置打开删除名片组件 DeleteSelect.vue。右击 src\components 文件夹,新建一个 DeleteSelect.vue,即删除名片组件。删除名片组件运行效果如图 12.25 所示。

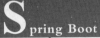

图 12.25 删除名片组件

DeleteSelect.vue 的代码如下：

```vue
<template>
  <div class="active">
    <NavMain></NavMain>
    <br>
    <table border="1" align="center">
      <tr>
        <th>ID</th>
        <th>姓名</th>
        <th>公司</th>
        <th>地址</th>
        <th>邮箱</th>
        <th>电话</th>
        <th>邮编</th>
        <th>照片</th>
        <th>操作</th>
      </tr>
      <tr v-for="card in cards">
        <td>{{ card.id }}</td>
        <td>{{ card.name }}</td>
        <td>{{ card.company }}</td>
        <td>{{ card.address }}</td>
        <td>{{ card.email }}</td>
        <td>{{ card.telephone }}</td>
        <td>{{ card.post }}</td>
        <td><img :src="card.logo"></td>
        <td><a style="cursor: pointer" @click="deleteCard(card.id)">删除</a></td>
      </tr>
    </table>
    <br>
  </div>
</template>
<script>
  import NavMain from '@/components/NavMain'
  export default {
    name: 'Main',
    components: {NavMain},
    data() {
      return {
        cards: []                           //保存数据
```

第12章 名片系统的设计与实现(Vue.js+JPA)

```
          }
        },
        created: function() {
          this.loadCards()
        },
        methods: {
          loadCards() {
            let _this = this
            this.$axios
              .get('/cards')
              .then(successResponse => {
                if(successResponse.data === "noLogin"){
                  alert("没有登录,请登录!")
                  _this.$router.replace({path: '/login'})
                }else{
                  _this.cards = successResponse.data
                }
              })
              .catch(failResponse => {
                alert("响应异常")
              })
          },
          deleteCard(cid) {
            if(window.confirm("真的删除该名片吗?id = " + cid)){
              this.$axios
                .post('/delete?cid = ' + cid)
                .then(successResponse => {
                  if(successResponse.data === "noLogin"){
                    alert("没有登录,请登录!")
                    this.$router.replace({path: '/login'})
                  }else{
                    alert("成功删除!")
                    this.$router.go(0)           //回到当前页面,并刷新
                  }
                })
                .catch(failResponse => {
                  alert("响应异常")
                })
            }
          }
        }
      }
</script>
<style>
  .active {
    background-image: url("../assets/bb.jpg");
  }
  table{
    font-size:12px;
    border-collapse:collapse
  }
  img{
    height: 20px;
    width: 20px;
```

```
}
</style>
```

在 DeleteSelect.vue 的 methods 中定义了删除的处理方法 deleteCard,即向后端/delete 接口发送数据,删除成功后,回到本页面。

7. 修改名片查询组件

单击图 12.23 中的"修改名片"超链接,根据路由配置打开修改名片查询组件 UpdateSelect.vue。右击 src\components 文件夹,新建一个 UpdateSelect.vue,即修改名片查询组件。修改名片查询组件运行效果如图 12.26 所示。

图 12.26 修改名片查询组件

UpdateSelect.vue 的代码与 DeleteSelect.vue 类似,不再赘述。

8. 修改名片组件

单击图 12.26 中的"修改"超链接,打开修改名片组件 EditCard.vue。修改名片组件的运行效果如图 12.27 所示。

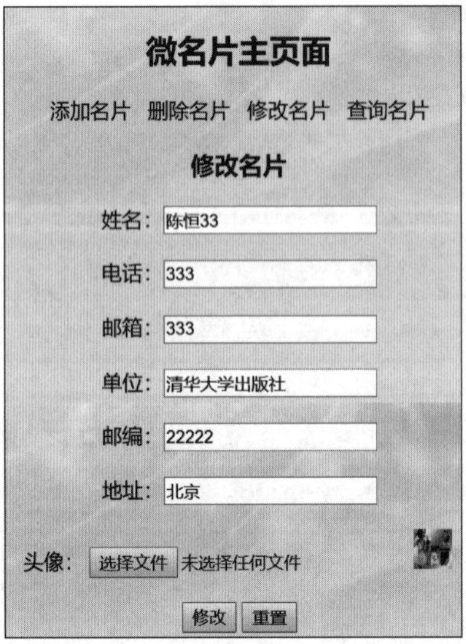

图 12.27 修改名片组件

EditCard.vue 的代码如下：

```html
<template>
  <div class="active">
    <NavMain></NavMain>
    <h3>修改名片</h3>
    <form>
      姓名:<input type="text" v-model="aCard.name" placeholder="请输入用户名"/><br><br>
      电话:<input type="text" v-model="aCard.telephone" placeholder="请输入电话"/><br><br>
      邮箱:<input type="email" v-model="aCard.email" placeholder="请输入邮箱"/><br><br>
      单位:<input type="text" v-model="aCard.company" placeholder="请输入单位"/><br><br>
      <!-- post 对应于实体属性名 -->
      邮编:<input type="text" v-model="aCard.post" placeholder="请输入邮编"/><br><br>
      地址:<input type="text" v-model="aCard.address" placeholder="请输入地址"/><br><br>
      头像:
      <input type="file" @change="getFile($event)"/><img :src="aCard.logo" alt="没有头像">
      <input type="hidden" v-model="aCard.logo"/><!-- 不选择图片时,使用隐藏域的图片路径,即没有修改图片 -->
      <input type="hidden" v-model="aCard.id"/>
      <br><br>
      <button type="button" @click="update($event)" :disabled="isDisable">修改</button>
      <button type="reset">重置</button>
    </form>
    <br>
  </div>
</template>
<script>
  import NavMain from '@/components/NavMain'
  export default {
    name: 'Edit',
    components: {NavMain},
    data() {
      return {
        isDisable:false,
        aCard:[]                              //存储返回结果
      }
    },
    created: function() {
      this.loadCard()
    },
    methods: {
      loadCard() {
        let _this = this
```

```js
      let cid = _this.$route.query.cid
      this.$axios
        .get('/aCard?cid=' + cid)
        .then(successResponse => {
          if(successResponse.data === "noLogin"){
            alert("没有登录,请登录!")
            _this.$router.replace({path: '/login'})
          }else{
            _this.aCard = successResponse.data
          }
        })
        .catch(failResponse => {
          alert("响应异常")
        })
    },
    //获得文件对象
    getFile(event){
      this.file = event.target.files[0];
    },
    update(event){
      this.isDisable = true;
      event.preventDefault();
      let formData = new FormData();
      formData.append('name', this.aCard.name);
      formData.append('telephone', this.aCard.telephone);
      formData.append('email', this.aCard.email);
      formData.append('company', this.aCard.company);
      formData.append('post', this.aCard.post);
      formData.append('address', this.aCard.address);
      formData.append('file', this.file);
      formData.append('logo', this.aCard.logo);
      formData.append('id', this.aCard.id);
      let config = {
        headers:{'Content-Type':'multipart/form-data'}
      };
      this.$axios
        .post('/add', formData, config)          //直接提交表单
        .then(successResponse => {
          if (successResponse.data === "ok") {
            alert("修改成功")
            let path = this.$route.query.redirect
            this.$router.replace({path: path === '/' || path === undefined ? '/toUpdateSelect': path})
          }else if(successResponse.data === "noLogin"){
            alert("没有登录,请登录!")
            this.$router.replace({path: '/login'})
          }else {
            alert("修改失败")
            this.isDisable = false
          }
        })
```

```
          .catch(failResponse => {
            alert("响应异常")
          })
      }
    }
  }
</script>
<style>
  .active{
    background-image: url("../assets/bb.jpg");
  }
  img{
    height: 30px;
    width: 30px;
  }
</style>
```

在 EditCard.vue 的 methods 中定义了页面信息初始化方法 loadCards,即从后端 /aCard 接口获得一个名片信息。另外,还定义了修改按钮处理方法 update,修改成功回到修改查询组件 UpdateSelect.vue。

12.3.7 配置页面路由

修改 src\router\index.js 的具体代码如下:

```
import Vue from 'vue'
import Router from 'vue-router'
//导入组件
import AppIndex from '@/components/AppIndex'
import Login from '@/components/Login'
import Register from '@/components/Register'
import Main from '@/components/Main'
import Add from '@/components/Add'
import DeleteSelect from '@/components/DeleteSelect'
import UpdateSelect from '@/components/UpdateSelect'
import EditCard from '@/components/EditCard'
//安装插件
Vue.use(Router)                            //挂载属性
//创建路由对象并配置路由规则
export default new Router({
  //将路由从默认的 hash 模式切换为 histroy 模式,运行项目,访问不加 # 号
  mode: 'history',
  routes: [
    //下面都是固定的写法,一个个 link 对象
    {
      path: '/login',
      name: 'Login',
      component: Login
    },
```

```
            {
                path: '/register',
                name: 'Register',
                component: Register
            },
            {
                path: '/main',
                name: 'Main',
                component: Main,
                meta:{auth:true}                          //需要登录权限验证
            },
            {
                path: '/toAdd',
                name: 'Add',
                component: Add,
                meta:{auth:true}                          //需要登录权限验证
            },
            {
                path: '/toDeleteSelect',
                name: 'DeleteSelect',
                component: DeleteSelect,
                meta:{auth:true}                          //需要登录权限验证
            },
            {
                path: '/toUpdateSelect',
                name: 'UpdateSelect',
                component: UpdateSelect,
                meta:{auth:true}                          //需要登录权限验证
            },
            {
                path: '/editCard',
                name: 'EditCard',
                component: EditCard,
                meta:{auth:true}                          //需要登录权限验证
            },
            {
                path: '/',
                name: 'AppIndex',
                component: AppIndex
            }
        ]
    })
```

12.3.8　设置反向代理

反向代理,即把前端的请求通过前端服务器转发给后端。设置反向代理,前端请求默认发送到 http://localhost:8443/cardmis,8443 与后端端口一致(见 12.2.3 节),cardmis 与控制器请求映射一致(见 12.2.7 节)。这里,因为使用了新的模块 axios,所以需要打开 IDEA 的 Terminal,在 Terminal 中,执行"npm install --save axios"命令安装该模块。

第12章 名片系统的设计与实现（Vue.js+JPA）

修改 src\main.js 代码实现反向代理，具体代码如下：

```javascript
import Vue from 'vue'
import App from './App'
import router from './router'
import store from './store'
//设置反向代理,前端请求默认发送到 http://localhost:8443/cardmis
let axios = require('axios')        //使用 axios 来完成 ajax 请求
//全局注册,之后可在其他组件中通过 this.$axios 发送数据
axios.defaults.baseURL = 'http://localhost:8443/cardmis'
//前端发送跨域请求时默认是不会携带 cookie 的,前端使用 axios.defaults.withCredentials
= true
//后端使用 res.setHeader("Access-Control-Allow-Credentials","true");
axios.defaults.withCredentials = true
Vue.prototype.$axios = axios
Vue.config.productionTip = false
//全局前置钩子函数:router.beforeEach(),它的作用就是在每次路由切换的时候调用
//这个钩子方法会接收三个参数:to、from、next
//to:Route:即将要进入的目标的路由对象
//from:Route:当前导航正要离开的路由
//next:Function:就是函数结束后执行什么
router.beforeEach((to,from,next) =>{
    //如果路由器需要验证
    if(to.matched.some(m => m.meta.auth)){
      //对路由进行验证
      if (store.state.isLogin == '0') {
        alert("您没有登录,无权访问!")
        //未登录则跳转到登录界面,query:{ redirect: to.fullPath}表示把当前路由信息传递过去
        //方便登录后跳转回来
        next({
          path: 'login',
          query: {redirect: to.fullPath}
        })
      } else {                            //已经登录
        next()                            //正常跳转到设置好的页面
      }
    }else{
      next()
    }
  }
)
new Vue({
  el: '#app',
  router,
  store,                                  //使用 store
  components: { App },
  template: '<App/>'
})
```

12.3.9 设置跨域支持

为了让后端能够访问到前端资源,需要配置跨域支持。在 config\index.js 中,找到 proxyTable 位置,修改为以下内容:

```
proxyTable: {
    '/cardmis': {
        target: 'http://localhost:8443',
        changeOrigin: true,
        pathRewrite: {
          '^/cardmis': ''
        }
    }
},
```

12.4 Vuex 与前端路由拦截器

实现前端路由拦截器,目的是在前端判断用户的登录状态。但登录状态应该被视为一个全局属性,而不应该只写在某一组件中。所以需要引入一个新的工具——Vuex,它是专门为 Vue 开发的状态管理方案。

12.4.1 引入 Vuex

首先,打开 IDEA 的 Terminal,在 Terminal 中,运行"npm install vuex --save"命令。然后,在 src 目录下新建一个文件夹 store,并在该目录下新建 index.js 文件,在该文件中引入 vue 和 vuex。具体代码如下:

```
import Vue from 'vue'
import Vuex from 'vuex'
Vue.use(Vuex)        //vuex是专为vue.js应用程序开发的状态管理模式
```

在 index.js 中设置前端系统需要的状态变量和方法。为实现前端路由拦截器,需要一个记录用户信息的变量 isLogin。同时,设置一个方法 changeLogin,触发该方法为用户变量 isLogin 赋值。具体代码如下:

```
import Vue from 'vue'
import Vuex from 'vuex'
Vue.use(Vuex)        //vuex是专为vue.js应用程序开发的状态管理模式
//提供仓库
export default new Vuex.Store({
    state:{
        //初始时给一个 isLogin = '0' 表示用户未登录
        isLogin:window.sessionStorage.getItem('user') == null ? '0' : window.sessionStorage.getItem('user')
    },
    mutations:{
```

```
      changeLogin(state, data) {
        state.isLogin = data;
        window.sessionStorage.setItem('user', data)
      }
    }
})
```

在登录组件中，登录成功后需要使用"this.$store.commit('changeLogin',this.loginForm.uname)"语句，触发 changeLogin 为用户变量 isLogin 赋值。

12.4.2 修改路由配置

为了区分组件是否需要路由拦截器拦截，需要修改一下 src\router\index.js，在需要拦截的路由中加一条元数据，设置一个 auth 字段，具体语句为"meta:{auth:true}"。完整的代码参见 12.3.7 节。

12.4.3 使用钩子函数判断是否登录

钩子函数，即在某些时机将被调用的函数。名片前端系统使用了 router.beforeEach() 钩子函数，意思是在访问每一个路由前调用。

在 src\main.js 中，首先使用"import store from './store'"语句添加对 store 的引用。其次，修改 Vue 对象中的内容。具体代码如下：

```
new Vue({
  el: '#app',
  router,
  store,                                  //使用 store
  components: { App },
  template: '<App/>'
})
```

最后，实现 beforeEach() 钩子函数，代码如下：

```
router.beforeEach((to,from,next)=>{
    //如果路由器需要验证
    if(to.matched.some(m=>m.meta.auth)){
      //对路由进行验证
      if (store.state.isLogin == '0') {
        alert("您没有登录,无权访问!")
        //未登录则跳转到登录界面,query:{ redirect: to.fullPath}表示把当前路由信息传递过去
        //方便登录后跳转回来
        next({
          path: 'login',
          query: {redirect: to.fullPath}
        })
      } else {                            //已经登录
        next()                            //正常跳转到设置好的页面
      }
    }else{
```

```
        next()
      }
    }
)
```

完整的 main.js 参见 12.3.8 节。

12.4.4　解决跨域请求 session 失效的问题

前端发送跨域请求时，默认是不携带 cookie 的，因此会话 session 将出现失效的问题，解决过程为：首先，在 src\main.js 中使用"axios.defaults.withCredentials=true"语句使每个请求带上 cookie 信息（参见 12.3.8 节的 main.js）；然后，在服务器端通过在响应 header 中设置 Access-Control-Allow-Credentials = true 来运行前端携带证书式访问，具体代码为 "response.setHeader("Access-Control-Allow-Credentials","true");"（参见 12.2.8 节的过滤器）。

12.5　测试运行

首先，运行后端系统 cardmis 的主类 CardmisApplication，启动 cardmis，如图 12.28 所示。

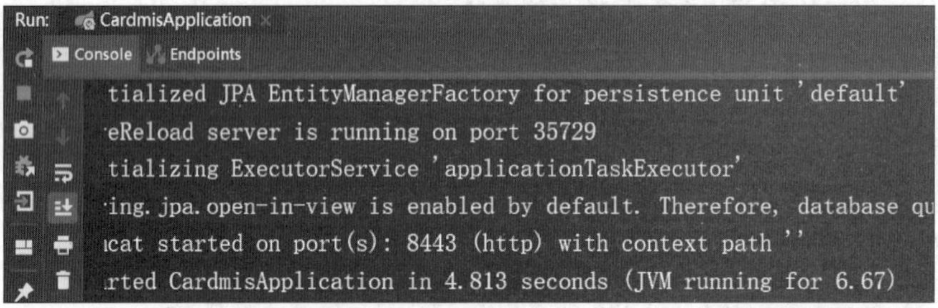

图 12.28　启动 cardmis

然后，启动前端系统 cardmis-vue，如图 12.29 所示。

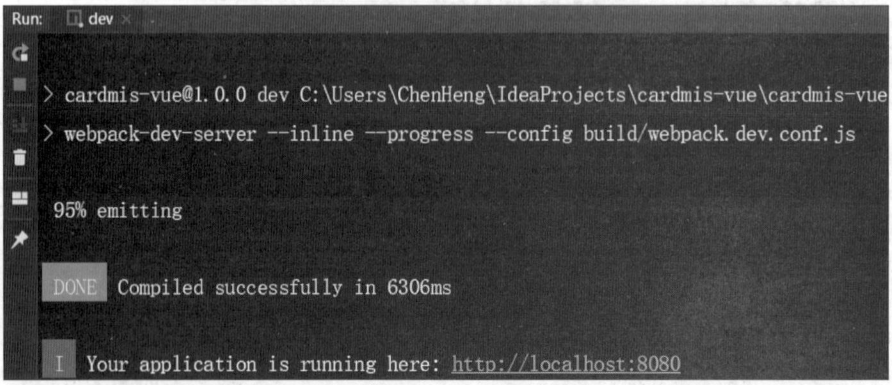

图 12.29　启动 cardmis-vue

前后端系统同时启动后，即可通过"http://localhost:8080/"测试运行。

12.6　小结

本章通过一个业务简单的名片系统，讲述了前后端分离开发的具体过程，旨在让读者了解 Vue.js＋Spring Boot 实现前后端分离开发的流程。

参 考 文 献

[1] 汪云飞.JavaEE开发的颠覆者：Spring Boot实战[M].北京：电子工业出版社,2016.
[2] 疯狂软件.Spring Boot 2企业应用实战[M].北京：电子工业出版社,2018.
[3] 王松.Spring Boot＋Vue全栈开发实战[M].北京：清华大学出版社,2018.
[4] 小马哥.Spring Boot编程思想(核心篇)[M].北京：电子工业出版社,2019.